Lecture Notes in Mathematics

Volume 2271

This series reports on new developments in all areas of mathematics and their applications - quickly, informally and at a high level. Mathematical texts analysing new developments in modelling and numerical simulation are welcome. The type of material considered for publication includes:

1. Research monographs
2. Lectures on a new field or presentations of a new angle in a classical field
3. Summer schools and intensive courses on topics of current research.

Texts which are out of print but still in demand may also be considered if they fall within these categories. The timeliness of a manuscript is sometimes more important than its form, which may be preliminary or tentative.

More information about this series at http://www.springer.com/series/304

Yoshishige Haraoka

Linear Differential Equations in the Complex Domain

From Classical Theory to Forefront

 Springer

Yoshishige Haraoka
Department of Mathematics
Kumamoto University
Kumamoto, Japan

ISSN 0075-8434   ISSN 1617-9692   (electronic)
Lecture Notes in Mathematics
ISBN 978-3-030-54662-5   ISBN 978-3-030-54663-2   (eBook)
https://doi.org/10.1007/978-3-030-54663-2

Mathematics Subject Classification: Primary: 34M03; Secondary: 34M35

This Springer imprint is published by the registered company Springer Nature Switzerland AG.
The registered company address is: Gewerbestrasse 11, 6330 Cham, Switzerland

Preface

Differential equations in the complex domain are fundamental objects in mathematics and physics. In physics, various fields such as mechanics, electromagnetism, and quantum mechanics use differential equations as fundamental tools. In particular, Bessel's differential equation, Legendre's differential equation, Gauss's hypergeometric differential equation, Kummer's confluent hypergeometric differential equation, and Hermite differential equation appear as fundamental equations in these fields. These differential equations are used in describing physical phenomena and hence are equations in real variables; however, we may regard them as equations in complex variables, and this viewpoint will bring deep understandings of these differential equations. In mathematics, linear ordinary differential equations in complex variables appear not only in analysis but also in number theory, algebraic geometry, differential geometry, representation theory, and so on and play substantial roles in each field. This diversity makes linear differential equations in the complex domain important and attractive.

On the other hand, apart from the relations to these various fields, the study of linear differential equations in the complex domain has been developed by itself. In particular, the theory of holonomic systems and the deformation theory of differential equations have made big progresses, and the results influenced many fields in mathematics and physics. Moreover, the book "Rigid Local Systems" by Nicolas M. Katz published in 1996 brought a decisive development to the theory of linear Fuchsian ordinary differential equations. We shall explain the development in detail.

A Fuchsian ordinary differential equation is a linear differential equation on the complex projective line $\mathbb{P}^1$ (or on a Riemann surface, in general) having only regular singular points as singular points. The local theory studies behaviors of solutions in a neighborhood of a regular singular point and has been completed already in the nineteenth century. On the other hand, the global theory, which studies analytic continuations of a solution specified at a regular singular point, is very difficult, and we have no uniform method of analysis in the global theory. However, there exists a special class of Fuchsian ordinary differential equations called rigid, for which we have explicit descriptions of global behaviors of solutions. Thus, as for the global theory on Fuchsian ordinary differential equations, everything is clearly described for rigid equations (and for some sporadic equations) and nothing is described for the other equations. Katz introduced the notions of the index of rigidity and the middle convolution, and clarified many structures of rigid equations and their monodromy representations. The methods in the Katz theory are also useful to the analysis of non-rigid equations, and then it turned out that there are various structures also for non-rigid equations. In particular, the spectral type, which represents the types of local behaviors at each singular point, plays a substantial role, and the space of the Fuchsian ordinary differential equations is classified on the basis of it. Equations in the same class can be transformed each other by the middle convolution, and we can evaluate the change of several analytic quantities by the transformation. Therefore, the study of the Fuchsian ordinary differential equations is reduced to the study of representatives of the classes. It was Toshio Oshima who completed this story after the Katz theory. At this point, Katz–Oshima theory is the most up-to-date understanding of the Fuchsian ordinary differential equations.

While the Katz theory made a striking breakthrough in the study of Fuchsian ordinary differential equations, the sources of it have been familiar to us. The middle convolution, which plays a central role in the Katz theory, is nothing but the Riemann–Liouville transform which is known to be a derivation in a complex time. The notion of rigidity is equivalent to being free from accessory parameters. Kenjiro Okubo derived a formula of counting the number of the accessory parameters and developed the study of equations free from accessory parameters. These facts do not reduce the significance of Katz's work, and on the contrary, I sincerely respect him for constructing such a solid scheme from these known materials. The Katz theory is translated by Dettweiler–Reiter in a language of linear algebra, which made a great contribution to circulate the theory. The main idea of Dettweiler–Reiter is to use a normal form of differential equations introduced by Okubo. Thus, Okubo gave a foundation of Katz–Oshima theory in idea and also in technique.[1]

In this book, we start from the basic theory of linear ordinary differential equations in the complex domain and then explain Katz–Oshima theory in detail. Moreover, we consider completely integrable systems in several variables and study the extension of Katz–Oshima theory to integrable systems.

The contents of this book are based on the knowledge and viewpoints on differential equations which the author deeply owe to the mathematicians with whom he communicated so far. Although it is difficult to list the names, the author would like to express his sincere gratitude to these mathematicians. On the other hand, the discussions with Masahiko Yoshinaga, Takeshi Abe, Hironobu Kimura, and Toshio Oshima helped the author in preparing particular subjects of this book. The author heartily thanks them. The author also thanks Shin Yokoyama who is the publisher of the Japanese version of this book. Without his effort, this book would

---

[1]Professor Kenjiro Okubo passed away in July 2014 when I was preparing the manuscript of the original version of the present book. Here I would like to express my sincere respect to his deep and wide perspective.

not have appeared. In preparing the English version, Donald A. Lutz gave the author many valuable and helpful comments, which improved this version very much. The author thanks him sincerely. Finally, the author genuinely thanks the referees who read the manuscript carefully and gave many useful advices.

Kumamoto, Japan   Yoshishige Haraoka
June 2020

Contents

Symbols

1. We denote the identity matrix by I. We also use I_n, where n denotes the size.
2. We denote the zero matrix by O. We also use O_n for the zero matrix of size $n \times n$, and O_{mn} for that of size $m \times n$.
3. We denote by E_{ij} the square matrix with 1 as the (i, j)-entry and 0 as the other entries. We call it a matrix element.
4. We denote by $J(\alpha; n)$ the Jordan cell of size n with eigenvalue α.
5. For a square matrix A and a scalar α, we write $A + \alpha I$ simply by $A + \alpha$.
6. We define the norm $||v||$ of a vector $v = {}^t(v_1, v_2, \ldots, v_n)$ by

$$||v|| = \max_{1 \leq i \leq n} |v_i|.$$

We also define the norm $||A||$ of a square matrix A by

$$||A|| = \sup_{v \neq 0} \frac{||Av||}{||v||}.$$

For vectors u, v, square matrices A, B, and a scalar c, we have

$$||cu|| = |c| \, ||u||, \quad ||u + v|| \leq ||u|| + ||v||,$$

$$||cA|| = |c| \, ||A||, \quad ||A + B|| \leq ||A|| + ||B||,$$

$$||Au|| \leq ||A|| \, ||u||, \quad ||AB|| \leq ||A|| \, ||B||.$$

Note that, if we denote $A = (a_{ij})_{1 \leq i, j \leq n}$, we have

$$\max_{1 \leq i, j \leq n} |a_{ij}| \leq ||A|| \leq n \max_{1 \leq i, j \leq n} |a_{ij}|.$$

7. For $a \in \mathbb{C}$ and $r > 0$, we denote by $B(a; r)$ the open disc with center a and of radius r:

$$B(a; r) = \{x \in \mathbb{C}; \ |x - a| < r\}.$$

Chapter 1
Introduction

In this book, we start from a basic theory of linear ordinary differential equations in the complex domain, arrive at Katz–Oshima theory, which is the forefront, and then explain the deformation theory and the theory of completely integrable systems in several variables so that the reader will get sufficient knowledge. In the following, we illustrate the contents of this book by using several examples.

The Gauss hypergeometric differential equation is a second order linear ordinary differential equation of the form

$$x(1-x)y'' + (\gamma - (\alpha + \beta + 1)x)y' - \alpha\beta y = 0, \tag{G}$$

where α, β, γ are constants. We rewrite this as

$$y'' + \frac{\gamma - (\alpha + \beta + 1)x}{x(1-x)}y' - \frac{\alpha\beta}{x(1-x)}y = 0,$$

where the poles $x = 0, 1$ appear in the coefficients of y' and y. By the coordinate change $t = 1/x$, we get

$$\frac{d^2 y}{dt^2} + \frac{(2-\gamma)t + \alpha + \beta - 1}{t(t-1)}\frac{dy}{dt} - \frac{\alpha\beta}{t^2(t-1)}y = 0,$$

where the poles $t = 0, 1$ appear in the coefficients of dy/dt and y. These poles are said to be singular points of the differential equation (G). In the coordinate $x, x = 0, 1, \infty$ are the singular points of (G). Let x_0 be a point different from $0, 1, \infty$. Then there exists holomorphic solutions of (G) in a neighborhood of $x = x_0$, and the whole set of such solutions makes a linear space of dimension two. Every holomorphic solution at x_0 can be analytically continued along any curve in $\mathbb{P}^1 \setminus \{0, 1, \infty\}$ with initial point x_0. Thus any solution is holomorphic at any point other than the singular points, and can be analytically continued to any

© The Editor(s) (if applicable) and The Author(s), under exclusive license to Springer Nature Switzerland AG 2020
Y. Haraoka, *Linear Differential Equations in the Complex Domain*, Lecture Notes in Mathematics 2271, https://doi.org/10.1007/978-3-030-54663-2_1

point other than the singular points. These are common properties for linear ordinary differential equations in the complex domain. In Chap. 3 we show these properties.

Then how do solutions behave around the singular points? The three singular points of (G) are regular singular points, which are characterized by the property that any solution is at most polynomial order of growth as x tends to the singular point. Usually a regular singular point is a branch point of a solution. Namely a solution is a multi-valued function which changes its value as x encircles a regular singular point. The multi-valuedness of the hypergeometric differential equation (G) at each regular singular point is described by a table

$$\left\{ \begin{array}{ccc} x = 0 & x = 1 & x = \infty \\ 0 & 0 & \alpha \\ 1 - \gamma & \gamma - \alpha - \beta & \beta \end{array} \right\},$$

which is called the Riemann scheme. The meaning of this table is as follows. Let ρ_1, ρ_2 be two numbers below a singular point, and t be a local coordinate at the singular point. Then there exists two linearly independent solutions having t^{ρ_1} and t^{ρ_2} as multiple-valued factors. We call ρ_1, ρ_2 the characteristic exponents. For example, the characteristic exponents at $x = 0$ are 0 and $1 - \gamma$, which means that there exist linearly independent solutions of the forms

$$y_1(x) = 1 + a_1 x + a_2 x^2 + \cdots,$$

$$y_2(x) = x^{1-\gamma} \left(1 + b_1 x + b_2 x^2 + \cdots \right)$$

in a neighborhood of $x = 0$. Among them, $y_1(x)$, which is a solution of exponent 0 at $x = 0$, is known to be the hypergeometric series

$$F(\alpha, \beta, \gamma; x) = \sum_{n=0}^{\infty} \frac{(\alpha, n)(\beta, n)}{(\gamma, n)n!} x^n,$$

where

$$(\alpha, n) = \frac{\Gamma(\alpha + n)}{\Gamma(\alpha)} = \begin{cases} 1 & (n = 0), \\ \alpha(\alpha + 1)(\alpha + 2) \cdots (\alpha + n - 1) & (n \geq 1). \end{cases}$$

The quantity that describes the change of solutions as x encircles a singular point is called the local monodromy. The local monodromies are determined by the Riemann scheme in general. For example, let μ be a loop which encircles $x = 0$ once in the positive direction. We denote the analytic continuation of a function $f(x)$ along the loop μ by $\mu_* f(x)$. Then we have

$$(\mu_* y_1(x), \mu_* y_2(x)) = (y_1(x), y_2(x)) \begin{pmatrix} 1 & 0 \\ 0 & e^{2\pi \sqrt{-1}(-\gamma)} \end{pmatrix},$$

where the matrix in the right hand side is the local monodromy at $x = 0$.

Compared with local monodromies, the global analysis which studies global analytic continuations is very difficult, although it is important. Usually we understand that the global analysis consists of the connection problem, which studies linear relations between local solutions specified by the Riemann scheme, and the monodromy problem, which describes analytic continuation along any loop. The monodromy is the quantity that describes the multi-valuedness, and essentially determines the Fuchsian ordinary differential equation. The connection problem has many important applications. For some differential equation appeared in physics, a solution holomorphic at two singular points simultaneously corresponds to a physical phenomenon, and, by solving the connection problem, we can specify when such solution exists. In order to solve these problems, we should compute the analytic continuation, and it is in general hard to get explicit results. On the other hand, if a differential equation has a particular structure, we may use it to solve the global problems. In the case of the hypergeometric differential equation, there is an integral representation of solutions

$$y(x) = \int_p^q t^{\beta-\gamma}(t-1)^{\gamma-\alpha-1}(t-x)^{-\beta}\, dt,$$

and we can use it to solve explicitly the connection problem and the monodromy problem as follows.

Let $y_1(x)$, $y_2(x)$ be the local solutions of the hypergeometric differential equation (G) at $x = 0$ given above, and define the local solutions $y_3(x)$, $y_4(x)$ at $x = 1$ by

$$y_3(x) = 1 + c_1(x-1) + c_2(x-1)^2 + \cdots,$$

$$y_4(x) = (1-x)^{\gamma-\alpha-\beta}(1 + d_1(x-1) + d_2(x-1)^2 + \cdots).$$

In the intersection $\{|x| < 1\} \cap \{|1-x| < 1\}$ of the domains of definition of $y_1(x), \ldots, y_4(x)$, linear relations among these solutions hold. For example, we have

$$y_1(x) = \frac{\Gamma(\gamma)\Gamma(\gamma-\alpha-\beta)}{\Gamma(\gamma-\alpha)\Gamma(\gamma-\beta)} y_3(x) + \frac{\Gamma(\gamma)\Gamma(\alpha+\beta-\gamma)}{\Gamma(\alpha)\Gamma(\beta)} y_4(x).$$

Such relations are called the connection relations, and we can derive them by using the integral representation. The coefficients of $y_3(x)$ and $y_4(x)$ in the right hand side are called the connection coefficients. From this relation, we can derive the condition for the existence of a solution holomorphic at $x = 0$ and $x = 1$ simultaneously. In fact, such solution exists only when the coefficient of $y_4(x)$ vanishes, and hence we have

$$\alpha \in \mathbb{Z}_{\leq 0} \text{ or } \beta \in \mathbb{Z}_{\leq 0}.$$

In this way, we can derive global properties of solutions if the connection coefficients are obtained explicitly.

The monodromy of the hypergeometric differential equation (G) can be obtained also by using the integral representation, or by using the rigidity. In general, the monodromy is a representation of the fundamental group. In the case of the hypergeometric differential equation, the monodromy is a representation of the fundamental group $\pi_1(\mathbb{P}^1 \setminus \{0, 1, \infty\})$, which has a presentation

$$\pi_1\left(\mathbb{P}^1 \setminus \{0, 1, \infty\}\right) = \langle \gamma_0, \gamma_1, \gamma_2 \mid \gamma_0 \gamma_1 \gamma_2 = 1 \rangle.$$

Then the monodromy is determined if we get the images M_j of the generators γ_j $(j = 0, 1, 2)$. We may assume that $\gamma_0, \gamma_1, \gamma_2$ are represented by loops encircling $x = 0, 1, \infty$, respectively, once in the positive direction. Then each M_j is similar to the local monodromy at each singular point. Moreover, we have $M_2 M_1 M_0 = I$ since the monodromy is a representation (precisely speaking, it is an anti-representation). By using these conditions, the equivalence class of (M_0, M_1, M_2) by simultaneous similar transformations is uniquely determined, and hence it should coincide with the monodromy. We call this property rigid: A representation of a fundamental group is said to be rigid if it is determined by the local monodromies uniquely up to isomorphisms. Naively speaking, to determine a representation is reduced to solve a system of algebraic equations, and then a representation is rigid if the numbers of the unknowns and the equations coincide. Hence, if a representation is rigid, we can get its explicit form. In this way, the monodromy of the hypergeometric differential equation (G) can be explicitly obtained by solving a system of algebraic equations.

A Fuchsian ordinary differential equation of the second order with four regular singular points specified by the Riemann scheme

$$\left\{ \begin{matrix} x = 0 & x = 1 & x = t & x = \infty \\ 0 & 0 & 0 & \alpha \\ \gamma_1 & \gamma_2 & \gamma_3 & \beta \end{matrix} \right\}$$

is called Heun's differential equation, where

$$\alpha + \beta + \gamma_1 + \gamma_2 + \gamma_3 = 2$$

is assumed. The last condition is called the Fuchs relation, which is a necessary condition for the existence of Fuchsian ordinary differential equations. Explicitly, Heun's differential equation is given by

$$y'' + \left(\frac{1 - \gamma_1}{x} + \frac{1 - \gamma_2}{x - 1} + \frac{1 - \gamma_3}{x - t} \right) y' + \frac{\alpha \beta x + q}{x(x - 1)(x - t)} y = 0,$$

where q is a parameter that cannot be determined by the Riemann scheme. In other words, for any special value of q, we have the same Riemann scheme. Such parameters are said to be accessory parameters. If a differential equation has accessory parameters, the equation cannot be determined by the local behaviors, and hence is not rigid. Heun's differential equation is the simplest example of non-rigid equations. The global analysis of non-rigid equations is very hard, and hence, except a few particular cases, we do not obtain the monodromy or the connection coefficients of Heun's differential equation.

We explain the above facts in general in Chaps. 4–7 and in Chap. 9. In Chap. 4, we construct local solutions at a regular singular point and describe the local monodromy. We do not assume any assumption for simplicity, and give the complete description. In Chap. 5, we study the monodromy representation. First we explain the general properties of the monodromy representations and the relation to the local monodromies, and then study the monodromy representations for Fuchsian ordinary differential equations in detail: What happens if the monodromy is reducible, solvable, or finite, and how we can know whether the monodromy is reducible or finite. For Fuchsian ordinary differential equations, the differential Galois group can be computed by the monodromy. Then we overview the differential Galois theory, so that the reader will grasp the outline of the theory. In Chap. 6, we treat the connection problem. We first give several examples to show how the connection problem concerns with problems in physics. Then we formulate the connection problem in general, describe the solution of the connection problem for the Gauss hypergeometric differential equation, and explain recent developments.

Chapter 7 is devoted to present the whole theory of Fuchsian ordinary differential equations. We explain from the classical theory to the modern Katz theory. The Katz theory is explained based on the two papers [34, 35] by Dettweiler–Reiter, so that the reader can understand by the knowledge of linear algebra. We consider the system

$$\frac{dY}{dx} = \left(\sum_{j=1}^{p} \frac{A_j}{x - a_j} \right) Y$$

of differential equations of the first order, where Y is a vector of unknown functions and A_j are constant matrices. This system is called a Fuchsian system of normal form, and behaves well with the deformation theory, which we discuss in Chap. 8, and with linear Pfaffian systems in several variables, which we study in Part II. On the other hand, Oshima [137] used a scalar differential equation

$$y^{(n)} + p_1(x)y^{(n-1)} + \cdots + p_n(x)y = 0$$

in developing his theory. A scalar differential equation is a normal form without any ambiguity, and then we can read various quantities directly. Both for Fuchsian systems of normal form and for scalar differential equations, the spectral types, which represents the multiplicities of the characteristic exponents, are defined. A spectral type is a tuple of partitions. Note that a Fuchsian differential equation does not necessarily exist for a prescribed tuple of partitions as the spectral type. It is a fundamental problem to characterize tuples of partitions which becomes the spectral type of a Fuchsian differential equation. In the last section of Chap. 7, we explain the solution of this problem by W. Crawley-Boevey relating it to the Katz theory.

Integral representations of solutions are explained in Chap. 9. We might need a big volume to explain the general theory of integral representation of solutions of linear differential equations. Then we treat a few illustrative examples, and study them in various aspects so that the reader will understand the outline of the theory. We explain how to construct a differential equation from an integral representation and how to compute the monodromy and the connection coefficients, in such a way that can be extended to the general case. As an application, the connection problem for Legendre's differential equation is solved by using the integral representation.

Chapter 8 deals with the deformation theory of Fuchsian ordinary differential equations. The deformation theory was originated by Riemann, and is one of the most profound theories in differential equations in the complex domain. By a deformation we mean to vary the accessory parameters of the differential equation so that the monodromy is invariant under the variation of the positions of the singular points. Such deformation is possible if and only if there exist partial differential equations in the positions of the singular points as variables compatible with the original ordinary differential equation. In other words, a deformation is possible if the ordinary differential equation is prolonged to a completely integrable system of partial differential equations. In this chapter, we derive Schlesinger's system which describes the deformation, explain the relation of the deformation theory and the Katz theory, and then clarify the Hamiltonian structure of the deformation, which are the fundamentals of the deformation theory.

The analysis of differential equations at an irregular singular point is rather complicated compared with the case of regular singular points. In the local analysis, a formal solution at an irregular singular point diverges in general. We can give an analytic meaning to the divergent formal solution by considering asymptotic expansions. Namely, there exists a sector centered at the irregular singular point and an analytic solution which is asymptotic to the formal solution in the sector. The analytic solution depends on the sector, and in distinct sectors, the analytic solutions asymptotic to the same formal solution in the respective sectors are distinct. This phenomenon is said to be the Stokes phenomenon. The Stokes phenomenon occurs in a neighborhood of an irregular singular point, however, it has a global analytic nature, and is very hard to compute. We overviewed these fundamental facts in Chap. 10.

These are the contents of Part I.

In Part II, we study completely integrable systems in several variables. To illustrate the contents, we give an example.

The power series

$$F_1\left(\alpha, \beta, \beta', \gamma; x, y\right) = \sum_{m,n=0}^{\infty} \frac{(\alpha, m+n)(\beta, m)(\beta', n)}{(\gamma, m+n)m!n!} x^m y^n$$

in two variables is called Appell's hypergeometric series F_1, where the symbol (α, n) has been already defined above. The function $z = F_1$ satisfies the system of partial differential equations

$$\begin{cases} x(1-x)z_{xx} + y(1-x)z_{xy} + (\gamma - (\alpha + \beta + 1)x)z_x - \beta y z_y - \alpha\beta z = 0, \\ x(1-y)z_{xy} + y(1-y)z_{yy} + (\gamma - (\alpha + \beta' + 1)y)z_y - \beta' x z_x - \alpha\beta' z = 0. \end{cases}$$
$$(F)$$

From this system, we can derive the system

$$\begin{cases} \dfrac{\partial u}{\partial x} = \left(\dfrac{A_0}{x} + \dfrac{A_1}{x-1} + \dfrac{C}{x-y} \right) u, \\[2mm] \dfrac{\partial u}{\partial y} = \left(\dfrac{B_0}{y} + \dfrac{B_1}{y-1} + \dfrac{C}{y-x} \right) u \end{cases}$$

of partial differential equations of the first order with the unknown vector $u = {}^t(z, x z_x, y z_y)$, where

$$A_0 = \begin{pmatrix} 0 & 1 & 0 \\ 0 & \beta' - \gamma + 1 & 0 \\ 0 & -\beta' & 0 \end{pmatrix}, \quad A_1 = \begin{pmatrix} 0 & 0 & 0 \\ -\alpha\beta & \gamma - \alpha - \beta - 1 & -\beta \\ 0 & 0 & 0 \end{pmatrix},$$

$$B_0 = \begin{pmatrix} 0 & 0 & 1 \\ 0 & 0 & -\beta \\ 0 & 0 & \beta - \gamma + 1 \end{pmatrix}, \quad B_1 = \begin{pmatrix} 0 & 0 & 0 \\ 0 & 0 & 0 \\ -\alpha\beta' & -\beta' & \gamma - \alpha - \beta' - 1 \end{pmatrix},$$

$$C = \begin{pmatrix} 0 & 0 & 0 \\ 0 & -\beta' & \beta \\ 0 & \beta' & -\beta \end{pmatrix}.$$

The last system can be written by using the total derivative in the form

$$du = \left(A_0 \frac{dx}{x} + A_1 \frac{dx}{x-1} + B_0 \frac{dy}{y} + B_1 \frac{dy}{y-1} + C \frac{d(x-y)}{x-y} \right) u. \qquad (P)$$

Total differential equations are also called Pfaffian systems. We see that the set of the singular points of the coefficients of the linear Pfaffian system (P) is

$$S = \{(x, y) \, ; \, x(x-1)y(y-1)(x-y) = 0\}.$$

Then we can show that, at any point in $\mathbb{C}^2 \setminus S$, the Pfaffian system (P) has the three dimensional space of holomorphic solutions, and that the holomorphic solutions can be continued analytically along any curve to the whole space $\mathbb{C}^2 \setminus S$. Thus any holomorphic solution is a holomorphic function on the universal covering space

$\widetilde{\mathbb{C}^2 \setminus S}$. We call S the singular locus of (P). The dimension of the solution space coincides with the size of the unknown vector of (P).

In this way, in rewriting the system of partial differential equations (F) into the linear Pfaffian system (P), we can readily see the dimension of the solution space and the singular locus. These are the advantage of linear Pfaffian systems as a normal form of completely integrable systems.

Next, as in the case of the hypergeometric differential equation, we look at the behavior of solutions at the singular locus S. The singular locus S is decomposed into five irreducible components:

$$S = \bigcup_{i=1}^{5} S_i,$$

$$S_1 = \{x = 0\}, \ S_2 = \{x = 1\}, \ S_3 = \{y = 0\}, \ S_4 = \{y = 1\}, \ S_5 = \{x = y\}.$$

We take a compactification of $\mathbb{C}^2$. For example, we may take $\mathbb{P}^2$ by adding the line S_0 at infinity. Then S_0 takes part in the set of the singular points. Hence the irreducible decomposition of the singular locus $\hat{S}$ in $\mathbb{P}^2$ becomes

$$\hat{S} = \bigcup_{i=0}^{5} S_i.$$

We find that, for each irreducible component $S_i = \{\varphi_i(x, y) = 0\}$, there exist complex numbers $\rho_{i1}, \rho_{i2}, \rho_{i3}$ such that we have a set of linearly independent local solutions of the forms

$$z_{i1}(x, y) = \varphi_i(x, y)^{\rho_{i1}} f_{i1}(x, y),$$

$$z_{i2}(x, y) = \varphi_i(x, y)^{\rho_{i2}} f_{i2}(x, y),$$

$$z_{i3}(x, y) = \varphi_i(x, y)^{\rho_{i3}} f_{i3}(x, y)$$

in a neighborhood of each point (x_0, y_0) in S_i, where $f_{ij}(x, y)$ denotes a function holomorphic at (x_0, y_0) satisfying $f_{ij}(x_0, y_0) \neq 0$. These complex numbers are said to be the characteristic exponents at S_i, and are the eigenvalues of the residue matrix at S_i in (P). For example, at $S_1 = \{x = 0\}$, the characteristic exponents are the eigenvalues $0, 0, \beta' - \gamma + 1$ of the residue matrix A_0, and hence there exists two dimensional solutions holomorphic in a neighborhood of $(0, y_0) \in S_1$ and a solution of the form

$$x^{\beta' - \gamma + 1} f(x, y),$$

where we assume $y_0 \neq 0, 1$. Therefore the local monodromy at S_1 is given by

$$
\begin{pmatrix}
1 & 0 & 0 \\
0 & 1 & 0 \\
0 & 0 & e^{2\pi\sqrt{-1}(\beta'-\gamma)}
\end{pmatrix}.
$$

We call the partition that represents the multiplicities of the eigenvalues of the local monodromy the spectral type. We see that, for the linear Pfaffian system (P) for F_1, the spectral type at every S_i becomes $(2, 1)$. Note that each irreducible component S_i of the singular locus $\hat{S}$ is not a point but a set with spread, while the characteristic exponents and the local monodromy are constant for all points on S_i. This is a remarkable property in linear Pfaffian systems. Thus, for linear Pfaffian systems, an irreducible component of the singular locus plays a similar role as a singular point of ordinary differential equations.

Thanks to this property, we can define the rigidity in a similar way as in the case of ordinary differential equations. The monodromy of (P), which describes the multi-valuedness of solutions, is a representation of the fundamental group $\pi_1(\mathbb{P}^2 \setminus \hat{S})$. Then we call a representation of the fundamental group rigid if it is uniquely determined by the local monodromies up to isomorphisms. In the case of ordinary differential equations, the domain of definition of an equation is a complement of a finite number of points, and the fundamental group is determined only by the number of the singular points. While in the case of linear Pfaffian systems, the configuration of the singular locus plays a decisive role to determine the fundamental group, so that we have various groups as fundamental groups, and in general, there hold many relations among the generators. Thus, naively saying, the monodromy of a linear Pfaffian system tends to be determined. As for the monodromy of the linear Pfaffian system (P) for F_1, by using the spectral type $(2, 1)$ of every local monodromy and a presentation of the fundamental group, we can show that it is (almost) rigid. In particular, we have an explicit description of the monodromy.

On the other hand, the system (F) of partial differential equations has integral representations of solutions. Explicitly, the following two representations are known:

$$
z(x, y) = \int_I t^{\alpha-1}(1 - t)^{\gamma-\alpha-1}(1 - xt)^{-\beta}(1 - yt)^{-\beta'} \, dt,
$$

$$
z(x, y) = \int_\Delta s^{\beta-1}t^{\beta'-1}(1 - s - t)^{\gamma-\beta-\beta'-1}(1 - xs - yt)^{-\alpha} \, ds \, dt.
$$

We may also obtain the monodromy by using these integrals.

Thus there are various similarities between ordinary differential equations and linear Pfaffian systems. Then in Part II, we analyze linear Pfaffian systems from the viewpoint of Katz–Oshima theory, which is the modern understanding of Fuchsian ordinary differential equations. In Chap. 11, we give fundamental theorems on the construction of local solutions at a regular point and on the analytic continuation of the local solution. The next Chap. 12 is devoted to the analysis at a regular singular point. As we have seen, every irreducible component of the singular locus plays a similar role as a singular point of ordinary differential equations. Then we construct local solutions in a neighborhood of a point of an irreducible component of the singular locus. We see that the characteristic exponents at the point do not depend on the choice of the point, and become constants on the irreducible component. On the other hand, an irreducible component of the singular locus is not a point but a variety, and hence we may restrict the linear Pfaffian system to the component. For example, for the linear Pfaffian system (P), there are double 0 among the characteristic exponents at the irreducible component $S_1 = \{x = 0\}$ of S, which means that there are two dimensional holomorphic solutions at S_1. Then we may take restrictions of the holomorphic solutions to S_1 by putting $x = 0$ into the solutions. The differential equation satisfied by these restrictions is the restriction of (P) to the singular locus S_1. As the restriction, we get an ordinary differential equation in the coordinate of S_1 (say, y), and it tuns out that the ordinary differential equation reduces to the Gauss hypergeometric differential equation. The restriction to a singular locus is a proper operation in completely integrable systems in several variables, and makes the structure of completely integrable systems fairly rich.

In Chap. 13, we study the monodromy of linear Pfaffian systems including the viewpoint of the Katz theory. As is seen above, a local monodromy is defined for each irreducible component of the singular locus. We explain this fact from the standpoint of topology. We clarify that, in extending the Katz theory to completely integrable systems in several variables, the definition of the local monodromy is crucial. On the other hand, since the monodromy is a representation of the fundamental group, we introduce a method due to Zariski-van Kampen to obtain a presentation of the fundamental group. By using the presentation of the fundamental group and the local monodromies, we may study the rigidity of the monodromy as in the case of ordinary differential equations. It turns out that the rigidity deeply depends on the spectral types of the local monodromies and also on the topology. We show as an example that the monodromy of the linear Pfaffian system (P) becomes almost rigid.

In Chap. 14, we define the middle convolution, which is another basic notion in the Katz theory, to a class of linear Pfaffian systems. Also in this case, the middle convolution is a basic operation, and we may expect many applications such as constructing new integrable systems, computing several analytic quantities, and so on. To analyze the moduli space of completely integrable systems, the middle convolution together with the restriction to a singular locus will play a substantial role.

Thus, in Part II, the basic theory of linear Pfaffian systems with regular singularities together with the recent development from the viewpoint of Katz–Oshima theory are stated. We expect this development will open a new fields of studies. On the other hand, the theory of holonomic systems has been highly developed. Completely integrable systems, in particular linear Pfaffian systems, are holonomic. Although linear Pfaffian system is a useful normal form, it does not completely describe its intrinsic quantities. For example, we can directly read the singular locus S for the linear Pfaffian system (P), however, it is not evident whether an actual singularity appears at each irreducible component of S. To determine the actual singular locus, we need the notion and theory of D-modules, and then the theory of holonomic systems works. It will be a good problem to incorporate the development brought by Katz–Oshima theory into the theory of holonomic systems.

One of the main aims of this book is to introduce Katz–Oshima theory for Fuchsian ordinary differential equations and to extend it to completely integrable systems. We believe that the extension of Katz–Oshima theory will become a substantial topic in future. However, we do not restrict to this story, but intend to provide a universal knowledge on linear differential equations in the complex domain. Here, by knowledge, we mean not only notions and theorems, but also ideas of proofs, viewpoints and ways of thinking. For the purpose, we give detailed proofs for the fundamental theorems, so that the reader may apply them in any exceptional case: these are construction of local solutions at a regular singular point, fundamental properties of the middle convolution, derivation of the Schlesinger equation in the deformation theory, and so on. Moreover, in order to get a broad perspective, we deal with the following various topics without detailed proofs—Gauss-Schwarz theory, differential Galois theory, the theory of twisted (co)homology, local analysis at an irregular singular point, theorems in topology on fundamental groups, and so on. We explain the main ideas and how to use them, so that it will help the reader when reading references on these topics.

Part I
Ordinary Differential Equations

Chapter 2
Scalar Differential Equations and Systems of Differential Equations

Linear ordinary differential equations are usually given in the scalar form

$$y^{(n)} + p_1(x)y^{(n-1)} + p_2(x)y^{(n-2)} + \cdots + p_n(x)y = 0 \qquad (2.1)$$

or the system of the first order equations

$$\frac{dY}{dx} = A(x)Y, \qquad (2.2)$$

where

$$Y = \begin{pmatrix} y_1 \\ y_2 \\ \vdots \\ y_n \end{pmatrix}, \qquad A(x) = (a_{ij}(x))_{1 \le i, j \le n}.$$

These two expressions are mathematically equivalent. Namely, we can transform the scalar equation (2.1) to the system (2.2) and vise versa. In order to transform the scalar equation to the system, we can take an unknown vector $Y = {}^t(y_1, y_2, \ldots, y_n)$ defined by

$$y_1 = y, \ y_2 = y', \ y_3 = y'', \ldots, \ y_n = y^{(n-1)}. \qquad (2.3)$$

Y. Haraoka, *Linear Differential Equations in the Complex Domain*, Lecture Notes in Mathematics 2271, https://doi.org/10.1007/978-3-030-54663-2_2

In this case, the Eq. (2.1) is transformed to the system

$$\frac{dY}{dx} = \begin{pmatrix} 0 & 1 & 0 & \cdots & 0 \\ 0 & 0 & 1 & \cdots & 0 \\ \vdots & \vdots & & \ddots & \vdots \\ 0 & 0 & \cdots\cdots & & 1 \\ -p_n(x) & -p_{n-1}(x) & \cdots\cdots & & -p_1(x) \end{pmatrix} Y. \tag{2.4}$$

We have many other way to define an unknown vector Y, and then the transformation is not unique.

In order to transform the system (2.2) to a scalar equation (2.1), we have to define the unknown function y of (2.1) in terms of the entries $y_1, y_2, \ldots, y_n$ of the unknown vector Y. Let us consider how to define y in the case $n = 2$. In this case, the system (2.2) is written as

$$\begin{cases} y_1' = a_{11}(x)y_1 + a_{12}(x)y_2, \\ y_2' = a_{21}(x)y_1 + a_{22}(x)y_2. \end{cases}$$

Differentiating both sides of the first equation, we get

$$y_1'' = a_{11}y_1' + a_{11}'y_1 + a_{12}y_2' + a_{12}'y_2$$
$$= a_{11}y_1' + (a_{11}' + a_{12}a_{21})y_1 + (a_{12}' + a_{12}a_{22})y_2$$

by the help of the second equation. If $a_{12}(x) \neq 0$, we obtain

$$y_2 = \frac{1}{a_{12}} y_1' - \frac{a_{11}}{a_{12}} y_1$$

from the first equation. Putting this, finally we get a scalar equation

$$y_1'' + p_1(x)y_1' + p_2(x)y_1 = 0$$

of order 2 with unknown function y_1. (Note that a zero of $a_{12}(x)$ may bring a new singularity to the scalar equation. Such singularity is called an apparent singularity, which will be mentioned in Sects. 5.4.1 and 7.3.) In the case $a_{12}(x) = 0$, if $a_{21}(x) \neq 0$, then in a similar way we get a scalar equation of order 2 with unknown function y_2. When $a_{12}(x) = a_{21}(x) = 0$, we define y by using auxiliary functions $u_1(x), u_2(x)$ as

$$y = u_1(x)y_1 + u_2(x)y_2.$$

If we can take $u_1(x)$, $u_2(x)$ satisfying

$$\begin{vmatrix} u_1 & u_2 \\ a_{11}u_1 + u_1' & a_{22}u_2 + u_2' \end{vmatrix} \neq 0,$$

we obtain a scalar second order equation with the unknown y. Note that, if $a_{11}(x) \neq a_{22}(x)$, the we can take $u_1 = u_2 = 1$, and if $a_{11}(x) = a_{22}(x)$, we can take $u_1 = 1$, $u_2 = x$.

In order to show that a similar argument holds for general n, we use a cyclic vector. The notion of cyclic vectors depends on a vector space over a field K. Since in this book we mainly consider differential equations with rational function coefficients, we take K to be the field of the rational functions $\mathbb{C}(x)$. Consider the n-dimensional vector space

$$V = \{v = (v_1, v_2, \ldots, v_n) \, ; \, v_i \in K \, (1 \leq i \leq n)\}$$

over $K = \mathbb{C}(x)$. We may consider a system of the first order equations

$$x \frac{dY}{dx} = A(x)Y \tag{2.5}$$

instead of (2.2), where the elements of $A(x)$ belong to $\mathbb{C}(x)$. (In fact, we obtain (2.5) from (2.2) by multiplying x to the both sides of (2.2) and denoting $xA(x)$ in the right hand side by $A(x)$.) Define the additive map ∇ from V to V by

$$\nabla : v \mapsto x \frac{dv}{dx} + vA(x).$$

Lemma 2.1 *([32]) There exists an element $u \in V$ such that*

$$u, \nabla u, \nabla^2 u, \ldots, \nabla^{n-1} u$$

are linearly independent.

Proof For each $v \in V$, there is a number $l = l(v)$ such that $v, \nabla v, \ldots, \nabla^{l-1} v$ are linearly independent and $v, \nabla v, \ldots, \nabla^{l-1} v, \nabla^l v$ are linearly dependent. We denote by m the maximum of $l(v)$, and take v satisfying $l(v) = m$. We assume $m < n$.

By this assumption, there exists an element $w \in V$ which is linearly independent with $v, \nabla v, \ldots, \nabla^{m-1} v$. Take any $c \in \mathbb{C}, k \in \mathbb{Z}$, and define u by

$$u = v + cx^k w.$$

Then we have

$$\nabla^i u = \nabla^i v + cx^k (\nabla + k)^i w \quad (i = 0, 1, 2, \ldots).$$

By the assumption, $u, \nabla u, \ldots, \nabla^m u$ are linearly dependent, so that the exterior product of these elements becomes 0:

$$u \wedge \nabla u \wedge \cdots \wedge \nabla^m u = 0.$$

The left hand side is a polynomial in c, which is identically zero, and in particular the coefficient of c^1 is 0. Then we have

$$\sum_{i=0}^{m} v \wedge \nabla v \wedge \cdots \wedge (\nabla + k)^i w \wedge \cdots \wedge \nabla^m v = 0.$$

The left hand side is a polynomial in k, which is also identically zero, and in particular the coefficient of the maximum degree is 0. Thus we have

$$v \wedge \nabla v \wedge \cdots \wedge \nabla^{m-1} v \wedge w = 0,$$

which contradicts to the definition of w. $\square$

We call the vector u in Lemma 2.1 a *cyclic vector*.

Now let u be a cyclic vector, and define P by

$$P = \begin{pmatrix} u \\ \nabla u \\ \vdots \\ \nabla^{n-1} u \end{pmatrix}.$$

Then P is invertible. By the gauge transformation

$$Z = PY,$$

the system (2.5) is transformed into the system

$$x \frac{dZ}{dx} = x \frac{dP}{dx} Y + Px \frac{dY}{dx}$$

$$= \left(x \frac{dP}{dx} + PA \right) Y$$

$$= \nabla P \cdot P^{-1} Z.$$

Thanks to Lemma 2.1, there exists $b_1, b_2, \ldots, b_n \in K$ such that

$$\nabla^n u = \begin{pmatrix} b_1 & b_2 & \cdots & b_n \end{pmatrix} P.$$

Then we have

$$\nabla P = \begin{pmatrix} \nabla u \\ \nabla^2 u \\ \vdots \\ \nabla^n u \end{pmatrix} = \begin{pmatrix} 0 & 1 & 0 & \cdots & 0 \\ 0 & 0 & 1 & \cdots & 0 \\ \vdots & \vdots & & \ddots & \\ 0 & 0 & \cdots\cdots & & 1 \\ b_1 & b_2 & \cdots\cdots & & b_n \end{pmatrix} P,$$

which implies that Z satisfies a system of the form (2.4). Thus the first entry z_1 of $Z = {}^t(z_1, z_2, \ldots, z_n)$ satisfies a scalar differential equation of order n with coefficients in $K = \mathbb{C}(x)$.

In this way, we see that a scalar equation (2.1) and a system (2.2) are equivalent. Then sometimes we call the both a differential equation. Each form will be more useful than the other for a particular problem.

Chapter 3
Analysis at a Regular Point

We consider a system of differential equations

$$\frac{dY}{dx} = A(x)Y \tag{3.1}$$

of the first order, where Y is an n-vector of unknowns and $A(x)$ an $n \times n$ matrix. Let D be a domain in $\mathbb{C}$, and we assume that $A(x)$ is holomorphic on D. Let a be any point in D, and take $r > 0$ so that $B(a; r) \subset D$. The following is the most fundamental theorem in the theory of linear ordinary differential equations in the complex domain.

Theorem 3.1 *For any $Y_0 \in \mathbb{C}^n$, there exist a unique solution $Y(x)$ of (3.1) satisfying the initial condition*

$$Y(a) = Y_0, \tag{3.2}$$

and the solution $Y(x)$ is holomorphic on $B(a; r)$.

Proof Without loss of generality, we may assume $a = 0$. Expand $A(x)$ in the Taylor series at $x = 0$:

$$A(x) = \sum_{m=0}^{\infty} A_m x^m.$$

This series converges in $|x| < r$.

First we shall construct a formal solution

$$Y(x) = \sum_{m=0}^{\infty} Y_m x^m. \tag{3.3}$$

© The Editor(s) (if applicable) and The Author(s), under exclusive
license to Springer Nature Switzerland AG 2020
Y. Haraoka, *Linear Differential Equations in the Complex Domain*, Lecture Notes
in Mathematics 2271, https://doi.org/10.1007/978-3-030-54663-2_3

Put (3.3) into the system (3.1) to obtain

$$\sum_{m=1}^{\infty} m Y_m x^{m-1} = \left(\sum_{m=0}^{\infty} A_m x^m \right) \left(\sum_{m=0}^{\infty} Y_m x^m \right).$$

We compare the coefficients of x^m in both sides, and get

$$(m+1)Y_{m+1} = \sum_{k+l=m} A_k Y_l \quad (m \geq 0). \tag{3.4}$$

By the recurrence relation (3.4), we see that Y_{m+1} is uniquely determined by $Y_0, Y_1,$ $\dots, Y_m$. Thus $\{Y_m\}_{m=0}^{\infty}$ is uniquely determined by Y_0.

Second we shall prove that the above formal solution is convergent. Take any r_1 satisfying $0 < r_1 < r$. Then every entry of $A(x)$ is bounded in $|x| \leq r_1$, and hence, by Cauchy's inequality, we see that there exists an $M > 0$ such that

$$\|A_m\| \leq \frac{M}{r_1{}^m} \quad (m \geq 0) \tag{3.5}$$

holds. Define a series $\{g_m\}_{m=0}^{\infty}$ by

$$g_0 = \|Y_0\|, \quad (m+1)g_{m+1} = \sum_{k+l=m} \frac{M}{r_1{}^k} g_l. \tag{3.6}$$

We show that the power series $\sum_{m=0}^{\infty} g_m x^m$ is a majorant series to $\sum_{m=0}^{\infty} Y_m x^m$ by induction. Assume that $\|Y_l\| \leq g_l$ holds for $1 \leq l \leq m$. Then we have

$$\|Y_{m+1}\| = \left\| \frac{1}{m+1} \sum_{k+l=m} A_k Y_l \right\|$$

$$\leq \frac{1}{m+1} \sum_{k+l=m} \|A_k\| \cdot \|Y_l\|$$

$$\leq \frac{1}{m+1} \sum_{k+l=m} \frac{M}{r_1{}^k} g_l$$

$$= g_{m+1},$$

which implies that $\|Y_m\| \leq g_m$ holds for all m. Thus, if we set

$$g(x) = \sum_{m=0}^{\infty} g_m x^m,$$

$Y(x)$ is convergent in a domain where $g(x)$ is convergent. We see that the right hand side of the recurrence formula (3.6) is the coefficient of x^m in the product of

$$\sum_{m=0}^{\infty} \frac{M}{r_1{}^m} x^m = \frac{M}{1 - \dfrac{x}{r_1}}$$

and $g(x)$, and the left hand side of (3.6) is the coefficient of x^m in $g'(x)$. This implies that $g(x)$ satisfies the differential equation

$$g' = \frac{M}{1 - \dfrac{x}{r_1}} g.$$

Moreover $g(x)$ satisfies the initial condition $g(0) = ||Y_0||$. Then we can expect that

$$g(x) = ||Y_0|| \left(1 - \frac{x}{r_1}\right)^{-r_1 M} \tag{3.7}$$

holds, where the branch of the power in the right hand side is so determined that the argument at $x = 0$ is 0. We can justify the above observation as follows. The coefficients in the Taylor expansion of the right hand side of (3.7) at $x = 0$ satisfy the recurrence relation (3.6), and, on the other hand, the series $\{g_m\}$ is uniquely determined by (3.6). Then $g(x)$ should coincide with the right hand side of (3.7).

Since $g(x)$ converges in $|x| < r_1$, $Y(x)$ converges in the same domain. Note that r_1 is an arbitrary positive number satisfying $r_1 < r$. Therefore we conclude that $Y(x)$ converges in $|x| < r$. □

From this theorem, we can derive two important properties of linear ordinary differential equations. One is the following result on the domain of definition of the solutions.

Theorem 3.2 *Any solution at a $\in D$ of the differential equation (3.1) can be continued analytically along any curve in D with the initial point a, and the result is also a solution of (3.1).*

Proof For the definition and basic properties of analytic continuations, please refer to the excellent description in Siegel [166].

Let C be a curve in D with the initial point a, and $\varphi(t)$ ($t \in [0, 1]$) its regular parametrization. Since C is a compact subset of the open set D, we can take a partition $0 = t_0 < t_1 < t_2 < \cdots < t_{n-1} < t_n = 1$ of $[0, 1]$ such that, by denoting $\varphi(t_i) = a_i$ ($0 \le i \le n$) and the part of C from a_j to a_{j+1} by C_j, there exists $r_i > 0$ for each i with the property $C_i \subset B(a_i; r_i) \subset D$. We set $B(a_i; r_i) = B_i$. It is enough to show that the direct continuation from B_i to B_{i+1} is possible for each i.

Let $Y_0(x)$ be a solution of (3.1) at $x = a$. Then $Y_0(x)$ is holomorphic in B_0. Let $Y_1(x)$ the solution of (3.1) at $x = a_1$ satisfying the initial condition

$$Y(a_1) = Y_0(a_1).$$

We know that such solution is unique and holomorphic in B_1. Since $Y_0(x)$ itself also satisfy the same initial condition, by the uniqueness we have

$$Y_0(x) = Y_1(x) \quad (x \in B_0 \cap B_1).$$

Thus $Y_1(x)$ is a direct continuation of $Y_0(x)$ to B_1.

In a similar way, we can show that, for each i, a solution in B_i can be continued to a solution in B_{i+1}, and hence $Y_0(x)$ can be analytically continued along C. The result of the analytic continuation does not depends on the partition of C. Since the analytic continuation is commutative with addition, multiplication and differentiation, we see that the result of the analytic continuation is also a solution of the differential equation (3.1).   □

By Theorem 3.2, we see that any solution of the system (3.1) is defined in the universal covering $\tilde{D}$ of D. In other words, it is a multi-valued holomorphic function on D.

Since any holomorphic function in a simply connected domain is single-valued, we get the following assertion.

Theorem 3.3 *Let a, b be two points in D. If two curves C_1, C_2 with the common initial point a and the common end point b are homotopic in D under fixing the initial point and the end point, the results of the analytic continuations of a solution of (3.1) at the point a along C_1 and that along C_2 coincide.*

We denote the result of the analytic continuation of a function $f(x)$ along a curve γ by $\gamma_* f(x)$.

The second important property derived from Theorem 3.1 is the following.

Theorem 3.4 *For the system of differential equation (3.1), the set of all solutions in $B(a; r)$ makes an n-dimensional vector space over $\mathbb{C}$.*

The assertion readily follows from the fact that any solution on $B(a; r)$ is uniquely determined by the initial value $Y_0 \in \mathbb{C}^n$.

In order to obtain a basis $Y_1(x), Y_2(x), \ldots, Y_n(x)$ of the vector space of the solutions, we take n linearly independent vectors $Y_{01}, Y_{02}, \ldots, Y_{0n} \in \mathbb{C}^n$ and, for each i, define $Y_i(x)$ by the initial condition

$$Y(a) = Y_{0i}.$$

By this construction, the vectors $Y_1(a), Y_2(a), \ldots, Y_n(a) \in \mathbb{C}^n$ are evidently linearly independent. Moreover, we can show that the values $Y_1(x), Y_2(x), \ldots, Y_n(x) \in \mathbb{C}^n$ at any point $x \in \tilde{D}$ are also linearly independent.

For n solutions $Y_1(x), Y_2(x), \ldots, Y_n(x)$ of the system (3.1), we consider the determinant

$$w(x) = |Y_1(x), Y_2(x), \ldots, Y_n(x)|.$$

This is an analogue of the Wronskian (3.10) which will be introduced later for a scalar equation. The following fact is substantial.

Lemma 3.5 *Let* $Y_1(x), Y_2(x), \ldots, Y_n(x)$ *be solutions of (3.1) in a common domain of definition. Then their determinant* $w(x)$ *satisfies the differential equation*

$$\frac{dw}{dx} = \operatorname{tr} A(x)\, w$$

of the first order.

Proof We write $A(x) = (a_{ij}(x))$, and, for each i, denote by u_i the i-th row of the $n \times n$-matrix $(Y_1(x), Y_2(x), \ldots, Y_n(x))$. Since each column $Y_j(x)$ is a solution of the system (3.1), we have

$$u_i' = \sum_{j=1}^{n} a_{ij} u_j.$$

Noting this, we get

$$w' = \begin{vmatrix} u_1' \\ u_2 \\ \vdots \\ u_n \end{vmatrix} + \begin{vmatrix} u_1 \\ u_2' \\ \vdots \\ u_n \end{vmatrix} + \cdots + \begin{vmatrix} u_1 \\ u_2 \\ \vdots \\ u_n' \end{vmatrix}$$

$$= a_{11} \begin{vmatrix} u_1 \\ u_2 \\ \vdots \\ u_n \end{vmatrix} + a_{22} \begin{vmatrix} u_1 \\ u_2 \\ \vdots \\ u_n \end{vmatrix} + \cdots + a_{nn} \begin{vmatrix} u_1 \\ u_2 \\ \vdots \\ u_n \end{vmatrix}$$

$$= \operatorname{tr} A\, w.$$

$\square$

Theorem 3.6 *Let* $Y_1(x), Y_2(x), \ldots, Y_n(x)$ *be solutions of the system (3.1) in a common domain of definition. If their determinant* $w(x)$ *vanishes at a point of the domain, it vanishes at any point of the domain.*

Proof By solving the differential equation in Lemma 3.5, we get

$$w(x) = w(a)e^{\int_a^x \operatorname{tr} A(t)\, dt}.$$

The assertion can be readily obtained from this expression.   □

This theorem assures that a tuple of n solutions linearly independent at a point of the domain of definition is linearly independent at any point of the domain $\tilde{D}$ of definition. Combining the above consideration, we obtain the following fundamental properties of the set of the solutions of the differential equation (3.1).

Theorem 3.7 *We consider the system differential equations (3.1) defined on the domain D with unknown vector of size n. The followings hold.*

(i) *The set of all solutions on the universal covering space $\tilde{D}$ of D makes an n dimensional vector space over $\mathbb{C}$.*

(ii) *For any simply connected domain D_1 of D, the set of all solutions on D_1 makes an n dimensional vector space over $\mathbb{C}$.*

The theorem asserts that the rank of the differential equation (3.1) is n. A tuple of n solutions linearly independent at a point of the domain of definition gives a basis of the space of the solutions on $\tilde{D}$ or D_1. We call such a tuple of solutions

$$\mathcal{Y}(x) = (Y_1(x), Y_2(x), \ldots, Y_n(x))$$

a *fundamental system of solutions*. The fundamental system of solutions can be regarded as an $n \times n$ matrix, which we also call a *fundamental matrix solution*.

For a scalar differential equation

$$y^{(n)} + p_1(x)y^{(n-1)} + p_2(x)y^{(n-2)} + \cdots + p_n(x)y = 0 \tag{3.8}$$

of higher order, we can apply the results for the system (3.1) by transforming it to the system (2.4) by the transformation (2.3). The condition that $A(x)$ is holomorphic on D corresponds to the condition that $p_1(x), p_2(x), \ldots, p_n(x)$ are holomorphic on D. The initial condition (3.2) corresponds to prescribing the values of $y(a), y'(a), \ldots, y^{(n-1)}(a)$. Then we have the following assertion.

Theorem 3.8 *We consider the scalar differential equation (3.8), where $p_1(x), p_2(x), \ldots, p_n(x)$ are holomorphic on a domain D. Let a be a point in D, and take $r > 0$ so that $B(a; r) \subset D$. For any values $y_0^0, y_1^0, \ldots, y_{n-1}^0 \in \mathbb{C}$, there exists a unique solution $y(x)$ of (3.8) satisfying the initial condition*

$$y(a) = y_0^0, \ y'(a) = y_1^0, \ldots, y^{(n-1)}(a) = y_{n-1}^0, \tag{3.9}$$

and the solution $y(x)$ is holomorphic on $B(a; r)$.

Similar assertions as Theorems 3.2, 3.3, and 3.4 can be obtained directly under the assumption that $p_1(x), p_2(x), \ldots, p_n(x)$ are holomorphic on D.

For a scalar differential equation of higher order, we can define the Wronskian which corresponds to the determinant of the fundamental matrix solution of the system (3.1). In general, for a set of functions $z_1(x), z_2(x), \ldots, z_k(x)$ with a common domain of definition, the determinant

$$
W(z_1, z_2, \ldots, z_k)(x) =
\begin{vmatrix}
z_1(x) & z_2(x) & \cdots & z_k(x) \\
z_1'(x) & z_2'(x) & \cdots & z_k'(x) \\
\vdots & \vdots & & \vdots \\
z_1^{(k-1)}(x) & z_2^{(k-1)}(x) & \cdots & z_k^{(k-1)}(x)
\end{vmatrix}
\tag{3.10}
$$

is called the *Wronskian* of $z_1(x), z_2(x), \ldots, z_k(x)$. By using the correspondence (2.3), we can derive the following assertion from Lemma 3.5.

Lemma 3.9 *Let $y_1(x), y_2(x), \ldots, y_n(x)$ be a set of solutions of (3.8) with a common domain of definition. Then the Wronskian*

$$
w(x) = W(y_1, y_2, \ldots, y_n)(x)
$$

satisfies the differential equation

$$
\frac{dw}{dx} + p_1(x)w = 0
$$

of the first order.

Thus we have similar assertions as Theorems 3.6 and 3.7. Namely, the Wronskian of a set of n solutions $y_1(x), y_2(x), \ldots, y_n(x)$ never vanish on the domain $\tilde{D}$ of definition or vanish identically on $\tilde{D}$. If it does not vanish, the set is said to be linearly independent. The set of all solutions on the universal covering space $\tilde{D}$ or on a simply connected domain D_1 in D makes a vector space of dimension n. A tuple of n linearly independent solutions

$$
\mathcal{Y}(x) = (y_1(x), y_2(x), \ldots, y_n(x)),
$$

which is an n-row vector, is called a *fundamental system of solutions*.

Chapter 4
Regular Singular Points

4.1 Definition of Regular Singularity

Definition 4.1 Let a be a point in $\mathbb{C}$, and $f(x)$ a (multi-valued) holomorphic function in a neighborhood of a except a. The point $x = a$ is said to be a *regular singular point* of $f(x)$ if $f(x)$ is not holomorphic at $x = a$, and if there exists a positive number N such that, for any $\theta_1 < \theta_2$,

$$|x - a|^N |f(x)| \to 0 \quad (|x - a| \to 0,\ \theta_1 < \arg(x - a) < \theta_2) \qquad (4.1)$$

holds. For a function $f(x)$ holomorphic in a neighborhood of ∞ of the Riemann sphere except ∞, the point $x = \infty$ is said to be a *regular singular point* of $f(x)$ if, for a local coordinate t at ∞, $t = 0$ is a regular singular point of $f(x(t))$. If $x = a$ is a regular singular point or a regular point, we say that $x = a$ is at most a regular singular point. A point which is not a regular point nor a regular singular point is said to be an *irregular singular point*.

Example 4.1 For any $\rho = \lambda + \sqrt{-1}\mu \in \mathbb{C} \setminus \mathbb{Z}_{\geq 0}$, the function $(x - a)^\rho$ is regular singular at $x = a$. In fact, by the equality

$$|(x - a)^\rho| = |e^{\rho \log(x-a)}|$$
$$= |e^{(\lambda + \sqrt{-1}\mu)(\log|x-a| + \sqrt{-1}\arg(x-a))}|$$
$$= e^{\lambda \log|x-a| - \mu \arg(x-a)}$$
$$= |x - a|^\lambda e^{-\mu \arg(x-a)},$$

(4.1) holds for $N > |\lambda|$ whenever $\arg(x - a)$ is bounded.

© The Editor(s) (if applicable) and The Author(s), under exclusive
license to Springer Nature Switzerland AG 2020
Y. Haraoka, *Linear Differential Equations in the Complex Domain*, Lecture Notes
in Mathematics 2271, https://doi.org/10.1007/978-3-030-54663-2_4

The function $\log(x - a)$ also is regular singular at $x = a$. In fact, by the equality

$$|\log(x - a)| = \sqrt{(\log|x - a|)^2 + (\arg(x - a))^2}.$$

(4.1) holds for any $N > 0$ whenever $\arg(x - a)$ is bounded.

The following fact is useful.

Theorem 4.1 *If $x = a$ is a regular singular point of $f(x)$, and if $f(x)$ is single-valued around $x = a$, then $x = a$ is a pole of $f(x)$.*

Proof In the inequality (4.1), by taking N bigger if necessary, we may assume that N is a positive integer. Then the function $(x-a)^N f(x)$ is single-valued and bounded at $x = a$, and hence holomorphic at $x = a$. The assertion immediately follows.   □

Theorem 4.2 *If $f(x)$, $g(x)$ are at most regular singular at $x = a$, then $f(x) \pm g(x)$, $f(x)g(x)$ and $f'(x)$ are at most regular singular at $x = a$.*

Proof The assertion for the sum and the product are easily obtained. We shall show the assertion for $f'(x)$.

Take a point x in a neighborhood of a, and fix a branch of $f(x)$ at x. Then we have

$$f'(x) = \frac{1}{2\pi\sqrt{-1}} \int_{|\zeta - x| = |x - a|/2} \frac{f(\zeta)}{(\zeta - x)^2} d\zeta.$$

By applying the inequality (4.1) to $f(\zeta)$ in the integrand, we obtain

$$|x - a|^{N+1}|f'(x)| \to 0$$

if $\arg(x - a)$ is bounded.   □

We call a singular point of the coefficients of a linear ordinary differential equation a *singular point* of the differential equation. In this book, we consider only poles as the singular points of the coefficients.

Definition 4.2 Let a be a point in $\mathbb{C}$, and $p_1(x)$, $p_2(x)$, $\ldots$, $p_n(x)$ meromorphic at $x = a$. The point $x = a$ is called a *regular singular point* of the differential equation

$$y^{(n)} + p_1(x)y^{(n-1)} + p_2(x)y^{(n-2)} + \cdots + p_n(x)y = 0, \tag{4.2}$$

if $x = a$ is a singular point of at least one of $p_1(x)$, $p_2(x)$, $\ldots$, $p_n(x)$ and if, for any solution $y(x)$ of the Eq. (4.2), $x = a$ is at most a regular singular point of $y(x)$.

We have a similar definition for a system of differential equations.

Definition 4.3 Let $x = a$ be a pole of a matrix function $A(x)$. The point $x = a$ is called a *regular singular point* of the system of differential equations

$$\frac{dY}{dx} = A(x)Y \tag{4.3}$$

if, for any solution $Y(x)$ of the system (4.3), $x = a$ is at most a regular singular point of every element of $Y(x)$.

4.2 Fuchs' Theorem and Analysis at Regular Singular Point

According to Definitions 4.2 and 4.3, it is necessary to know the behaviors of all solutions to show the regular singularity. However, for a scalar differential equation, we have an effective criterion for the regular singularity.

Theorem 4.3 (Fuchs' Theorem) *A singular point $x = a$ of a scalar differential equation (4.2) is a regular singular point if and only if, for each j $(1 \le j \le n)$, $x = a$ is a pole of $p_j(x)$ of order at most j.*

Proof "only if" part: Assume that $x = a$ is at most a regular singular point for any solution of (4.2). Take $r > 0$ so small that $p_1(x), p_2(x), \ldots, p_n(x)$ are holomorphic in $D = \{0 < |x - a| < r\}$. Let b be a point in D, and U a disk centered at b and contained in D. Take a fundamental system of solutions $\mathcal{Y}(x) = (y_1(x), y_2(x), \ldots, y_n(x))$ of (4.2) in U. For the curve

$$\gamma : x = a + (b - a)e^{\sqrt{-1}\theta} \quad (0 \le \theta \le 2\pi),$$

the analytic continuation $\gamma_* \mathcal{Y}(x)$ along γ becomes again a fundamental system of solutions. Then there exists $M \in GL(n, \mathbb{C})$ such that $\gamma_* \mathcal{Y}(x) = \mathcal{Y}(x)M$ holds (cf. Theorem 4.10). Let g be an eigenvalue of M, and v an eigenvector for g. If we set

$$y_0(x) = \mathcal{Y}(x)v,$$

we have

$$\gamma_* y_0(x) = \gamma_* \mathcal{Y}(x)v = \mathcal{Y}(x)Mv = \mathcal{Y}(x)gv = gy_0(x).$$

Thus $y_0(x)$ is a solution holomorphic on D with this multi-valued property. Set

$$\rho = \frac{1}{2\pi\sqrt{-1}} \log g,$$

where the branch of $\log g$ will be specified soon later. If we set

$$z(x) = (x - a)^{-\rho} y_0(x),$$

$z(x)$ is regular singular at $x = a$, and moreover is single-valued around $x = a$ because

$$\gamma_* z(x) = e^{-2\pi\sqrt{-1}\rho}(x - a)^{-\rho} g y_0(x) = z(x)$$

holds. Then, thanks to Theorem 4.1, $x = a$ is at most a pole of $z(x)$. If we change the branch of $\log g$, an integer is added to ρ. Then we can choose a branch of $\log g$ so that the order of $z(x)$ at $x = a$ is 0. Thus we may assume that $z(x)$ is holomorphic at $x = a$ and $z(a) \neq 0$. Summing up the above, we have a solution of (4.2) of the form

$$y_0(x) = (x - a)^{\rho} z(x), \tag{4.4}$$

where $\rho \in \mathbb{C}$ and $z(x)$ is holomorphic on $|x - a| < r$ and satisfies $z(a) \neq 0$.

Now we shall show the "only if" part of the assertion by induction on n.

For the case $n = 1$, by using the solution (4.4), we have

$$p_1(x) = -\frac{y_0'(x)}{y_0(x)} = -\frac{\rho}{x - a} - \frac{z'(x)}{z(x)}.$$

Then $x = a$ is a pole of $p_1(x)$ of order at most 1 since $z(a) \neq 0$.

We assume that the "only if" part of the assertion holds for any differential equations of order at most $n - 1$. Consider the solution $y_0(x)$ of (4.2) given by (4.4). For a solution $y(x)$ of the differential equation (4.2), the function $u(x) = y(x)/y_0(x)$ satisfies an n-th order linear ordinary differential equation, which has $x = a$ as at most a regular singular point. We know that $u = 1$ is a solution of the differential equation, and hence the differential equation can be written as

$$u^{(n)} + q_1(x)u^{(n-1)} + \cdots + q_{n-1}(x)u' = 0. \tag{4.5}$$

Then we obtain the differential equation

$$v^{(n-1)} + q_1(x)v^{(n-2)} + \cdots + q_{n-1}(x)v = 0 \tag{4.6}$$

satisfied by $v = u'$, and $x = a$ is also a regular singular point of this equation. Then by the assumption of the induction, for each $1 \leq i \leq n - 1$, $x = a$ is a pole of $q_i(x)$ of order at most i. We shall write $p_j(x)$ in terms of $q_i(x)$. From $u = y/y_0(x)$, we obtain

$$u^{(i)} = \sum_{k=0}^{i} \binom{i}{k} \left(y_0(x)^{-1}\right)^{(i-k)} y^{(k)}.$$

Putting these into (4.5) and collecting the coefficients of $y^{(j)}$, we get

$$p_j(x) = y_0(x) \sum_{i=n-j}^{n} \binom{i}{n-j} q_{n-i}(x)(y_0(x)^{-1})^{(i-n+j)}.$$

By the definition (4.4) of $y_0(x)$, we see that $x = a$ is a pole of $y_0(x)(y_0(x)^{-1})^{(k)}$ of order k. Since we know that $x = a$ is a pole of $q_i(x)$ of order at most i, $x = a$ is a pole of $p_j(x)$ of order at most j. This completes the proof of the "only if" part.

The "if" part of the assertion will be shown by using the following two theorems. We give the proof after the proof of Theorem 4.5. $\qquad\square$

For $a \in \mathbb{C}$ and $r > 0$, we denote by $\mathcal{M}_{a,r}$ the ring of the holomorphic functions on $0 < |x - a| < r$ which are meromorphic at $x = a$.

Theorem 4.4 *Let $A(x)$ be an $n \times n$ matrix function with entries meromorphic at $x = a$. The system of differential equations*

$$\frac{dY}{dx} = A(x)Y \tag{4.7}$$

is regular singular at $x = a$ if and only if, for a sufficiently small $r > 0$, there exists $P(x) \in \mathrm{GL}(n, \mathcal{M}_{a,r})$ such that, by the transformation

$$Y = P(x)Z, \tag{4.8}$$

the coefficient $B(x)$ of the transformed system

$$\frac{dZ}{dx} = B(x)Z \tag{4.9}$$

has a pole $x = a$ of order 1.

Proof First we show the "only if" part. We set $Y = {}^t(y_1, y_2, \ldots, y_n)$. By using a cyclic vector, we can take a linear combination y of $y_1, y_2, \ldots, y_n$ so that the system (4.7) is transformed into a scalar differential equation (4.2) for y. We see that the linear transformation from ${}^t(y_1, y_2, \ldots, y_n)$ to ${}^t(y, y', \ldots, y^{(n-1)})$ belongs to $\mathrm{GL}(n, \mathcal{M}_{a,r})$ if we take $r > 0$ sufficiently small. In particular, the coefficients $p_j(x)$ $(1 \leq j \leq n)$ are holomorphic in $0 < |x - a| < r$. Since $x = a$ is a regular singular point of the system (4.7), it is also a regular singular point of the scalar equation (4.2). Then by the "only if" part of Theorem 4.3 which we have proved, $x = a$ is a pole of $p_j(x)$ of order at most j. Set

$$P_j(x) = (x - a)^j p_j(x) \quad (1 \leq j \leq n).$$

Then $P_j(x)$ is holomorphic at $x = a$, and the Eq. (4.2) can be written as

$$(x - a)^n y^{(n)} + P_1(x)(x - a)^{n-1} y^{(n-1)} + \cdots + P_{n-1}(x)(x - a)y' + P_n(x)y = 0.$$
(4.10)

Now we define a vector $Z = {}^t(z_1, z_2, \ldots, z_n)$ of new unknowns by

$$\begin{cases} z_1 = y, \\ \\ z_2 = (x - a)y', \\ \\ \vdots \\ \\ z_j = (x - a)^{j-1} y^{(j-1)}, \\ \\ \vdots \\ \\ z_n = (x - a)^{n-1} y^{(n-1)}. \end{cases}$$
(4.11)

By this definition, we get

$$z_1' = \frac{1}{x - a} z_2,$$

$$z_j' = \frac{j - 1}{x - a} z_j + \frac{1}{x - a} z_{j+1} \quad (2 \le j \le n - 1).$$

Moreover, by using the Eq. (4.10), we obtain

$$z_n' = (n - 1)(x - a)^{n-2} y^{(n-1)} + (x - a)^{n-1} y^{(n)}$$

$$= \frac{n - 1}{x - a} z_n - \frac{1}{x - a} (P_1 z_n + \cdots + P_{n-1} z_2 + P_n z_1).$$

Thus Z satisfies the system of the form (4.9), where the entries of the coefficient matrix $B(x)$ have poles at $x = a$ of order at most 1. Since the transformation from $(y_1, y_2, \ldots, y_n)$ to $(y, y', \ldots, y^{(n-1)})$ belongs to $GL(n, \mathcal{M}_{a,r})$ and the transformation (4.11) also belongs to $GL(n, \mathcal{M}_{a,r})$, the transformation matrix $P(x)$ from Y to Z belongs to $GL(n, \mathcal{M}_{a,r})$. This completes the proof of the "only if" part.

The "if" part is proved by showing that any solution of the system (4.9) is regular singular at $x = a$ if $B(x)$ has a pole at $x = a$ of order 1. Then it is enough to construct a fundamental system of solutions of the system (4.9) which are regular singular at $x = a$. This will be done in the next theorem. □

In the next theorem we give a fundamental system of solutions of the system (4.9) at a regular singular point. To describe the result, we prepare the following notion.

Definition 4.4 Let A be a square matrix. A is said to be *non-resonant* if there is no integral difference among distinct eigenvalues, and is said to be *resonant* otherwise. Namely, A is resonant if it has eigenvalues λ and $\lambda + k$ with $k \in \mathbb{Z} \setminus \{0\}$.

For example, the matrices A and B satisfying

$$A \sim \begin{pmatrix} \lambda & & \\ & \lambda & \\ & & \mu \end{pmatrix}, \quad B \sim \begin{pmatrix} \lambda & 1 & \\ & \lambda & \\ & & \mu \end{pmatrix}$$

with $\lambda - \mu \notin \mathbb{Z}$ are non-resonant, while the matrix C satisfying

$$C \sim \begin{pmatrix} \lambda & & \\ & \lambda + 1 & \\ & & \mu \end{pmatrix}$$

is resonant.

Theorem 4.5 *Let $B(x)$ be an $n \times n$ matrix function which are meromorphic at $x = a$ and has a pole at $x = a$ of order 1. We expand $B(x)$ into the Laurent series*

$$B(x) = \frac{1}{x - a} \sum_{m=0}^{\infty} B_m (x - a)^m \tag{4.12}$$

at $x = a$. We consider a system of differential equations

$$\frac{dZ}{dx} = B(x)Z. \tag{4.13}$$

(i) *Assume that B_0 is non-resonant. Then the system (4.13) has a fundamental matrix solution of the form*

$$Z(x) = F(x)(x - a)^{B_0}, \quad F(x) = I + \sum_{m=1}^{\infty} F_m (x - a)^m. \tag{4.14}$$

   The series $F(x)$ converges in the domain where the Laurent series of $B(x)$ converges.

(ii) *For a general B_0, we divide the eigenvalues of B_0 into classes such that the eigenvalues in the same class are congruent modulo integers. For each class, we take the eigenvalue ρ with minimal real part as the representative. Then the eigenvalues of the class can be given by*

$$\{\rho + k_0, \rho + k_1, \rho + k_2, \ldots, \rho + k_d\},$$

where $0 = k_0 < k_1 < k_2 < \cdots < k_d$ are integers. We denote the Jordan canonical form of B_0 by

$$B_0^J = \bigoplus_\rho B_{0,\rho}, \qquad (4.15)$$

where ρ stands for a representative of a class, and $B_{0,\rho}$ denotes the direct sum of Jordan cells with the eigenvalues contained in the class. Namely we have

$$B_{0,\rho} = \bigoplus_{i=0}^d B_{0,\rho,i}, \qquad (4.16)$$

$$B_{0,\rho,i} = \bigoplus_j J(\rho + k_i; p_{ij}),$$

where p_{ij} is a positive integer. For each i we set

$$p_i = \max_j p_{ij},$$

and set

$$q = \sum_{i=0}^d p_i - 1.$$

Then the system (4.13) has a fundamental system of solutions consisting of the solutions of the form

$$(x - a)^\rho \sum_{j=0}^q (\log(x - a))^j \sum_{m=0}^\infty Z_{jm}(x - a)^m. \qquad (4.17)$$

We decompose a vector $v \in \mathbb{C}^n$ into a direct sum corresponding to the direct sum decomposition (4.16) of B_0^J, and denote the component corresponding to $B_{0,\rho,i}$ by $[v]_{\rho,i}$. Let $P \in GL(n, \mathbb{C})$ be a matrix such that $P^{-1} B_0 P = B_0^J$. Then the solution (4.17) is uniquely determined by prescribing

$$\left[P^{-1} Z_{00} \right]_{\rho,0}, \left[P^{-1} Z_{0k_1} \right]_{\rho,1}, \ldots, \left[P^{-1} Z_{0k_d} \right]_{\rho,d},$$

and depends on these data linearly. The series $\sum_{m=0}^\infty Z_{jm}(x - a)^m$ $(0 \le j \le q)$ which appear in the solution (4.17) converge in the domain where the Laurent series of $B(x)$ converges.

Definition 4.5 For the system of differential equations (4.13), the eigenvalues of B_0 are called the *characteristic exponents* at $x = a$. For the solution (4.17) of the system (4.13), $\rho + k_i$ is called the *characteristic exponent* if

$$\left[P^{-1} Z_{0k_j} \right]_{\rho,j} = 0 \ (0 \le j < i), \quad \left[P^{-1} Z_{0k_i} \right]_{\rho,i} \ne 0$$

hold.

Remark 4.1 The characteristic exponents defined by Definition 4.5 are not intrinsic quantities. We may translate eigenvalues of B_0 by integers by operating a linear transformation with coefficients meromorphic at $x = a$ to the system (4.13). The equivalence class of the eigenvalues of B_0 modulo integers are intrinsic, and then it may be natural to define the characteristic exponents by the equivalence classes. In a similar way, it is sometimes possible to transform a resonant B_0 to non-resonant one.

Proof of Theorem 4.5 Without loss of generality, we may assume that $a = 0$.

We prove (i). Let F_m $(m = 1, 2, \dots)$ be an $n \times n$ matrix of constant entries, and set

$$F(x) = \sum_{m=0}^{\infty} F_m x^m.$$

We are going to construct the matrix solution of the form

$$Z(x) = F(x) x^{B_0}. \tag{4.18}$$

Putting (4.18) into the system (4.13), we get

$$F'(x) x^{B_0} + \frac{1}{x} F(x) B_0 x^{B_0} = B(x) F(x) x^{B_0}.$$

Multiply x^{-B_0} to both sides from the right. If we can differentiate $F(x)$ term by term, we have

$$\sum_{m=1}^{\infty} m F_m x^{m-1} + \sum_{m=0}^{\infty} F_m B_0 x^{m-1} = \left(\sum_{m=0}^{\infty} B_m x^{m-1} \right) \left(\sum_{m=0}^{\infty} F_m x^m \right).$$

We shall compare the coefficients of x^m in the both sides. First we compare the coefficients of x^{-1} to get

$$F_0 B_0 = B_0 F_0.$$

This can be fulfilled by taking

$$F_0 = I.$$

Next we compare the coefficients of x^{m-1} $(m \geq 1)$ to get

$$m F_m + F_m B_0 = B_0 F_m + \sum_{k=0}^{m-1} B_{m-k} F_k.$$

We rewrite this equation as

$$F_m (B_0 + m) - B_0 F_m = \sum_{k=0}^{m-1} B_{m-k} F_k, \qquad (4.19)$$

where we regard $F_0, F_1, \ldots, F_{m-1}$ as given ones and F_m as an unknown. Since B_0 is non-resonant, there is no common eigenvalue of $B_0 + m$ and B_0. Then, owing to Lemma 2.7 which we shall show later, the Eq. (4.19) has a unique solution F_m. This shows that the formal series $F(x)$ can be determined, and thus we obtain a formal solution (4.18).

Now we show the convergence of the formal series $F(x)$ by constructing a majorant series. Let r be the radius of convergence of the series $\sum_{m=0}^{\infty} B_m x^m$, and take any r_1 satisfying $0 < r_1 < r$. Then there exists a constant $M > 0$ such that

$$\|B_m\| \leq \frac{M}{r_1^m} \quad (m \geq 0)$$

holds. By the property of the norm of matrices, we have

$$\|F_m B_0 - B_0 F_m\| \leq 2\|B_0\| \, \|F_m\|.$$

Then, for any $\varepsilon > 0$,

$$\frac{\|F_m B_0 - B_0 F_m\|}{m\|F_m\|} < \varepsilon$$

holds for sufficiently large m if $\|F_m\| \neq 0$. Then we have the inequality

$$\|F_m (B_0 + m) - B_0 F_m\| = m\|F_m\| \left\| \frac{F_m}{\|F_m\|} + \frac{F_m B_0 - B_0 F_m}{m\|F_m\|} \right\|$$

$$\geq m\|F_m\|(1 - \varepsilon)$$

for sufficiently large m. Note that the inequality among the left hand side and the right hand side holds also for the case $||F_m|| = 0$. Thus we see that there exist a positive integer m_0 and a constant $c > 0$ such that

$$||F_m(B_0 + m) - B_0 F_m|| \geq cm||F_m|| \tag{4.20}$$

holds for any $m \geq m_0$. Now we define a sequence $\{g_m\}$ by

$$g_m = ||F_m|| \qquad (0 \leq m \leq m_0),$$

$$cmg_m = \sum_{k=0}^{m-1} \frac{M}{r_1^{m-k}} g_k \qquad (m \geq m_0).$$

Then by using (4.20) and (4.19), we can show that

$$||F_m|| \leq g_m$$

holds for any m. Thus, if we set

$$g(x) = \sum_{m=0}^{\infty} g_m x^m,$$

$g(x)$ is a majorant series of $F(x)$. We shall show the convergence of $g(x)$. Compute

$$\sum_{m=1}^{\infty} cmg_m x^{m-1}$$

in two ways by using the definition of g_m. Then we obtain the differential equation

$$g'(x) = \frac{\frac{M}{cr_1}}{1 - \frac{x}{r_1}} g(x) + h(x),$$

where $h(x)$ is a rational function which has the only pole at $x = r_1$. The general solution of the differential equation is given by

$$g(x) = \left(1 - \frac{x}{r_1}\right)^{-\frac{M}{c}} \int \left(1 - \frac{x}{r_1}\right)^{\frac{M}{c}} h(x)\, dx,$$

and is holomorphic in $|x| < r_1$. This implies that $F(x)$ is convergent in $|x| < r_1$. Since $r_1 < r$ is arbitrary, we find that $F(x)$ is convergent in $|x| < r$.

Note that, since we have shown the convergence of $F(x)$, the derivation of the formal series using term by term differentiation is justified.

Next we prove (ii). By a transformation

$$Z = P Z_1$$

with $P \in \mathrm{GL}(n, \mathbb{C})$, we have a system of differential equations

$$\frac{d Z_1}{d x} = P^{-1} B(x) P Z_1$$

for Z_1. Then we may assume that B_0 is already of the Jordan canonical form B_0^J.

Take the representative ρ of a class of eigenvalues of B_0. For the sake of simplicity, in the direct sum decomposition of $B_0^J = B_0$, the component $B_{0,\rho}$ is assumed to be in the first position. Then we can set

$$B_0 = B_{0,\rho} \oplus B_0'$$

$$= B_{0,\rho,0} \oplus B_{0,\rho,1} \oplus \cdots \oplus B_{0,\rho,l} \oplus B_0',$$

where B_0' denotes the direct sum of the other components. We decompose $v \in \mathbb{C}^n$ according to this decomposition, and denote $[v]_{\rho,i}$ simply by $[v]_i$. We also denote the component of v corresponding to B_0' by $[v]'$. Thus we have

$$v = \begin{pmatrix} [v]_0 \\ [v]_1 \\ \vdots \\ [v]_d \\ [v]' \end{pmatrix}.$$

Since we put $a = 0$, the expression (4.17) of a solution becomes

$$Z(x) = x^\rho \sum_{j=0}^{q} (\log x)^j \sum_{m=0}^{\infty} Z_{jm} x^m. \tag{4.21}$$

We put it to the system (4.13). The left hand side becomes

$$\frac{d Z}{d x} = \sum_{j=0}^{q} (\log x)^j \sum_{m=0}^{\infty} (\rho + m) Z_{jm} x^{\rho+m-1} + \sum_{j=1}^{q} j (\log x)^{j-1} \sum_{m=0}^{\infty} Z_{jm} x^{\rho+m-1}.$$

Comparing the coefficients of the both sides of (4.13) for each degree of $\log x$ and each exponent of x, we get the following relations. First, the comparison of terms without $\log x$ yields

$$(B_0 - \rho)Z_{00} = Z_{10}, \tag{4.22}$$

$$(B_0 - (\rho + m))Z_{0m} = Z_{1m} - \sum_{t=0}^{m-1} B_{m-t} Z_{0t} \quad (m > 0). \tag{4.23}$$

Next, the comparison of terms with $(\log x)^j$ for $1 \le j \le q - 1$ yields

$$(B_0 - \rho)Z_{j0} = (j + 1)Z_{j+1,0}, \tag{4.24}$$

$$(B_0 - (\rho + m))Z_{jm} = (j + 1)Z_{j+1,m} - \sum_{t=0}^{m-1} B_{m-t} Z_{jt} \quad (m > 0). \tag{4.25}$$

Finally we get

$$(B_0 - \rho)Z_{q0} = 0, \tag{4.26}$$

$$(B_0 - (\rho + m))Z_{qm} = - \sum_{t=0}^{m-1} B_{m-t} Z_{qt} \quad (m > 0) \tag{4.27}$$

by the comparison of terms with $(\log x)^q$.

Take any value of $[Z_{00}]_0$, and set

$$Z_{00} = \begin{pmatrix} [Z_{00}]_0 \\ 0 \end{pmatrix}.$$

Define Z_{j0} $(1 \le j \le q)$ by

$$Z_{j0} = \frac{1}{j!}(B_0 - \rho)^j Z_{00},$$

then the relations (4.22) and (4.24) hold, and moreover, noting $(B_{0,\rho,0} - \rho)^{p_0} = O$ and $q \ge p_0 - 1$, the relation (4.26) also holds. We define Z_{jm}, Z_{0m} by the relations (4.27), (4.25), and (4.23) by using Z_{00}, Z_{j0} $(1 \le j \le q)$ obtained above. For $1 \le m < k_1$, we have

$$\det(B_0 - (\rho + m)) \ne 0,$$

and then Z_{jm}, Z_{0m} are uniquely determined for these m. We show by the flowchart the procedure of the determination of these coefficients.

$$
\begin{array}{ccccccc}
Z_{00} & \to & Z_{10} & \to \cdots \to & Z_{q-1,0} & \to & Z_{q0} \\
 & & & & & & \downarrow \\
Z_{01} & \leftarrow & Z_{11} & \leftarrow \cdots \leftarrow & Z_{q-1,1} & \leftarrow & Z_{q1} \\
 & & & & & & \downarrow \\
Z_{02} & \leftarrow & Z_{12} & \leftarrow \cdots \leftarrow & Z_{q-1,2} & \leftarrow & Z_{q2} \\
 & & & & & & \downarrow \\
 & & & & & & \vdots \\
 & & & & & & \downarrow \\
Z_{0,k_1-1} & \leftarrow & Z_{1,k_1-1} & \leftarrow \cdots \leftarrow & Z_{q-1,k_1-1} & \leftarrow & Z_{q,k_1-1}
\end{array}
$$

Note that, from $Z_{j0} = 0$ ($j \geq p_0$) it follows

$$
Z_{jm} = 0 \quad (j \geq p_0,\ 0 \leq m < k_1). \tag{4.28}
$$

We consider $m = k_1$. For $i \neq 1$, $B_{0,\rho,i} - (\rho + k_1)$ and $B_0' - (\rho + k_1)$ are invertible, so that

$$
[Z_{qk_1}]_i \quad (i \neq 1), \quad [Z_{qk_1}]'
$$

are uniquely determined by the relation (4.27) with $m = k_1$. Then by using (4.25) and (4.23) with $m = k_1$ recursively, we see that

$$
[Z_{jk_1}]_i, \quad [Z_{jk_1}]' \quad (0 \leq j \leq q,\ i \neq 1)
$$

are uniquely determined. Take any $[Z_{0k_1}]_1$, and define

$$
Z_{0k_1} = \begin{pmatrix} [Z_{0k_1}]_0 \\ [Z_{0k_1}]_1 \\ \vdots \\ [Z_{0k_1}]_d \\ [Z_{0k_1}]' \end{pmatrix}.
$$

By using this Z_{0k_1}, we define

$$
[Z_{1k_1}]_1 = (B_{0,\rho,1} - (\rho + k_1))[Z_{0k_1}]_1 + \left[\sum_{t=0}^{k_1-1} B_{k_1-t} Z_{0t} \right]_1,
$$

and recursively define $[Z_{jk_1}]_1$ $(1 < j \leq q)$ by

$$[Z_{j+1,k_1}]_1 = \frac{1}{j+1} \left\{ (B_{0,\rho,1} - (\rho + k_1))[Z_{jk_1}]_1 + \left[\sum_{t=0}^{k_1-1} B_{k_1-t} Z_{jt} \right]_1 \right\}.$$

(4.29)

Thus we defined Z_{jk_1} $(0 \leq j \leq q)$ so that the relations (4.23) and (4.25) with $m = k_1$ hold. However, we should check that Z_{qk_1} satisfies the relation (4.27) with $m = k_1$. By the equality (4.28), we have

$$\sum_{t=0}^{k_1-1} B_{k_1-t} Z_{jt} = 0$$

for $j \geq p_0$. Then, for $j \geq p_0$, the definition (4.29) becomes

$$[Z_{j+1,k_1}]_1 = \frac{1}{j+1} (B_{0,\rho,1} - (\rho + k_1))[Z_{jk_1}]_1,$$

which induces

$$[Z_{jk_1}]_1 = \frac{p_0!}{j!} (B_{0,\rho,1} - (\rho + k_1))^{j-p_0} [Z_{p_0k_1}]_1 \quad (j \geq p_0).$$

Hence we get $[Z_{jk_1}]_1 = 0$ for $j \geq p_0 + p_1$, and also have

$$(B_{0,\rho,1} - (\rho + k_1))[Z_{p_0+p_1-1,k_1}]_1 = 0,$$

which shows

$$(B_0 - (\rho + k_1))Z_{qk_1} = 0$$

since $q \geq p_0 + p_1 - 1$. Thus the both sides of (4.27) with $m = k_1$ becomes 0, so that it holds.

We continue a similar process. Since $B_0 - (\rho + m)$ is invertible for $k_1 < m < k_2$, Z_{jm} can be determined uniquely by (4.27), (4.25), and (4.23). For $m = k_2$, we take any $[Z_{0k_2}]_2$. Then Z_{jk_2} can be determined uniquely so that the relations (4.23), (4.25), and (4.27) with $m = k_2$ hold.

It is similar for the case $m > k_2$, and then we see that the formal solution (4.21) can be determined uniquely by prescribing any values of

$$[Z_{00}]_0, [Z_{0k_1}]_1, \ldots, [Z_{0k_d}]_d.$$

Now we are going to show that the formal solution (4.21) converges in a neighborhood of $x = 0$. For $0 \leq j \leq q$, we set

$$Z_j(x) = \sum_{m=0}^{\infty} Z_{jm} x^m.$$

In order to show the convergence of $Z_j(x)$, we use Lemma 4.8 which will be given later. First we consider $Z_q(x)$. By rewriting (4.27) as

$$((B_0 - \rho) - m) W_{qm} = -\sum_{t=0}^{k_l - 1} B_{m-1-t} W_{qt} - \sum_{t=k_l}^{m-1} B_{m-1-t} W_{qt},$$

we see that the lemma can be applied by setting

$$A = B_0 - \rho, \quad C_m = -\sum_{t=0}^{k_l - 1} B_{m-1-t} Z_{qt}.$$

Then $Z_q(x)$ converges. For $Z_j(x)$ $(0 \leq j < q)$, thanks to (4.25), we can apply the lemma by setting

$$A = B_0 - \rho, \quad C_m = (j+1) Z_{j+1,m} - \sum_{t=0}^{k_l - 1} B_{m-1-t} Z_{jt}$$

if we have shown the convergence of $Z_{j+1}(x)$. Thus we have shown that the formal solution (4.21) converges in a neighborhood of $x = 0$.

Here we consider the radius of convergence of $Z_j(x)$ $(0 \leq j \leq q)$. Let r be the radius of convergence of the series

$$\sum_{m=0}^{\infty} B_m x^m = \tilde{B}(x).$$

Since

$$Z(x) = x^{\rho} \sum_{j=0}^{q} (\log x)^j Z_j(x)$$

is a solution of the differential equation (4.13), the function

$$U(x) = x^{-\rho} Z(x)$$

satisfies the differential equation

$$\frac{dU}{dx} = \frac{\tilde{B}(x) - \rho}{x} U.$$

Put $U(x) = \sum_{j=0}^{q} (\log x)^j Z_j(x)$ into the differential equation and compare the coefficients of $(\log x)^q$ of the both sides. Then we get

$$\frac{dZ_q}{dx} = \frac{\tilde{B}(x) - \rho}{x} Z_q.$$

The coefficient of this differential equation is holomorphic in the domain $0 < |x| < r$. Since $Z_q(x)$ is a solution holomorphic in a neighborhood of $x = 0$, owing to Theorem 3.2, it can be continued analytically to the domain $0 < |x| < r$ where the coefficient is holomorphic, and hence is holomorphic in $|x| < r$. Thus the radius of convergence of $Z_q(x)$ is at least r. The function $Z_{q-1}(x)$ satisfies the linear inhomogeneous differential equation

$$\frac{dZ_{q-1}}{dx} = \frac{\tilde{B}(x) - \rho}{x} Z_{q-1} - \frac{q}{x} Z_q.$$

Solving this by the method of variation of constants, we see that the solution $Z_{q-1}(x)$, which is holomorphic in a neighborhood of $x = 0$, can be continued analytically to the domain $0 < |x| < r$ where the coefficient $(\tilde{B}(x) - \rho)/x$ and the inhomogeneous term $q Z_q/x$ are holomorphic, and hence is holomorphic in $|x| < r$. Thus the radius of convergence of $Z_{q-1}(x)$ is also at least r. Continuing similar arguments, we see that the radius of convergence of every $Z_j(x)$ $(0 \leq j \leq q)$ is at least r.

Note that the dimension of the datum $[Z_0]_0, [Z_{k_1}]_1, \ldots, [Z_{k_d}]_d$, which is used to determine the formal solution, coincides with the size of $B_{0,\rho}$. Then, by constructing formal solutions for all ρ, we obtain n linearly independent solutions of the form (4.21), and hence a fundamental system of solutions. $\qquad \square$

Corollary 4.6 *Suppose that B_0 is non-resonant. Let Λ be the Jordan canonical form of B_0, and P a non-singular matrix such that $P^{-1} B_0 P = \Lambda$. Then there exists a fundamental matrix solution of the form*

$$Z(x) = G(x)(x - a)^{\Lambda}, \quad G(x) = P + \sum_{m=1}^{\infty} G_m (x - a)^m$$

of the differential equation (4.13).

Proof Let $Z_0(x)$ the fundamental matrix solution given in Theorem 4.5, (i). We have

$$Z_0(x) = F(x)(x-a)^{B_0}$$

$$= \left(I + \sum_{m=1}^{\infty} F_m(x-a)^m \right) (x-a)^{P \Lambda P^{-1}}$$

$$= \left(P + \sum_{m=1}^{\infty} F_m P(x-a)^m \right) (x-a)^{\Lambda} P^{-1}.$$

Then the matrix

$$Z(x) = Z_0(x)P$$

satisfies the condition.   □

Lemma 4.7 *Let A be an $n \times n$-matrix, and B an $m \times m$-matrix. The linear map*

$$X \mapsto XA - BX$$

from $\mathrm{M}(m \times n; \mathbb{C})$ to itself, which is then an endomorphism, is an isomorphism if and only if the matrices A, B have no common eigenvalue.

Proof Note that an endomorphism is an isomorphism if and only if the kernel is 0.

Assume first that A, B have a common eigenvalue λ. Let ${}^t u$ be a row λ-eigenvector of A, and v a column λ-eigenvector of B. Thus we have

$${}^t u A = \lambda\, {}^t u, \quad Bv = \lambda v,$$

and ${}^t u \neq 0$, $v \neq 0$. Set $X = v\, {}^t u$. Then $X \neq O$, and we have

$$XA - BX = v\, {}^t u A - Bv\, {}^t u = \lambda v\, {}^t u - \lambda v\, {}^t u = O,$$

which implies that the kernel is not 0.

Assume next that the matrices A, B have no common eigenvalue. Let X be a matrix in the kernel of the linear map. Then we have

$$XA = BX.$$

Let λ be any eigenvalue of A, and v a generalized λ-eigenvector of A. Namely $v \neq 0$, and

$$(A - \lambda)^k v = 0$$

holds for some integer $k \geq 1$. Now we have

$$0 = X(A - \lambda)^k v = (B - \lambda)^k X v,$$

and hence λ is an eigenvalue of B unless $Xv = 0$. By the assumption the eigenvalue λ of A is not an eigenvalue of B, and then we have $Xv = 0$. Since the generalized eigenvectors of A make a basis of $\mathbb{C}^n$, we conclude that $X = O$. $\qquad \square$

Lemma 4.8 *Let k be a non-negative integer, A an $n \times n$-matrix such that $k+1, k+2, \ldots$ are not eigenvalues of A. Let $\{B_m\}_{m=0}^{\infty}$ be a sequence of $n \times n$-matrices, and $\{C_m\}_{m=k+1}^{\infty}$ a sequence of vectors, such that*

$$\|B_m\| \leq K\beta^m, \quad \|C_m\| \leq K\gamma^m$$

hold by some constants $K > 0, \beta > 0, \gamma > 0$. Define a sequence $\{Z_m\}_{m=k}^{\infty}$ of vectors by

$$(A - m)Z_m = C_m + \sum_{t=k}^{m-1} B_{m-1-t} Z_t \quad (m > k)$$

with arbitrarily given initial data Z_k. Then the series

$$W(x) = \sum_{m=k}^{\infty} W_m x^m$$

converges in a neighborhood of $x = 0$.

Proof By the assumption, $\|(A - m)^{-1}\|$ is bounded for $m > k$. Then there exists a constant $L > 0$ such that

$$\|(A - m)^{-1}\| \leq L.$$

Noting the equality

$$W_m = (A - m)^{-1} C_m + (A - m)^{-1} \sum_{t=k}^{m-1} B_{m-1-t} W_t \quad (m > k),$$

we define a sequence $\{g_m\}_{m=k}^{\infty}$ by

$$g_k = \|W_k\|,$$

$$g_m = LK\gamma^m + LK \sum_{t=k}^{m-1} \beta^{m-1-t} g_t \quad (m > k).$$

Then we can show

$$\|W_m\| \leq g_m$$

for any $m \geq k$. Thus the power series

$$g(x) = \sum_{m=k}^{\infty} g_m x^m$$

is a majorant series for $W(x)$. On the other hand, we have

$$\sum_{m=k+1}^{\infty} g_m x^m = LK \sum_{m=k+1}^{\infty} (\gamma x)^m + LK \sum_{m=k+1}^{\infty} \sum_{t=k}^{m-1} \beta^{m-1-t} g_t x^m$$

$$= LK \cdot \frac{(\gamma x)^{k+1}}{1-\gamma x} + LK \sum_{t=k}^{\infty} g_t x^t \sum_{m=t+1}^{\infty} \beta^{m-1-t} x^{m-t}$$

$$= LK \cdot \frac{(\gamma x)^{k+1}}{1-\gamma x} + LK \cdot \frac{x}{1-\beta x} g(x),$$

from which we obtain

$$g(x) = g_k x^k + LK \cdot \frac{(\gamma x)^{k+1}}{1-\gamma x} + LK \cdot \frac{x}{1-\beta x} g(x).$$

Solving this linear equation, we get

$$g(x) = \frac{1-\beta x}{1-(\beta+LK)x} \left(g_k + \frac{LK\gamma^{k+1} x}{1-\gamma x} \right) x^k.$$

Then $g(x)$ converges in the domain

$$|x| < \min\left(\frac{1}{\beta + LK}, \frac{1}{\gamma} \right),$$

which shows that $W(x)$ converges in the same domain. $\square$

Remark 4.2 For any constant matrix B_0, the differential equation

$$\frac{dZ}{dx} = \frac{B_0}{x-a} Z$$

has a fundamental matrix solution

$$\mathcal{Z}(x) = (x - a)^{B_0}. \tag{4.30}$$

In this case, we do not need to assume that B_0 is non-resonant. The differential equation (4.13) in Theorem 4.5 can be regarded as obtained by adding to the coefficient B_0 the perturbation $\sum_{m=1}^{\infty} B_m (x - a)^m$. In the proof of Theorem 4.5, we a priori assumed the forms (4.14) and (4.17) of the solution, which are understood as obtained by adding the effect of the perturbation to the solution (4.30).

Since we have proved Theorem 4.5, the proof of the "if" part of Theorem 4.4 is completed. Moreover, we complete the proof of the "if" part of Theorem 4.3.

Proof of Theorem 4.3 (continued). Assume that, in the scalar differential equation (4.2), $x = a$ is a pole of $p_j(x)$ of order at most j for $1 \le j \le n$. Then, as is shown in the proof of Theorem 4.4, we can transform the scalar equation (4.2) to a system (4.13) with single pole at $x = a$ in the coefficient. Then thanks to Theorem 4.5, any solution of the system (4.13) is regular singular at $x = a$, and hence so is any solution of the scalar equation (4.13). This completes the proof of the "if" part of Theorem 4.3.   □

For a scalar differential equation, we can construct a fundamental set of solutions⋅ at a regular singular point via a system of differential equations of the first order, however, we can also construct it directly. Let $x = a$ be a regular singular point of the scalar differential equation

$$y^{(n)} + p_1(x)y^{(n-1)} + p_2(x)y^{(n-2)} + \cdots + p_n(x)y = 0. \tag{4.31}$$

By Fuchs' theorem, we see that $x = a$ is a pole of $p_j(x)$ of order at most j for each j. Then we can set

$$p_j(x) = \frac{1}{(x - a)^j} \sum_{m=0}^{\infty} p_{jm}(x - a)^m.$$

We consider a formal solution of the form

$$y(x) = (x - a)^{\rho} \sum_{m=0}^{\infty} y_m (x - a)^m \quad (y_0 \ne 0). \tag{4.32}$$

Putting it into the Eq. (4.10), we get

$$\sum_{m=0}^{\infty} (\rho + m)(\rho + m - 1) \cdots (\rho + m - n + 1) y_m (x - a)^{\rho + m - n}$$

$$+ \left(\sum_{m=0}^{\infty} p_{1m} (x - a)^m \right) \left(\sum_{m=0}^{\infty} (\rho + m) \cdots (\rho + m - n + 2) y_m (x - a)^{\rho + m - n} \right)$$

$$+ \cdots$$

$$+ \left(\sum_{m=0}^{\infty} p_{n-1,m} (x - a)^m \right) \left(\sum_{m=0}^{\infty} (\rho + m) y_m (x - a)^{\rho + m - n} \right)$$

$$+ \left(\sum_{m=0}^{\infty} p_{nm} (x - a)^m \right) \left(\sum_{m=0}^{\infty} y_m (x - a)^{\rho + m - n} \right) = 0.$$

By comparing the coefficients of the lowest term $(x - a)^{\rho - n}$, we get

$$f(\rho) y_0 = 0,$$

where we set

$$f(\rho) = \rho(\rho - 1) \cdots (\rho - n + 1) + p_{10}\rho(\rho - 1) \cdots (\rho - n + 2)$$

$$+ \cdots + p_{n-1,0}\rho + p_{n0}.$$

Since we assumed $y_0 \neq 0$, we have

$$f(\rho) = 0. \tag{4.33}$$

This is an algebraic equation of degree n for ρ, which we call the *characteristic equation*.

Definition 4.6 For a scalar differential equation (4.31), the roots of the characteristic equation (4.33) are called the *characteristic exponents*.

We continue to construct a formal solution. Letting the coefficient of the term $(x - a)^{\rho + m - n}$ for $m > 0$ be 0, we get an equation of the form

$$f(\rho + m) y_m = F_m(\rho; y_1, y_2, \ldots, y_{m-1}),$$

where $F_m(\rho; y_1, y_2, \ldots, y_{m-1})$ is a linear combination of $y_0, y_1, \ldots, y_{m-1}$ with polynomials in ρ as coefficients. If $f(\rho + m) \neq 0$ for any $m > 0$, y_m's are uniquely determined, and then we get a formal solution (4.32). In particular, if there is no integral difference among the characteristic exponents, $\rho + m$ $(m > 0)$ cannot be a characteristic exponent for any characteristic exponent ρ, and hence we can get n linearly independent solutions of the form (4.32). Even if there are integral differences among the characteristic exponents $\rho_1, \rho_2, \ldots, \rho_k$, for the exponent ρ_i with the maximum real part, there exists a formal solution of the form (4.32) because $f(\rho_i + m) \neq 0 \ (m > 0)$ holds.

Consider the case that there are characteristic exponents ρ_1, ρ_2 with an integral difference. If $\rho_1 = \rho_2$, there is a solution (4.32) of characteristic exponent ρ_1 unless $\rho_1 + m \ (m > 0)$ are not characteristic exponents, however, we have only 1-dimensional solution space in this way. If $\rho_2 = \rho_1 + k \ (k > 0)$, in constructing a solution with characteristic exponent ρ_1, we come to the equation

$$f(\rho_1 + k)y_k = F_k(\rho_1; y_0, y_1, \ldots, y_{k-1}),$$

whose left hand side becomes 0. If $F_k(\rho_1; y_0, y_1, \ldots, y_{k-1})$ happens to be 0, this equation holds for any y_k, and then we can proceed to the next step by setting any value of y_k. If $F_k(\rho_1; y_0, y_1, \ldots, y_{k-1}) \neq 0$, the equation cannot holds, and hence there is no solution of the form (4.32) with the characteristic exponent ρ_1.

Thus we understand that a fundamental system of solutions consisting of solutions of the form (4.32) may not be obtained only when the characteristic equation (4.33) has roots with integral differences. In order to describe a fundamental system of solutions for such case, we classify the roots of the characteristic equation into the classes consisting of roots with integral differences each other. Then each class can be described by using a complex number ρ and a sequence $0 = k_0 < k_1 < \cdots < k_d$ of integers as

$$\rho + k_0, \rho + k_1, \ldots, \rho + k_d.$$

We denote the multiplicities of the above roots by $e_0, e_1, \ldots, e_d$, respectively, and set

$$q = \sum_{i=0}^{d} e_i - 1.$$

We have the following assertion.

Theorem 4.9 *Let $x = a$ be a regular singular point of the scalar differential equation (4.10). Then there exists a fundamental system of solutions consisting of solutions of the forms*

$$(x-a)^\rho \left[\sum_{m=0}^{\infty} y_m (x-a)^m + \sum_{j=1}^{q} (\log(x-a))^j \sum_{m=0}^{\infty} z_{jm} (x-a)^m \right]. \qquad (4.34)$$

This theorem can be proved in a similar manner as Theorem 4.5. It is also proved by the Frobenius method, that is to obtain the solution (4.34) from the solution (4.32) by differentiating with respect to ρ. Refer to Coddington–Levinson [27, Chapter 4, §8] for the Frobenius method.

Example 4.2 We shall describe the assertion of Theorem 4.9 for the second order scalar differential equation

$$y'' + p(x)y' + q(x)y = 0. \tag{4.35}$$

Let $x = a$ a regular singular point, and set

$$p(x) = \frac{1}{x-a} \sum_{m=0}^{\infty} p_m (x-a)^m, \quad q(x) = \frac{1}{(x-a)^2} \sum_{m=0}^{\infty} q_m (x-a)^m.$$

Then we have the characteristic equation

$$f(\rho) := \rho(\rho - 1) + p_0\rho + q_0 = 0$$

at $x = a$. Denote by ρ_1, ρ_2 the roots of this equation. We have the following three cases.

(i) $\rho_1 - \rho_2 \notin \mathbb{Z}$
(ii) $\rho_1 - \rho_2 \in \mathbb{Z} \setminus \{0\}$
(iii) $\rho_1 = \rho_2$

For the case (i), we have linearly independent solutions of the form

$$y_1(x) = (x-a)^{\rho_1} \sum_{m=0}^{\infty} y_{1m}(x-a)^m \quad (y_{10} \neq 0),$$

$$y_2(x) = (x-a)^{\rho_2} \sum_{m=0}^{\infty} y_{2m}(x-a)^m \quad (y_{20} \neq 0).$$

For the case (ii), we assume that $\operatorname{Re}\rho_1 < \operatorname{Re}\rho_2$ and set $\rho_2 = \rho_1 + k \ (k \in \mathbb{Z}_{>0})$. Then we have a solution of the form

$$y_2(x) = (x-a)^{\rho_2} \sum_{m=0}^{\infty} y_{2m}(x-a)^m \quad (y_{20} \neq 0).$$

In constructing a formal solution

$$y_1(x) = (x-a)^{\rho_1} \sum_{m=0}^{\infty} y_{1m}(x-a)^m$$

with the characteristic exponent ρ_1, we come to the equation

$$f(\rho_1 + k)y_{1k} = F_k(\rho_1; y_{10}, y_{11}, \ldots, y_{1,k-1}). \tag{4.36}$$

The left hand side becomes 0 by $f(\rho_1+k) = f(\rho_2) = 0$. We consider the following two sub-cases.

(ii-i) $F_k(\rho_1; y_{10}, y_{11}, \ldots, y_{1,k-1}) \neq 0$
(ii-ii) $F_k(\rho_1; y_{10}, y_{11}, \ldots, y_{1,k-1}) = 0$

For the case (ii-i), the Eq. (4.36) never holds for any value of y_{1k}. If we set

$$y(x) = (x - a)^{\rho_1} \left[\sum_{m=0}^{\infty} y_m(x - a)^m + \log(x - a) \sum_{m=k}^{\infty} z_m(x - a)^m \right]$$

•
$$= u(x) + \log(x - a)v(x)$$

and put it into the differential equation, we get

$$u'' + \log(x - a)v'' + \frac{2}{x - a} v' - \frac{1}{(x - a)^2} v$$

$$+ p(x)\left(u' + \log(x - a)v' + \frac{1}{x - a} v \right) + q(x)(u + \log(x - a)v) = 0. \tag{4.37}$$

From this we obtain

$$v'' + p(x)v' + q(x)v = 0,$$

which implies that $v(x)$ is a solution of (4.35) of characteristic exponent $\rho_1+k = \rho_2$. Hence $v(x)$ is a scalar multiple of $y_2(x)$. Thus in the case (ii-i) we have, together with $y_2(x)$, a solution

$$y(x) = (x - a)^{\rho_1} \sum_{m=0}^{\infty} y_m(x - a)^m + y_2(x) \log(x - a) \quad (y_0 \neq 0)$$

which is linearly independent with $y_2(x)$.

For the case (ii-ii), the Eq. (4.36) holds for any y_{1k}, and then we have a solution of the form

$$y_1(x) = (x - a)^{\rho_1} \sum_{m=0}^{\infty} y_{1m}(x - a)^m \quad (y_{10} \neq 0).$$

The solutions $y_1(x)$, $y_2(x)$ make a fundamental system of solutions. The ambiguity of y_{1k} corresponds to replacing $y_1(x)$ by a linear combination

$$y_1(x) + cy_2(x)$$

for a constant c.

For the case (iii), we have a solution

$$y_1(x) = (x-a)^{\rho_1} \sum_{m=0}^{\infty} y_{1m}(x-a)^m \quad (y_{10} \neq 0).$$

In order to obtain another linearly independent solution, we set

$$y(x) = (x-a)^{\rho_1} \left[\sum_{m=0}^{\infty} y_m(x-a)^m + \log(x-a) \sum_{m=0}^{\infty} z_m(x-a)^m \right]$$

$$= u(x) + \log(x-a)v(x),$$

and put it into the differential equation. Similarly as in the case (ii-i), we see that we can take $v(x) = y_1(x)$. Then we have a solution

$$y(x) = (x-a)^{\rho_1} \sum_{m=0}^{\infty} y_m(x-a)^m + y_1(x)\log(x-a) \quad (y_0 \neq 0),$$

and it, together with $y_1(x)$, makes a fundamental system of solutions.

It is sometimes said that (i) is a generic case, (ii-i) and (iii) logarithmic cases, and (ii-ii) an apparent case. The generic case is most generic, and among non-generic cases the logarithmic case is generic. The apparent case is most exceptional. Note that, in the most generic case and in the most exceptional case, we have a fundamental system of solutions consisting of solutions of the same form.

4.3 Local Monodromy

Let r be a positive number, and define a domain D by

$$D = \{x \in \mathbb{C} \, ; \, 0 < |x-a| < r\}.$$

We consider a system of differential equations

$$\frac{dY}{dx} = A(x)Y \tag{4.38}$$

Fig. 4.1 Domain U and
loop Γ

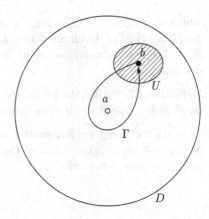

of rank n with the coefficient $A(x)$ holomorphic on the domain D. Take a simply
connected domain U in D, and fix a point $b \in U$. Let Γ be a loop in the domain D
with the base point b which encircles a once in the positive direction (Fig. 4.1).

Theorem 4.10 *Let $\mathcal{Y}(x)$ be any fundamental matrix solution on U of the differential equation (4.38). The analytic continuation $\Gamma_*\mathcal{Y}(x)$ of $\mathcal{Y}(x)$ along the loop Γ becomes again a fundamental matrix solution on U. Therefore there exists $L \in \mathrm{GL}(n, \mathbb{C})$ such that*

$$\Gamma_*\mathcal{Y}(x) = \mathcal{Y}(x)L \tag{4.39}$$

holds.

Proof As is shown in Theorem 3.6, determinants of results of analytic continuations
of $\mathcal{Y}(x)$ do not vanish whenever the determinant of $\mathcal{Y}(x)$ does not vanish at a point.
Then $\Gamma_*\mathcal{Y}(x)$ becomes a fundamental matrix solution in U.

Since $\mathcal{Y}(x)$ is a fundamental matrix solution, there exists an $n \times n$-matrix L
satisfying (4.39). By taking the determinant of the both sides of (4.39), we see that
$\det L \neq 0$, which implies $L \in \mathrm{GL}(n, \mathbb{C})$.   □

The matrix L appeared in Theorem 4.10 is called the *circuit matrix* (or the
monodromy matrix) for Γ with respect to the fundamental matrix solution $\mathcal{Y}(x)$
of the differential equation (4.38). The circuit matrix L depends on the choice of the
fundamental matrix solution as follows. Let $\tilde{\mathcal{Y}}(x)$ be another fundamental matrix
solution. Then there exists an $n \times n$-matrix C such that $\tilde{\mathcal{Y}}(x) = \mathcal{Y}(x)C$. We see
$C \in \mathrm{GL}(n, \mathbb{C})$ in a similar manner as in the proof of Theorem 4.10. Then we have

$$\Gamma_*\tilde{\mathcal{Y}}(x) = \Gamma_*\mathcal{Y}(x)C$$

$$= \mathcal{Y}(x)LC$$

$$= \tilde{\mathcal{Y}}(x)C^{-1}LC,$$

which shows that $C^{-1}LC$ is the circuit matrix with respect to $\tilde{\mathcal{Y}}(x)$. Therefore the conjugacy class of the circuit matrix L in $\mathrm{GL}(n, \mathbb{C})$ does not depend on the choice of the fundamental matrix solution. Taking this into account, we have the following definition.

Definition 4.7 We call the conjugacy class of the circuit matrix L in $\mathrm{GL}(n, \mathbb{C})$ the *local monodromy* at $x = a$ of the differential equation (4.38).

In the following we assume that the singular point $x = a$ of the differential equation (4.38) is regular singular. Then, by Theorem 4.4, we have the fundamental matrix solution of the form

$$\mathcal{Y}(x) = P(x)\mathcal{Z}(x),$$

where $P(x)$ is a matrix function holomorphic and single-valued in D, and $\mathcal{Z}(x)$ is a fundamental matrix solution of the differential equation of the form

$$\frac{dZ}{dx} = B(x)Z, \quad B(x) = \frac{1}{x-a} \sum_{m=0}^{\infty} B_m (x-a)^m. \tag{4.40}$$

Note that r may be replaced by a smaller one. Since $P(x)$ is single-valued in D, we have

$$\Gamma_* \mathcal{Y}(x) = P(x)\Gamma_* \mathcal{Z}(x).$$

Then the local monodromy of (4.38) coincides with the local monodromy of (4.40). Hence it is enough to obtain the latter.

Let U, Γ be as above. The next lemma is fundamental.

Lemma 4.11 *Let ρ be a complex number, and A a square matrix with entries in the complex numbers. We fix any branch of the functions $\log(x - a)$, $(x - a)^\rho$ and $(x - a)^A$ on U. Then the result of the analytic continuation of these functions along Γ are given by*

$$\Gamma_* \log(x - a) = \log(x - a) + 2\pi\sqrt{-1}, \tag{4.41}$$

$$\Gamma_*(x - a)^\rho = (x - a)^\rho e^{2\pi\sqrt{-1}\rho}, \tag{4.42}$$

$$\Gamma_*(x - a)^A = (x - a)^A e^{2\pi\sqrt{-1}A}. \tag{4.43}$$

Proof By the analytic continuation along Γ, $\arg(x - a)$ increases by 2π. The assertions (4.42) and (4.41) immediately follow from the definitions of the functions.

We recall the definition

$$e^A = \sum_{m=0}^{\infty} \frac{A^m}{m!}$$

of the exponential function of matrix argument. Note that it follows from this definition that

$$e^{A+B} = e^A e^B$$

holds if $AB = BA$. The function $(x - a)^A$ is defined by using the exponential function as

$$(x - a)^A = e^{A \log(x-a)}.$$

Then it follows from (4.41)

$$
\begin{aligned}
\Gamma_*(x - a)^A &= e^{A(\log(x-a)+2\pi \sqrt{-1})} \\
&= e^{A \log(x-a)+2\pi \sqrt{-1}A} \\
&= e^{A \log(x-a)} e^{2\pi \sqrt{-1}A} \\
&= (x - a)^A e^{2\pi \sqrt{-1}A}.
\end{aligned}
$$

□

Theorem 4.12 *For the system of differential equations (4.40), we assume that B_0 is non-resonant. Then the local monodromy at $x = a$ of the differential equation is given by*

$$e^{2\pi \sqrt{-1}B_0}.$$

Proof By Theorem 4.5, (i), we have a fundamental matrix solution of the form

$$\mathcal{Z}(x) = F(x)(x - a)^{B_0},$$

where $F(x)$ is holomorphic at $x = a$. Then the assertion follows directly from (4.43) in Lemma 4.11. □

Next we consider the general case, namely the case where B_0 is not necessarily non-resonant. The linear transformation which sends B_0 to its Jordan canonical form does not change the local monodromy of the differential equation (4.40), so that we may assume that B_0 is already B_0^J given in (4.15). By Theorem 4.5(ii), there exist a solution of the form (4.17), which is uniquely determined by prescribing the data

$$[Z_0]_{\rho,0}, [Z_{k_1}]_{\rho,1}, \ldots, [Z_{k_d}]_{\rho,d}.$$

We rewrite the solution

$$Z(x) = (x - a)^{\rho} \sum_{j=0}^{q} (\log(x - a))^j \sum_{m=0}^{\infty} Z_{jm}(x - a)^m$$

given by (4.17). We shall consider the analytic continuation of this solution. Set

$$v_0 = [Z_0]_{\rho,0}, v_1 = [Z_{k_1}]_{\rho,1}, \ldots, v_d = [Z_{k_d}]_{\rho,d}$$

and

$$v = \begin{pmatrix} v_0 \\ v_1 \\ \vdots \\ v_d \end{pmatrix}.$$

The solution $Z(x)$ is uniquely determined by the constant vector v, so that we shall denote

$$Z(x) = Z(x; v).$$

By Theorem 3.2, $\Gamma_* Z(x; v)$ is also a solution of the same differential equation (4.40), and by Lemma 4.11, the exponent after the analytic continuation is congruent to ρ modulo $\mathbb{Z}$. Then $\Gamma_* Z(x; v)$ is also a solution of the form (4.17). It follows that there exists a constant vector v' such that

$$\Gamma_* Z(x; v) = Z(x; v').$$

We denote this vector v' by $\Gamma_* v$; namely, we have

$$\Gamma_* Z(x; v) = Z(x; \Gamma_* v).$$

We are going to calculate $\Gamma_* v$.

As is described in Lemma 4.11, we have

$$(x - a)^{\rho} \rightsquigarrow e^{2\pi\sqrt{-1}\rho}(x - a)^{\rho},$$

$$\log(x - a) \rightsquigarrow \log(x - a) + 2\pi\sqrt{-1}$$

as results of the analytic continuation along Γ. Noting that

$$(\log(x - a))^j \rightsquigarrow \left(\log(x - a) + 2\pi\sqrt{-1}\right)^j,$$

we have

$$Z_{0m} \rightsquigarrow e^{2\pi\sqrt{-1}\rho} \left(Z_{0m} + \sum_{j=1}^{q} \left(2\pi\sqrt{-1} \right)^{j} Z_{jm} \right). \tag{4.44}$$

In particular, we have

$$v_i = [Z_{0k_i}]_{\rho,i} \rightsquigarrow e^{2\pi\sqrt{-1}\rho} \left(v_i + \sum_{j=1}^{q} \left(2\pi\sqrt{-1} \right)^{j} [Z_{jk_i}]_{\rho,i} \right),$$

from which we shall obtain the description of the change of v. We use results and notations in the proof of Theorem 4.5. We use $[\ ]_i$ for $[\ ]_{\rho,i}$ as in the proof of the theorem.

First we see the change of $v_0 = [Z_{00}]_0$. From

$$Z_{j0} = \frac{1}{j!}(B_0 - \rho)^j Z_{00},$$

we obtain

$$[Z_{j0}]_0 = \frac{1}{j!}(B_{0,\rho,0} - \rho)^j [Z_{00}]_0 = \frac{1}{j!}(B_{0,\rho,0} - \rho)^j v_0.$$

Putting it to (4.44) and noting $(B_{0,\rho,0} - \rho)^{p_0} = O$, we get

$$v_0 \rightsquigarrow e^{2\pi\sqrt{-1}\rho} \left(v_0 + \sum_{j=1}^{p_0-1} (2\pi\sqrt{-1})^j \frac{1}{j!}(B_{0,\rho,0} - \rho)^j v_0 \right)$$

$$= e^{2\pi\sqrt{-1}\rho} \sum_{j=0}^{p_0-1} \frac{(2\pi\sqrt{-1}(B_{0,\rho,0} - \rho))^j}{j!} v_0.$$

Next we see the change of v_1. By the proof of Theorem 4.5, we have

$$[Z_{1k_1}]_1 = (B_{0,\rho,1} - (\rho + k_1))[Z_{0k_1}]_1 + \left[\sum_{t=0}^{k_1-1} B_{k_1-t} Z_{0t} \right]_1,$$

$$[Z_{jk_1}]_1 = \frac{1}{j} \left((B_{0,\rho,1} - (\rho + k_1))[Z_{j-1,k_1}]_1 + \left[\sum_{t=0}^{k_1-1} B_{k_1-t} Z_{j-1,t} \right]_1 \right).$$

The vectors Z_{0t}, $Z_{j-1,t}$ in the right hand sides are determined linearly by v_0, because $t < k_1$. Then there exists a matrix C_j such that

$$[Z_{jk_1}]_1 = \frac{1}{j!}(B_{0,\rho,1} - (\rho + k_1))^j v_1 + C_j v_0.$$

Hence, by noting $(B_{0,\rho,1} - (\rho + k_1))^{p_1} = O$, we have

$$v_1 \rightsquigarrow e^{2\pi\sqrt{-1}\rho}\left(v_1 + \sum_{j=1}^{q}(2\pi\sqrt{-1})^j[Z_{jk_1}]_1\right)$$

$$= e^{2\pi\sqrt{-1}\rho}\left(\sum_{j=0}^{p_1-1}\frac{(2\pi\sqrt{-1}(B_{0,\rho,1} - (\rho + k_1)))^j}{j!}v_1 + C_{10}v_0\right),$$

where we set $\sum_{j=1}^{q}(2\pi\sqrt{-1})^j C_j = C_{10}$.

Continuing a similar consideration, we obtain

$$v_h \rightsquigarrow e^{2\pi\sqrt{-1}\rho}\left(\sum_{j=0}^{p_h-1}\frac{\left(2\pi\sqrt{-1}(B_{0,\rho,h} - (\rho + k_h))\right)^j}{j!}v_h + \sum_{h'=0}^{h-1}C_{hh'}v_{h'}\right)$$

for $h = 1, 2, \ldots, d$. Thus, summing up the above, we have

$$\Gamma_* v = L_\rho v, \tag{4.45}$$

$$L_\rho = e^{2\pi\sqrt{-1}\rho}\begin{pmatrix} D_0 & & & \\ C_{10} & D_1 & & \\ \vdots & \ddots & \ddots & \\ C_{d0} & \cdots & C_{d,d-1} & D_d \end{pmatrix}, \tag{4.46}$$

where we set

$$D_h = \sum_{j=0}^{p_h-1}\frac{(2\pi\sqrt{-1}(B_{0,\rho,h} - (\rho + k_h)))^j}{j!} \quad (0 \le h \le d).$$

Note that $C_{hh'}$ in off-diagonal blocks depend on ρ and on coefficients of higher order terms of the Laurent series expansion of $B(x)$. We see that all the eigenvalues of L_ρ are $e^{2\pi\sqrt{-1}\rho}$. In fact, by the similar transformation with the matrix

$$Q = \begin{pmatrix} & & I \\ & I & \\ & \iddots & \\ I & & \end{pmatrix},$$

L_ρ is sent to the block upper triangular matrix

$$Q^{-1}L_\rho Q = e^{2\pi\sqrt{-1}\rho} \begin{pmatrix} D_d & C_{d,d-1} & \cdots & C_{d0} \\ & \ddots & \ddots & \vdots \\ & & D_1 & C_{10} \\ & & & D_0 \end{pmatrix},$$

which is an upper triangular matrix because so are the diagonal blocks. All the eigenvalues of each diagonal block are $e^{2\pi\sqrt{-1}\rho}$, and hence L_ρ has the only eigenvalue $e^{2\pi\sqrt{-1}\rho}$.

Now we translate the above result to the description of the local monodromy. Let e_i the unit vector of the same size as v which has the only non-zero entry at the i-th position. We set

$$Z(x; e_i) = Z_i(x).$$

Since $Z(x; v)$ is linear in v, we have

$$Z(x; v) = (Z_1(x), Z_2(x), \ldots)v.$$

Using (4.45), we get

$$\begin{aligned} \Gamma_* Z_i(x) &= \Gamma_* Z(x; e_i) \\ &= Z(x; \Gamma_* e_i) \\ &= Z(x; L_\rho e_i) \\ &= (Z_1(x), Z_2(x), \ldots)L_\rho e_i. \end{aligned}$$

Collecting the above result for $i = 1, 2, \ldots$, we obtain

$$\begin{aligned} \Gamma_*(Z_1(x), Z_2(x), \ldots) &= (Z_1(x), Z_2(x), \ldots)L_\rho I \\ &= (Z_1(x), Z_2(x), \ldots)L_\rho. \end{aligned}$$

Thus we have described the analytic continuation for the solutions (4.17) corresponding to the class of the eigenvalue ρ of B_0.

Combining the results for all the classes of the eigenvalues of B_0, we obtain the following assertion.

Theorem 4.13 *Consider a system of differential equations*

$$\frac{dZ}{dx} = B(x)Z, \quad B(x) = \frac{1}{x-a}\sum_{m=0}^{\infty} B_m(x-a)^m,$$

where B_0 is assumed to coincide with B_0^J given in (4.15), (4.16). Then there exists a fundamental matrix solution $\mathcal{Z}(x)$ consisting of the solutions of the form (4.17) such that

$$\Gamma_* \mathcal{Z}(x) = \mathcal{Z}(x)L,$$
$$L = \bigoplus_\rho L_\rho . \tag{4.47}$$

holds, where L_ρ is given by (4.46) and the direct sum is taken over the equivalence classes of the eigenvalues of B_0.

Theorems 4.12 and 4.13 show by which the local monodromy of the system of differential equations (4.40) is determined. If the residue matrix B_0 is non-resonant, the local monodromy is determined only by B_0. If B_0 is resonant, it may depend on some coefficients of the Laurent series expansion of $B(x)$ other than B_0. In any case, the local monodromy is determined by a finite number of coefficients.

We can derive the description of the local monodromies for scalar differential equations from the above description for systems of differential equations. Consider a scalar differential equation

$$y^{(n)} + p_1(x)y^{(n-1)} + \cdots + p_n(x)y = 0 \tag{4.48}$$

with a regular singular point at $x = a$. Let

$$\mathcal{Y}(x) = (y_1(x), y_2(x), \ldots, y_n(x))$$

be a fundamental system of solutions on the simply connected domain U. Then there exists a matrix $L \in \mathrm{GL}(n, \mathbb{C})$ such that

$$\Gamma_* \mathcal{Y}(x) = \mathcal{Y}(x)L.$$

The matrix L is called the circuit matrix for Γ. We define the local monodromy of the Eq. (4.48) at $x = a$ by the conjugacy class in $\mathrm{GL}(n, \mathbb{C})$ of the circuit matrix. Any scalar differential equation can be transformed to a system of differential equations of the first order, keeping $x = a$ regular singular, by a linear transformation with univalent functions as coefficients. This can be realized by the transformation (2.3), in which case we have a fundamental matrix solution for the system with the fundamental system of solutions $\mathcal{Y}(x)$ as the first row. Thus, in this case, the circuit matrix L for the fundamental matrix solution is just the circuit matrix for the fundamental system of solutions $\mathcal{Y}(x)$.

Chapter 5
Monodromy

5.1 Definition

In the last section we have considered the local monodromy, which describes the change of the fundamental system of solutions caused by the analytic continuation along a loop encircling a regular singular point. In this chapter we study the (global) monodromy, which describes the changes caused by global analytic continuations.

Let D be a domain in the Riemann sphere $\mathbb{P}^1 = \mathbb{C} \cup \{\infty\}$. We consider a system of linear differential equations

$$\frac{dY}{dx} = A(x)Y \tag{5.1}$$

of the first order, or a scalar differential equation

$$y^{(n)} + p_1(x)y^{(n-1)} + \cdots + p_n(x)y = 0, \tag{5.2}$$

both of which have holomorphic functions in D as coefficients. (We call the both a differential equation, if there is no possibility of confusion.) Let the rank of the system (5.1) be n. Take a point $b \in D$ and a simply connected neighborhood U in D of b. We fix a fundamental system of solutions $\mathcal{Y}(x)$ of (5.1) or (5.2) on U. If we consider the system (5.1), $\mathcal{Y}(x)$ is an $n \times n$-matrix, and if we consider the scalar equation (5.2), it is an n-row vector.

Let Γ be a loop in D with the base point b. By the analytic continuation of $\mathcal{Y}(x)$ along Γ, we have a matrix $M \in \mathrm{GL}(n, \mathbb{C})$ such that

$$\Gamma_* \mathcal{Y}(x) = \mathcal{Y}(x)M,$$

© The Editor(s) (if applicable) and The Author(s), under exclusive
license to Springer Nature Switzerland AG 2020
Y. Haraoka, *Linear Differential Equations in the Complex Domain*, Lecture Notes
in Mathematics 2271, https://doi.org/10.1007/978-3-030-54663-2_5

which can be shown in a similar way as Theorem 4.8. Thanks to Theorem 3.3, the matrix M is uniquely determined by the homotopy class $[\Gamma] = \gamma$ of Γ in D. We denote $M = M_\gamma$. Thus we have the map

$$\rho : \pi_1(D, b) \to \mathrm{GL}(n, \mathbb{C})$$
$$\gamma \mapsto M_\gamma. \tag{5.3}$$

We shall see that a product in $\pi_1(D, b)$ is mapped to a product in $\mathrm{GL}(n, \mathbb{C})$. We define the product $\Gamma_1\Gamma_2$ of two loops Γ_1, Γ_2 with base point b to be a loop obtained by encircling Γ_1 first and then encircling Γ_2. Setting $[\Gamma_1] = \gamma_1$, $[\Gamma_2] = \gamma_2$, we have $[\Gamma_1\Gamma_2] = \gamma_1\gamma_2$. Now we obtain

$$(\Gamma_1\Gamma_2)_*\mathcal{Y}(x) = (\Gamma_2)_*((\Gamma_1)_*\mathcal{Y}(x))$$
$$= (\Gamma_2)_*\mathcal{Y}(x)M_{\gamma_1}$$
$$= \mathcal{Y}(x)M_{\gamma_2}M_{\gamma_1},$$

which implies

$$M_{\gamma_1\gamma_2} = M_{\gamma_2}M_{\gamma_1}.$$

Thus we have proved

$$\rho(\gamma_1\gamma_2) = \rho(\gamma_2)\rho(\gamma_1). \tag{5.4}$$

This shows that the map ρ is an anti-homomorphism of groups.

Definition 5.1 The anti-homomorphism ρ is called a *monodromy representation* of the differential equation (5.1) (or (5.2)) with respect to the fundamental system of solutions $\mathcal{Y}(x)$. The image of ρ, which is a subgroup of $\mathrm{GL}(n, \mathbb{C})$, is called the *monodromy group* of (5.1) (or (5.2)). The image $\rho(\gamma) = M_\gamma$ of $\gamma \in \pi_1(D, b)$ is said to be a *monodromy matrix* or a *circuit matrix* for γ.

Remark 5.1 Since ρ is an anti-homomorphism, we may have to call it a monodromy *anti*-representation, however, conventionally we call it a monodromy representation. In some literature, the product in $\pi_1(D, b)$ is defined in converse order, in which case ρ is actually a representation.

We consider how the monodromy representation depends on the fundamental system of solutions $\mathcal{Y}(x)$. This is the same argument as in the case of local monodromies. Take another fundamental system of solutions $\tilde{\mathcal{Y}}(x)$. Then there exists $C \in \mathrm{GL}(n, \mathbb{C})$ such that

$$\tilde{\mathcal{Y}}(x) = \mathcal{Y}(x)C$$

holds in U. Let γ be any element in $\pi_1(D, b)$, and take a representative Γ of γ. Then we have

$$\Gamma_* \tilde{\mathcal{Y}}(x) = \Gamma_*(\mathcal{Y}(x)C)$$
$$= \mathcal{Y}(x)M_\gamma C$$
$$= \tilde{\mathcal{Y}}(x)C^{-1}M_\gamma C.$$

Thus, if we denote the monodromy representation with respect to $\tilde{\mathcal{Y}}(x)$ by

$$\tilde{\rho} : \pi_1(D, b) \to \mathrm{GL}(n, \mathbb{C})$$

$$\gamma \mapsto \tilde{M}_\gamma,$$

we have

$$\tilde{M}_\gamma = C^{-1}M_\gamma C.$$

Hence $\tilde{\rho}$ is conjugate to ρ. To sum up, we get the following assertion.

Theorem 5.1 *Let $\mathcal{Y}(x), \tilde{\mathcal{Y}}(x)$ be fundamental systems of solutions on U of the differential equation (5.1) (or (5.2)), and ρ and $\tilde{\rho}$ the monodromy representations with respect to $\mathcal{Y}(x)$ and $\tilde{\mathcal{Y}}(x)$, respectively. Then we have*

$$\tilde{\rho}(\gamma) = C^{-1}\rho(\gamma)C \quad (\gamma \in \pi_1(D, b)),$$

where the matrix $C \in \mathrm{GL}(n, \mathbb{C})$ is determined by

$$\tilde{\mathcal{Y}}(x) = \mathcal{Y}(x)C.$$

In particular, the conjugacy class of monodromy representations is uniquely determined by the differential equation (5.1) (or (5.2)).

Next we consider the dependence of the monodromy representation on the base point b. Take another point $b' \in D$ and a path L in D which connect b and b'. We set b the initial point of L. Then we have an isomorphism

$$L_\# : \pi_1(D, b) \to \pi_1(D, b')$$

$$[\Gamma] \mapsto [L^{-1}\Gamma L]$$

among the fundamental groups. For a fundamental system of solutions $\mathcal{Y}(x)$ on U, $L_*\mathcal{Y}(x)$ becomes a fundamental system of solutions on a simply connected neighborhood of b' in D. The monodromy representation ρ with respect to $\mathcal{Y}(x)$

and the monodromy representation ρ' with respect to $L_* \mathcal{Y}(x)$ are related so that the following diagram becomes commutative:

$$
\begin{array}{ccc}
\pi_1(D, b) & \xrightarrow{\rho} & \mathrm{GL}(n, \mathbb{C}) \\
\downarrow{\scriptstyle L_\#} & & \downarrow{\scriptstyle \mathrm{id.}} \\
\pi_1(D, b') & \xrightarrow{\rho'} & \mathrm{GL}(n, \mathbb{C}).
\end{array}
$$

More explicitly, from the relation

$$
(L^{-1} \Gamma L)_* (L_* \mathcal{Y}(x)) = (\Gamma L)_* \mathcal{Y}(x) = L_* \mathcal{Y}(x) M_{[\Gamma]},
$$

we obtain

$$
\rho'([L^{-1} \Gamma L]) = \rho([\Gamma]).
$$

Thus the conjugacy class of the image of a monodromy representation does not depend on the fundamental system of solutions $\mathcal{Y}(x)$ nor on the base point b. It is determined uniquely by the differential equation. Then we call the conjugacy class the monodromy group of (5.1) (or (5.2)).

5.2 Global Monodromy and Local Monodromy

We shall consider the relation between the (global) monodromy and the local monodromy. In order to make the argument simple, we consider a domain D which is obtained from a simply connected domain by subtracting a finite number of points. Namely, we assume that there exist a simply connected domain $\hat{D} \subset \mathbb{P}^1$ and a finite subset S of $\hat{D}$ such that

$$
D = \hat{D} \setminus S. \tag{5.5}
$$

Take a point $b \in D$, and consider a loop Γ in D with base point b. Let a be a point in S. When $a = \infty$, we may take a chart in $\mathbb{P}^1$ around ∞. Hence, without loss of generality, we may assume $a \neq \infty$. If a point x moves along Γ from the initial point to the end point, the argument $\arg(x - a)$ increases or decreases by an integer multiple of 2π. We call the integer the *winding number* of Γ with respect to a (Fig. 5.1).

Analytically, the winding number is given by the integral

$$
\frac{1}{2\pi \sqrt{-1}} \int_\Gamma \frac{dx}{x - a}.
$$

Fig. 5.1 A loop in D

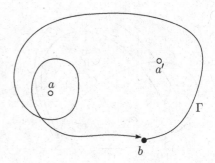

By this expression, we see that the winding number is invariant if we vary Γ in D with keeping the base point fixed. Then the winding number can be defined for the homotopy class $[\Gamma]$ in $\pi_1(D, b)$. We denote it by $n([\Gamma], a)$.

Now take a point $c \in D$ so close to a that the circle K with center a and radius $|c - a|$ is contained in D and any other $a' \in S$ is in the exterior of K. We regard c as the initial and the end point of K, and attach the positive direction to K. Take any curve L in D with initial point b and end point c. The loop LKL^{-1} (or the element of $\pi_1(D, b)$ represented by LKL^{-1}) is called a *(+1)-loop*[1] for a. For a (+1)-loop γ for a, we have $n(\gamma, a) = 1$ and $n(\gamma, a') = 0$ ($a' \in S \setminus \{a\}$), however, these conditions on the winding numbers are not sufficient for the definition of (+1)-loop.

Proposition 5.2 *Any two (+1)-loops for a are conjugate in $\pi_1(D, b)$.*

Proof Let $\gamma = [LKL^{-1}], \gamma' = [L'KL'^{-1}]$ be (+1)-loops for a. If we set $\mu = [L'L^{-1}]$, we have

$$\mu\gamma\mu^{-1} = \left[\left(L'L^{-1}\right)\left(LKL^{-1}\right)\left(LL'^{-1}\right)\right]$$
$$= \left[L'KL'^{-1}\right]$$
$$= \gamma',$$

which shows the assertion (Fig. 5.2).   □

We consider a monodromy representation ρ of the differential equation (5.1) (or (5.2)). Take a point $a \in S$. Let $\gamma \in \pi_1(D, b)$ be a (+1)-loop for a. The image $\rho(\gamma)$ depends on γ, however, thanks to Proposition 5.2, the conjugacy class $[\rho(\gamma)]$ does not depend on γ and is uniquely determined by a. Moreover, the conjugacy class does not depend on the base point b nor on the choice of the fundamental system of solutions which is used to define ρ. Thus the conjugacy class $[\rho(\gamma)]$ depends only on the differential equation and the point a. This shows that $[\rho(\gamma)]$ coincides with the local monodromy at a. Thus we obtain the following assertion.

---

[1]It is also called a *monodromy* for a.

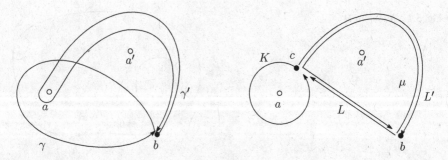

Fig. 5.2   (+1)-loops are conjugate

Theorem 5.3 *Let D be the domain defined by (5.5), and a be a point in S. Consider a differential equation (5.1) (or (5.2)) with coefficients holomorphic in D, and let ρ be its monodromy representation. Then, for any (+1)-loop γ for a, the conjugacy class [ρ(γ)] coincides with the local monodromy of (5.1) (or (5.2)) at a.*

So far we have distinguished a closed curve Γ from its homotopy class [Γ]. Since there is no possibility of confusion except the argument in Chap. 8, in the following except Chap. 8 we will not distinguish a closed curve from its homotopy class.

5.3   Monodromy of Differential Equations with Rational Coefficients

We consider a differential equation (5.1) (or (5.2)) whose coefficients are rational functions. In this case, the singular points makes a finite set. Let $a_0, a_1, \ldots, a_p$ be the singular points. One of them can be ∞. Then the domain where the differential equation is defined is

$$D = \mathbb{P}^1 \setminus \{a_0, a_1, \ldots, a_p\}. \tag{5.6}$$

Take a base point $b \in D$. The fundamental group $\pi_1(D, b)$, which is the domain of a monodromy representation, has the following presentation

$$\pi_1(D, b) = \langle \gamma_0, \gamma_1, \ldots, \gamma_p \mid \gamma_0 \gamma_1 \cdots \gamma_p = 1 \rangle, \tag{5.7}$$

where, for each j ($0 \le j \le p$), γ_j is a (+1)-loop for a_j located as in the following figure (Fig. 5.3).

Let U be a simply connected neighborhood of b in D, $\mathcal{Y}(x)$ a fundamental system of solutions of (5.1) (or (5.2)) on U, and

$$\rho : \pi_1(D, b) \to \mathrm{GL}(n, \mathbb{C})$$

Fig. 5.3 γ_j's

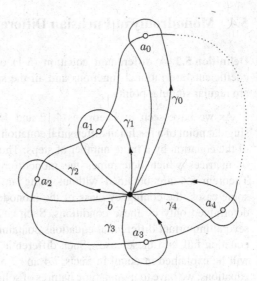

the monodromy representation with respect to $\mathcal{Y}(x)$. We set

$$\rho(\gamma_j) = M_j \quad (0 \le j \le p).$$

Then we have the following assertion.

Theorem 5.4

(i) *The monodromy representation ρ is uniquely determined by the tuple $(M_0, M_1, \ldots, M_p)$ of matrices.*

(ii) *For the tuple of the matrices $(M_0, M_1, \ldots, M_p)$, the relation*

$$M_p \cdots M_1 M_0 = I$$

holds.

(iii) *For each j, we denote the local monodromy at $x = a_j$ by $\mathcal{L}_j$. Then we have*

$$M_j \in \mathcal{L}_j.$$

Proof The assertions (i) and (ii) follow directly from the presentation (5.7). The assertion (iii) follows from Theorem 5.3.   □

5.4   Monodromy of Fuchsian Differential Equations

Definition 5.2 A differential equation (5.1) or (5.2) is called *Fuchsian*[2] if all coefficients are rational functions and all the singular points $x = a_0, a_1, \ldots, a_p$ are regular singular points.

As we have seen in Theorems 4.12 and 4.13, the local monodromy at each singular point of a Fuchsian differential equation is computable from the coefficients of the equation by a finite number of steps. Thus, for the tuple $(M_0, M_1, \ldots, M_p)$ of matrices which determines the monodromy, the conditions (ii) and (iii) in Theorem 5.4 are obtained without using any analytic nature of solutions. In some cases, the conjugacy class of the monodromy representation happens to be determined only by these conditions. Such cases are very exceptional, however, several important differential equations containing the hypergeometric differential equation fall into these cases. Such differential equations are called rigid, which will be explained in detail in Sects. 7.4 and 7.5. For the other generic differential equations, we have to use analytic natures of solutions to determine the monodromy, because the monodromy is a collection of results of analytic continuations of solutions. Thus, in general, it is very difficult to determine the monodromy. On the other hand, once the monodromy of a Fuchsian differential equation is determined, we can derive many properties of solutions of the equation. We are going to explain how to derive properties of solutions from the monodromy.

In the following, we use the same notations as in Theorem 5.4. In particular, we note that the monodromy group G is a group generated by the tuple $(M_0, M_1, \ldots, M_p)$ of matrices:

$$G = \langle M_0, M_1, \ldots, M_p \rangle.$$

5.4.1   Equivalence of Fuchsian Differential Equations and Monodromies

A singular point of a Fuchsian differential equation is called *apparent* if the local monodromy at the point is trivial, i.e. the conjugacy class of the identity matrix. The other singular points are called *branch points*. A singular point is called a *generalized apparent singular point* if the local monodromy at the point is a

---

[2]A system (5.1) is often called a *Fuchsian system* if the order of every pole of $A(x)$ is one. In this book, we call such system a Fuchsian system of normal form (see Sect. 7.3). We use "Fuchsian" in more intrinsic meaning.

conjugacy class of a scalar matrix. If $x = a$ is a generalized apparent singular point, and if the local monodromy is given by αI ($\alpha \neq 0$), we take a branch of $\log \alpha$, set

$$\lambda = \frac{\log \alpha}{2\pi \sqrt{-1}}$$

and transform the differential equation by the gauge transformation of multiplying solutions by $(x - a)^{-\lambda}$. Then, for the transformed equation, the singular point $x = a$ is apparent.

Let $\{a_0, a_1, \dots, a_p\}$ be the set of the singular points of a Fuchsian differential equation, and $\{a_0, a_1, \dots, a_q\}$ the subset of the branch points, where $0 \leq q \leq p$. Then the monodromy representation is reduced to the representation of the fundamental group $\pi_1(\mathbb{P}^1 \setminus \{a_0, a_1, \dots, a_q\})$.

Theorem 5.5 *For two Fuchsian differential equations with the same set of the branch points and of the same rank, the monodromy representations are isomorphic if and only if the equations are transformed from one to the other by a gauge transformation with coefficients in rational functions.*

Proof Let X be the space obtained from $\mathbb{P}^1$ by subtracting the common branch points and the apparent singular points of both differential equations. Take a point $b \in X$. Then the monodromy representations of both equations can be regarded as representations of $\pi_1(X, b)$.

First we assume that both differential equations are given by systems of the first order

$$\frac{dY}{dx} = A(x)Y, \quad \frac{dZ}{dx} = B(x)Z.$$

If the monodromy representations are isomorphic, we can choose a fundamental system of solutions $\mathcal{Y}(x)$, $\mathcal{Z}(x)$ for each equation so that the monodromy representation with respect to $\mathcal{Y}(x)$ and one with respect to $\mathcal{Z}(x)$ coincide. Then, for each $\gamma \in \pi_1(X, b)$, there exists a common matrix $M_\gamma \in GL(n, \mathbb{C})$ such that

$$\gamma_* \mathcal{Y}(x) = \mathcal{Y}(x)M_\gamma, \quad \gamma_* \mathcal{Z}(x) = \mathcal{Z}(x)M_\gamma$$

hold. Then we have

$$\gamma_* \left(\mathcal{Z}(x)\mathcal{Y}(x)^{-1} \right) = \mathcal{Z}(x)\mathcal{Y}(x)^{-1},$$

which implies that $P(x) = \mathcal{Z}(x)\mathcal{Y}(x)^{-1}$ is single valued on X. We see that $P(x)$ is holomorphic on X, because the determinant of $\mathcal{Y}(x)$ does not vanish on X as we have shown in Theorem 3.6. Moreover, at each point in $\mathbb{P}^1 \setminus X$, $P(x)$ is at most regular singular. Then, thanks to Theorem 4.1, we see that any entry of $P(x)$ is meromorphic on $\mathbb{P}^1$, which is nothing but a rational function. Thus we have

$$\mathcal{Z}(x) = P(x)\mathcal{Y}(x), \quad P(x) \in GL(n, \mathbb{C}(x)),$$

which implies that the two equations are related by the transformation

$$Z = P(x)Y.$$

On the other hand, it is clear that, for two equations which are related by this transformation, their monodromy representations are isomorphic.

For scalar equations, by associating a fundamental system of solutions $(y_1(x), y_2(x), \ldots, y_n(x))$ with the Wronsky matrix

$$\mathcal{Y}(x) = \begin{pmatrix} y_1(x) & y_2(x) & \cdots & y_n(x) \\ y_1'(x) & y_2'(x) & \cdots & y_n'(x) \\ & \cdots & \cdots & \\ y_1^{(n-1)}(x) & y_2^{(n-1)}(x) & \cdots & y_n^{(n-1)}(x) \end{pmatrix},$$

we can reduce the proof to the above.   □

5.4.2 Reducibility

Definition 5.3 A subgroup G of $GL(n, \mathbb{C})$ is said to be *reducible* if there exists a non-trivial G-invariant subspace W of $\mathbb{C}^n$. Namely, G is reducible if there exists a subspace $W \subset \mathbb{C}^n$ such that $W \neq \mathbb{C}^n$, $W \neq \{0\}$ satisfying

$$gW \subset W \qquad (\forall g \in G).$$

G is said to be *irreducible* if it is not reducible.

For Fuchsian differential equations, a differential equation is reduced to differential equations of lower ranks if the monodromy group is reducible. We shall explain this fact for scalar equations and for systems of the first order.

First, we consider a scalar Fuchsian differential equation (5.2). Let $a_0, a_1, \ldots, a_p$ be the regular singular points, and define D by (5.6). By using the differential operator

$$\partial = \frac{d}{dx},$$

the differential operator which defines the left hand side of (5.2) is written as

$$L = \partial^n + p_1(x)\partial^{n-1} + \cdots + p_{n-1}(x)\partial + p_n(x), \tag{5.8}$$

and then the Eq. (5.2) can be given by $Ly = 0$.

Theorem 5.6 *Let L be the differential operator defined by (5.8), and assume that the differential equation $Ly = 0$ is Fuchsian. If the monodromy group G of the differential equation $Ly = 0$ is reducible, there exist two differential operators*

$$
\begin{aligned}
K &= \partial^k + q_1(x)\partial^{k-1} + \cdots + q_k(x), \\
M &= \partial^m + r_1(x)\partial^{m-1} + \cdots + r_m(x)
\end{aligned}
\tag{5.9}
$$

with rational coefficients such that

$$
L = MK
\tag{5.10}
$$

holds, where k, m are integers satisfying $1 \le m < n$, $k + m = n$, and $q_1(x), \ldots, q_k(s), r_1(x), \ldots, r_m(x)$ are rational functions.

Conversely, if the decomposition (5.10) exists for L, the monodromy group of $Ly = 0$ is reducible.

Proof Let $\mathcal{Y}(x) = (y_1(x), y_2(x), \ldots, y_n(x))$ be a fundamental system of solutions such that G is the monodromy groups with respect to it. Since G is reducible, there exists a non-trivial G-invariant subspace $W \subset \mathbb{C}^n$. Set dim $W = k$. Then we have $1 \le k < n$. Take a basis $w_1, w_2, \ldots, w_k$ of W, and extend it to a basis of $\mathbb{C}^n$ by adding $v_{k+1}, \ldots, v_n$. Since $gW \subset W$ holds for any $g \in G$, we have $gw_j \in W$ $(1 \le j \le k)$. Then, by setting

$$
P = (w_1, w_2, \ldots, w_k, v_{k+1}, \ldots, v_n),
$$

we get

$$
gP = P \left(\begin{array}{c|c} * & * \\ \hline O & * \end{array} \right),
$$

where the right matrix in the right hand side is partitioned into $(k, m) \times (k, m)$-blocks. Note that $P \in \mathrm{GL}(n, \mathbb{C})$. By Theorem 3.1, we see that the monodromy group with respect to the fundamental system of solutions $\mathcal{Y}(x)P$ is given by

$$
P^{-1}GP = \left\{ P^{-1}gP \; ; \; g \in G \right\},
$$

which consists of block upper-triangular matrices.

Then, by taking $\mathcal{Y}(x)P$ as the original $\mathcal{Y}(x) = (y_1(x), y_2(x), \ldots, y_n(x))$, we may assume that G consists of block upper-triangular matrices. Namely, for any $\gamma \in \pi_1(D, b)$, we have

$$
\gamma_* \mathcal{Y}(x) = \mathcal{Y}(x) \left(\begin{array}{c|c} M_\gamma^{11} & * \\ \hline O & * \end{array} \right),
$$

where $M_\gamma^{11} \in GL(k, \mathbb{C})$. Therefore we get

$$\gamma_*(y_1(x), \ldots, y_k(x)) = (y_1(x), \ldots, y_k(x))M_\gamma^{11}.$$

Then moreover we have

$$\gamma_* \left(y_1^{(i)}(x), \ldots, y_k^{(i)}(x) \right) = \left(y_1^{(i)}(x), \ldots, y_k^{(i)}(x) \right) M_\gamma^{11} \tag{5.11}$$

for any $i = 0, 1, 2, \ldots$.

Now let y be a differential indeterminate, and define $\Delta_0, \Delta_1, \ldots, \Delta_k$ by

$$\begin{vmatrix} y & y_1 & \cdots & y_k \\ y' & y_1' & \cdots & y_k' \\ \vdots & \vdots & & \vdots \\ y^{(k)} & y_1^{(k)} & \cdots & y_k^{(k)} \end{vmatrix} = \Delta_0 y^{(k)} + \Delta_1 y^{(k-1)} + \cdots + \Delta_k y.$$

In particular, Δ_0 is the Wronskian

$$\Delta_0 = W(y_1, y_2, \ldots, y_k)(x).$$

If Δ_0 is identically zero, $y_1(x), \ldots, y_k(x)$ are linearly dependent over $\mathbb{C}$, which will be shown later in Theorem 5.16. This contradicts that $y_1(x), \ldots, y_n(x)$ is a fundamental system of solutions, and hence Δ_0 is not identically zero. Let N be the set of zeros in D of Δ_0. For $i = 1, 2, \ldots, k$, we set

$$q_i(x) = \frac{\Delta_k}{\Delta_0}.$$

Then $q_i(x)$ is holomorphic on $D \setminus N$, and the points in N are at most poles. Take any $\gamma \in \pi_1(D, b)$, and assume that the representative of γ is taken as a loop in the domain $D \setminus N$. Then we obtain

$$\gamma_* q_i(x) = \gamma_* \left(\frac{\Delta_i}{\Delta_0} \right) = \frac{\Delta_i |M_\gamma^{11}|}{\Delta_0 |M_\gamma^{11}|} = q_i(x)$$

from (5.11). Thus $q_i(x)$ is holomorphic and single valued in $D \setminus N$. Moreover, since $q_i(x)$ is a ratio of differential polynomials in $y_1(x), \ldots, y_k(x)$, $q_i(x)$ is at most regular singular at the points $a_0, a_1, \ldots, a_p$. Hence, owing to Theorem 4.1, $q_i(x)$ is a rational function. In particular, we see that the set N is a finite set. Define the differential operator K by

$$K = \partial^k + q_1(x)\partial^{k-1} + \cdots + q_k(x).$$

Then we see that $(y_1(x), \ldots, y_k(x))$ makes a fundamental system of solutions of the differential equation $Ky = 0$.

The orders of the differential operators L, K are n, k, respectively, with $n > k$, and then we can divide L by K. If we set

$$L_1 = L - \partial^{n-k} K,$$

the order of the operator L_1 is at most $n - 1$. Then we can write

$$L_1 = r_1(x)\partial^{n-1} + \cdots,$$

where $r_1(x)$ is a rational function (may be 0). Next set

$$L_2 = L_1 - r_1(x)\partial^{n-k-1} K.$$

Then the order of L_2 is at most $n - 2$, and we can write

$$L_2 = r_2(x)\partial^{n-2} + \cdots$$

with a rational function $r_2(x)$. Continuing these processes, we come to

$$L - (\partial^{n-k} + r_1(x)\partial^{n-k-1} + \cdots + r_{n-k}(x))K = R,$$

where the order of R is at most $k-1$. Now $y_1(x), \ldots, y_k(x)$ are common solutions of $Ly = 0$ and $Ky = 0$, and then they are solutions of $Ry = 0$. Since $y_1(x), \ldots, y_k(x)$ are linearly independent and the order of R is at most $k - 1$, we have $R = 0$. Thus, by setting

$$M = \partial^{n-k} + r_1(x)\partial^{n-k-1} + \cdots + r_{n-k}(x),$$

we get (5.10). This completes the proof.

The converse assertion is clear. □

When L is decomposed into a product of differential operators as in (5.10), we can solve the differential equation $Ly = 0$ as follows. First we solve the differential equation $Ky = 0$ of order k to obtain a fundamental system of solutions $y_1(x), \ldots, y_k(x)$. Since the solutions are also solutions of $Ly = 0$, we have obtained k linearly independent solutions of $Ly = 0$. Next we solve the differential equation $Mz = 0$ of order $m = n - k$ to obtain a fundamental system of solutions $z_1(x), \ldots, z_m(x)$. If y is a solution of $Ly = 0$, we have

$$0 = Ly = M(Ky),$$

and hence $z = Ky$ is a solution of $Mz = 0$. Then, by the method of variation of constants, we solve the inhomogeneous linear differential equation

$$Ky = z_i(x)$$

to get a solution $y_{k+i}(x)$ for $i = 1, 2, \ldots, m$. We see that m solutions $y_{k+1}(x), \ldots, y_n(x)$ are linearly independent solutions of the equation $Ly = 0$. Thus we obtain n solutions $y_1(x), \ldots, y_k(x), y_{k+1}(x), \ldots, y_n(x)$ of $Ly = 0$, and it is easy to see that these solutions are linearly independent. In this way, we can obtain a fundamental system of solutions by solving lower order differential equations.

The following fact is a direct consequence of the proof of Theorem 5.6.

Corollary 5.7 *Notations being as in Theorem 5.6 and its proof. If we set*

$$G_1 = \left\{ M_\gamma^{11} ; \gamma \in \pi_1(D, b) \right\},$$

G_1 *is the monodromy group of the differential equation*

$$Ky = 0$$

with respect to the fundamental system of solutions $(y_1(x), y_2(x), \ldots, y_k(x))$.

We remark that the zeros of Δ_0 are singular points of the differential equation $Ky = 0$, and are not singular points of $Ly = 0$, which implies that $y_1, \ldots, y_k$ are single valued around these points. Then the zeros of Δ_0 are apparent singular points of $Ky = 0$.

Next we consider systems of differential equations of the first order. By using a cyclic vector, we can transform a system to a scalar differential equation, and then we can apply Theorem 5.6. Thus we obtain the following resut.

Theorem 5.8 *Assume that a system of differential equations of the first order*

$$\frac{dY}{dx} = A(x)Y \tag{5.12}$$

is Fuchsian. If the monodromy group of the system is reducible, the system (5.12) can be transformed, by a gauge transformation

$$Y = P(x)Z, \quad P(x) \in \mathrm{GL}(n, \mathbb{C}(x))$$

with rational coefficients, to a system

$$\frac{dZ}{dx} = B(x)Z, \quad B(x) = \begin{pmatrix} B_{11}(x) & B_{12}(x) \\ O & B_{22}(x) \end{pmatrix}$$

with a block upper-triangular matrix coefficient.

Proof By using a cyclic vector, we can transform the system (5.12) to a scalar differential equation. Namely, there exists $Q(x) \in GL(n, \mathbb{C}(x))$ such that the gauge transformation

$$Y = Q(x)U$$

sends (5.12) to

$$\frac{dU}{dx} = \begin{pmatrix} 0 & 1 & & & \\ & \ddots & \ddots & & \\ & & \ddots & \ddots & \\ & & & 0 & 1 \\ -p_n(x) & \cdots & \cdots & -p_2(x) & -p_1(x) \end{pmatrix} U, \tag{5.13}$$

where $p_i(x) \in \mathbb{C}(x)$ $(1 \le i \le n)$. Write $U = {}^t(u_1, u_2, \ldots, u_n)$ and set $u = u_1$. Then we have $u_i = u^{(i-1)}$, and the system (5.13) is equivalent to a scalar differential equation $Lu = 0$ with the differential operator (5.8). Since a gauge transformation with coefficients in rational functions does not change the monodromy group, the monodromy group of $Lu = 0$ is still reducible. Then, owing to Theorem 5.6, there exist differential operators K, M of the form (5.9) such that

$$L = MK.$$

If we set $Ku = v$, u is a solution of $Lu = 0$ if and only if v is a solution of $Mv = 0$. Thus the differential equation $Lu = 0$ is equivalent to

$$\begin{cases} Ku = v, \\ Mv = 0. \end{cases}$$

Now we set

$$Z = {}^t\left(u, u', \ldots, u^{(k-1)}, v, v', \ldots, v^{(m-1)}\right).$$

Then the differential equation $Lu = 0$ is transformed to the system

$$\frac{dZ}{dx} = \begin{pmatrix} 0 & 1 & & & & & & \\ & \ddots & \ddots & & & & & \\ & & 0 & 1 & & & & \\ -q_k & \cdots & \cdots & -q_1 & 1 & & & \\ \hline & & & & 0 & 1 & & \\ & & O & & & \ddots & \ddots & \\ & & & & & & 0 & 1 \\ & & & & -r_m & \cdots & \cdots & -r_1 \end{pmatrix} Z$$

with a block upper-triangular matrix coefficient. We look at the transformation from U to Z. We have

$$v = Ku = u_{k+1} + q_1(x)u_k + \cdots + q_k(x)u_1,$$

and by differentiating this, we get

$$v^{(i)} = u_{i+k+1} + \varphi_i,$$

where φ_i is a linear combination of $u_1, \ldots, u_{i+k}$ with rational function coefficients. Thus the transformation from U to Z is a non-singular gauge transformation with rational function coefficients:

$$Z = R(x)U, \quad R(x) \in \mathrm{GL}(n, \mathbb{C}(x)).$$

Hence, by setting

$$P(x) = Q(x)R(x)^{-1},$$

we obtain the assertion of the theorem.   □

We have shown that a Fuchsian differential equation with a reducible monodromy group is reduced to lower order differential equations. However, our proof shows only the existence of lower order differential equations, and is not constructive. To find the lower order differential equations is another problem.

Now we introduce one criterion for irreducibility of a group $G \subset \mathrm{GL}(n, \mathbb{C})$ by using generators of the group. The following is a key lemma.

Lemma 5.9 *Let $M \in \mathrm{M}(n \times n, \mathbb{C})$ be a semi-simple (diagonalizable) matrix, and*

$$\mathbb{C}^n = \bigoplus_\lambda V_\lambda$$

be the direct sum decomposition of $\mathbb{C}^n$ into the eigenspaces of M. Here λ denotes an eigenvalue of M, V_λ the λ-eigenspace of M, and λ runs through all distinct eigenvalues of M. For each λ, we denote the projection onto the component V_λ by π_λ:

$$\pi_\lambda : \mathbb{C}^n \to V_\lambda.$$

If W is an M-invariant subspace of $\mathbb{C}^n$, we have $\pi_\lambda(W) \subset W$ for each λ.

Proof Let $\lambda_1, \lambda_2, \ldots, \lambda_l$ be the distinct eigenvalues of M. Take any $w \in W$, and set $\pi_{\lambda_i}(w) = w_i$. Then we have

$$w = w_1 + w_2 + \cdots + w_l. \tag{5.14}$$

Since $w_i \in V_{\lambda_i}$, we have $M w_i = \lambda_i w_i$. By noting this, we operate M^j to both sides of (5.14) to obtain

$$M^j w = \lambda_1{}^j w_1 + \lambda_2{}^j w_2 + \cdots + \lambda_l{}^j w_l.$$

Collecting these equalities for $j = 0, 1, 2, \ldots, l-1$, we get

$$\left(w, Mw, M^2 w, \ldots, M^{l-1} w \right) = (w_1, w_2, \ldots, w_l) \begin{pmatrix} 1 & \lambda_1 & \lambda_1{}^2 & \cdots & \lambda_1{}^{l-1} \\ 1 & \lambda_2 & \lambda_2{}^2 & \cdots & \lambda_2{}^{l-1} \\ \vdots & \vdots & \vdots & & \vdots \\ 1 & \lambda_l & \lambda_l{}^2 & \cdots & \lambda_l{}^{l-1} \end{pmatrix}.$$

Since each column of the left hand side is in W and the right matrix of the right hand side is non-singular by $\lambda_i \neq \lambda_j$ ($i \neq j$), we conclude $w_1, w_2, \ldots, w_l \in W$.   □

To study the irreducibility of a group $G \subset \mathrm{GL}(n, \mathbb{C})$ is to study the existence of a non-trivial G-invariant subspace. We can show the irreducibility by showing, for any G-invariant subspace W, $W = \mathbb{C}^n$ or $W = \{0\}$.

We consider the case that G is generated by a finite number of semi-simple matrices $M_1, M_2, \ldots, M_m$:

$$G = \langle M_1, M_2, \ldots, M_m \rangle.$$

Let W be any G-invariant subspace. For each M_j, take a projection π_λ to an eigenspace. Then we have $\pi_\lambda(W) \subset W$ by Lemma 5.9. We use this relation for various M_j and λ to determine W. Then, in some cases, we may show the irreducibility of G. We shall explain this procedure by an example.

Example 5.1 Let m be an integer greater than 1, and set $n = 2m$. Take non-zero complex numbers λ_i ($1 \leq i \leq m$), μ_i ($1 \leq i \leq m-1$), ν, ρ_1, ρ_2 satisfying

$$\prod_{i=1}^{m} \lambda_i \cdot \prod_{i=1}^{m-1} \mu_i \cdot \nu = \rho_1{}^m \rho_2{}^m,$$

$$\lambda_i \neq \lambda_j \ (i \neq j), \ \mu_i \neq \mu_j \ (i \neq j), \ \rho_1 \neq \rho_2$$

$$\lambda_i \neq 1 \ (1 \leq i \leq m), \ \mu_i \neq 1 \ (1 \leq i \leq m-1), \ \nu \neq 1. \qquad (5.15)$$

By using these quantities, we define $n \times n$ non-singular matrices M_1, M_2, M_3 as follows:

$$M_1 = \begin{pmatrix} \lambda_1 & & & \xi_{11} & \cdots & \xi_{1m} \\ & \ddots & & \vdots & & \vdots \\ & & \lambda_m & \xi_{m1} & \cdots & \xi_{mm} \\ \hline & O_m & & & I_m & \end{pmatrix},$$

$$M_2 = \begin{pmatrix} & I_m & & O_{m,m-1} & O_{m,1} \\ \hline \eta_{11} & \cdots & \eta_{1m} & \mu_1 & & \eta_{1n} \\ \vdots & & \vdots & & \ddots & \vdots \\ \eta_{m-1,1} & \cdots & \eta_{m-1,m} & & \mu_{m-1} & \eta_{m-1,n} \\ \hline & O_{1,m} & & O_{1,m-1} & 1 \end{pmatrix}, \qquad (5.16)$$

$$M_3 = \begin{pmatrix} I_{n-1} & O_{n-1,1} \\ \hline \zeta_1 \cdots \zeta_{n-1} & \nu \end{pmatrix},$$

where

$$\xi_{ij} = -\frac{(\lambda_i - \rho_1)(\lambda_i - \rho_2)}{(\rho_1\rho_2)^{m-1}} \prod_{\substack{1 \le k \le m \\ k \ne i}} \frac{\lambda_k\mu_j - \rho_1\rho_2}{\lambda_k - \lambda_i} \qquad (1 \le i \le m, 1 \le j \le m-1),$$

$$\xi_{im} = -\frac{(\lambda_i - \rho_1)(\lambda_i - \rho_2)}{(\rho_1\rho_2)^{m-1}} \prod_{\substack{1 \le k \le m \\ k \ne i}} \frac{\lambda_k}{\lambda_k - \lambda_i} \qquad (1 \le i \le m),$$

$$\eta_{ij} = \prod_{\substack{1 \le l \le m-1 \\ l \ne i}} \frac{\lambda_j\mu_l - \rho_1\rho_2}{\mu_l - \mu_i} \qquad (1 \le i \le m-1, 1 \le j \le m),$$

$$\eta_{in} = \prod_{\substack{1 \le l \le m-1 \\ l \ne i}} \frac{1}{\mu_i - \mu_l} \qquad (1 \le i \le m-1),$$

$$\zeta_j = \frac{\nu}{\rho_1\rho_2} \prod_{l=1}^{m-1} (\rho_1\rho_2 - \lambda_j\mu_l) \qquad (1 \le j \le m),$$

$$\zeta_{m+j} = -\frac{1}{\mu_j{}^m} \prod_{k=1}^{m} \frac{\lambda_k\mu_j - \rho_1\rho_2}{\lambda_k} \qquad (1 \le j \le m-1).$$

We consider the irreducibility of the group

$$G = \langle M_1, M_2, M_3 \rangle.$$

Later we will explain how this group is obtained.

Proposition 5.10 *The group G is irreducible if and only if*

$$\lambda_i \neq \rho_k \qquad (1 \le i \le m, \ k = 1, 2),$$
$$\lambda_i \mu_j \neq \rho_1 \rho_2 \qquad (1 \le i \le m, \ 1 \le j \le m - 1), \qquad (5.17)$$
$$\rho_1 \neq 1, \ \rho_2 \neq 1$$

holds.

Proof It is easy to see that the matrices M_1, M_2, M_3 are semi-simple, and hence we can apply Lemma 5.9. For each M_i, we calculate the direct sum decomposition into the eigenspaces and the projections onto the direct sum components. Denote by e_i the unit vector in $\mathbb{C}^n$ with the only non-zero entry 1 in i-th position.

We see that the eigenvalues of M_1 are 1 of multiplicity m and λ_i $(1 \le i \le m)$ of multiplicity free. Let X_0 be the 1-eigenspace of M_1, X_i the λ_i-eigenspace of M_1, and take a direct sum decomposition

$$\mathbb{C}^n = X_0 \oplus X_1 \oplus \cdots \oplus X_m.$$

Denote by

$$p_i : \mathbb{C}^n \to X_i \qquad (0 \le i \le m)$$

the projection onto each component. Then we have, for $v = {}^t(v_1, \ldots, v_n) \in \mathbb{C}^n$,

$$p_0(v) = \begin{pmatrix} x_1(v) \\ \vdots \\ x_m(v) \\ v_{m+1} \\ \vdots \\ v_n \end{pmatrix}, \quad p_i(v) = (v_i - x_i(v))e_i \quad (1 \le i \le m),$$

where

$$x_i(v) = \frac{\sum_{k=1}^{m} \xi_{ik} v_{m+k}}{1 - \lambda_i} \qquad (1 \le i \le m).$$

Note that $X_i = \langle e_i \rangle$ for $1 \le i \le m$.

The eigenvalues of M_2 are 1 of multiplicity $m + 1$ and μ_i $(1 \le i \le m - 1)$ of multiplicity free. Let Y_0 be the 1-eigenspace of M_2, Y_i the μ_i-eigenspace of M_2, and take a direct sum decomposition

$$\mathbb{C}^n = Y_0 \oplus Y_1 \oplus \cdots \oplus Y_{m-1}.$$

Denote by

$$q_i : \mathbb{C}^n \to Y_i \qquad (0 \le i \le m - 1)$$

the projection onto each component. Then we have, for $v = {}^t(v_1, \ldots, v_n) \in \mathbb{C}^n$,

$$q_0(v) = \begin{pmatrix} v_1 \\ \vdots \\ v_m \\ y_{m+1}(v) \\ \vdots \\ y_{n-1}(v) \\ v_n \end{pmatrix}, \quad q_i(v) = (v_{m+i} - y_{m+i}(v))e_{m+i} \quad (1 \le i \le m - 1),$$

where

$$y_{m+i}(v) = \frac{\sum_{k=1}^m \eta_{ik} v_k + \eta_{in} v_n}{1 - \mu_i} \qquad (1 \le i \le m - 1).$$

Note that $Y_i = \langle e_{m+i} \rangle$ for $1 \le i \le m - 1$.

The eigenvalues of M_3 are 1 of multiplicity $n - 1$ and v of multiplicity free. Let Z_0 be the 1-eigenspace of M_3, Z_1 the v-eigenspace of M_3, and take a direct sum decomposition

$$\mathbb{C}^n = Z_0 \oplus Z_1.$$

Denote by

$$r_i : \mathbb{C}^n \to Z_i \qquad (i = 0, 1)$$

the projection onto each component. Then we have, for $v = {}^t(v_1, \ldots, v_n) \in \mathbb{C}^n$,

$$r_0(v) = \begin{pmatrix} v_1 \\ \vdots \\ v_{n-1} \\ z_n(v) \end{pmatrix}, \quad r_1(v) = (v_n - z_n(v))e_n,$$

where

$$z_n(v) = \frac{\sum_{k=1}^{n-1} \zeta_k v_k}{1 - v}.$$

Note that $Z_1 = \langle e_n \rangle$.

Assume that $G = \langle M_1, M_2, M_3 \rangle$ is reducible, and let W be a non-trivial G-invariant subspace. From the explicit forms of X_i $(1 \le i \le m)$, Y_i $(1 \le i \le m-1)$ and Z_1, we obtain the direct sum decomposition

$$\mathbb{C}^n = \bigoplus_{i=1}^{m} X_i \oplus \bigoplus_{i=1}^{m-1} Y_i \oplus Z_1$$

into 1-dimensional subspaces. Since we assume $W \ne \mathbb{C}^n$, at least one intersection of W and one of the direct sum components is $\{0\}$. First assume that

$$W \cap X_i = \{0\}$$

holds for some i $(1 \le i \le m)$. We fix this i.

Take any $v \in W \setminus \{0\}$ and set $y_j = q_j(v)$ $(0 \le j \le m-1)$. Then, owing to Lemma 5.9, we have $y_j \in W$. If $y_j \ne 0$ holds for some j $(1 \le j \le m-1)$, we get $W \cap Y_j \ne \{0\}$, which implies $e_{m+j} \in W$. Then, again thanks to Lemma 5.9, we have $p_i(e_{m+j}) \in W$. Since $W \cap X_i = \{0\}$, we have

$$p_i(e_{m+j}) = \frac{\xi_{ij}}{\lambda_i - 1} e_i = 0,$$

from which we obtain

$$\xi_{ij} = 0.$$

If $y_j = 0$ holds for any j $(1 \le j \le m-1)$, we have $y_0 \ne 0$ by $v \ne 0$. Then we get $y_0 \in W \setminus \{0\}$. Set $r_0(y_0) = z_0, r_1(y_0) = z_1$. Then, by Lemma 5.9, we get $z_k \in W \cap Z_k$ $(k = 0, 1)$. If $z_1 \ne 0$, by a similar argument as above, we get

$$\xi_{im} = 0.$$

If $z_1 = 0$, we have

$$v = y_0 = z_0 \in Y_0 \cap Z_0.$$

In this case, we consider $w = M_1 v \in W$ instead of v. Then, in a similar argument, we get $\xi_{ij} = 0, \xi_{im} = 0$, or

$$w = M_1 v \in Y_0 \cap Z_0.$$

If the last condition holds, we have $v \in Y_0$, $v \in Z_0$, $M_1 v \in Y_0$ and $M_1 v \in Z_0$, which yields the linear equation

$$
\begin{cases}
\displaystyle\sum_{l=1}^{m} \eta_{kl} v_l + (\mu_k - 1)v_{m+k} + \eta_{kn} v_n = 0 & (1 \le k \le m-1), \\[2ex]
\displaystyle\sum_{l=1}^{n-1} \zeta_l v_l + (v - 1)v_n = 0, \\[2ex]
\displaystyle\sum_{l=1}^{m} \eta_{kl}\left((\lambda_l - 1)v_l + \sum_{p=1}^{m} \xi_{lp} v_{m+p} \right) = 0 & (1 \le k \le m-1), \\[2ex]
\displaystyle\sum_{l=1}^{m} \zeta_l \left((\lambda_l - 1)v_l + \sum_{p=1}^{m} \xi_{lp} v_{m+p} \right) = 0.
\end{cases}
$$

Denote the coefficient matrix of this equation by Q. Then the linear equation is written as $Qv = 0$. We see that the determinant of Q is expressed as follows ([49]):

$$
\det Q = \pm \frac{\displaystyle\prod_{1 \le i < j \le m} \left(\frac{\rho_1 \rho_2}{\lambda_i} - \frac{\rho_1 \rho_2}{\lambda_j} \right)}{\displaystyle\prod_{1 \le k < l \le m-1} (\mu_l - \mu_k)} (\rho_1 - 1)^m (\rho_2 - 1)^m.
$$

Since $v \ne 0$, we get $\det Q = 0$, which implies

$$
\rho_1 = 1, \text{ or } \rho_2 = 1
$$

under the assumption (5.15).

If we assume $W \cap Y_i = \{0\}$ or $W \cap Z_1 = \{0\}$, we get the same condition. Thus, the condition (5.17) implies that G is irreducible.

Conversely, we assume that the condition (5.17) does not hold. Set $V_i = \langle e_i \rangle$ $(1 \le i \le n)$. If $\lambda_i = \rho_1$ or $\lambda_i = \rho_2$ holds, $\bigoplus_{j \ne i} V_j$ is a G-invariant subspace. If $\lambda_i \mu_j = \rho_1 \rho_2$ holds, $V_i \oplus V_{m+j}$ is a G-invariant subspace. If $\rho_1 = 1$ or $\rho_2 = 1$ holds, we have $\det Q = 0$, and in this case the 0-eigenspace of Q becomes G-invariant. Thus, if the condition (5.17) does not hold, G becomes reducible.   □

Here we explain the origin of the above group G. As will be explained in Chap. 7, we define a numerical data called the spectral type for each Fuchsian differential equation. Roughly speaking, the spectral data describes the multiplicities of the eigenvalues of the local monodromies. The Fuchsian differential equation having the spectral data

$$
\left((m, 1^m), \left(m+1, 1^{m-1} \right), (2m-1, 1), (m, m) \right)
$$

becomes rigid, which means that the monodromy is uniquely determined by the local monodromies. Notational remark: we denote

$$1^m = \overbrace{1, 1, \ldots, 1}^{m}.$$

The group G is obtained as the monodromy group of the Fuchsian differential equation with this spectral type. To be accurate, the group G is the monodromy group of the differential equation if it is irreducible (namely if the condition (5.17) holds). If G is reducible, G may be the monodromy group, however, sometimes not.

Please refer to [48, 49, 186] for more detail of this example.

5.4.3 Finite Monodromy

We consider a scalar Fuchsian differential equation

$$y^{(n)} + p_1(x)y^{(n-1)} + \cdots + p_n(x)y = 0. \tag{5.18}$$

Let $x = a_0, a_1, \ldots, a_p$ the set of the singular points.

Theorem 5.11 *The monodromy group of the Fuchsian differential equation (5.18) is a finite group if and only if any solution of the differential equation is an algebraic function.*

Proof Take any point $x_0 \in D = \mathbb{P}^1(\mathbb{C}) \setminus \{a_0, a_1, \ldots, a_p\}$, a simply connected neighborhood U of x_0 in D, and a fundamental system of solutions $\mathcal{Y}(x)$ of (5.18) on U. Suppose that the monodromy group with respect to $\mathcal{Y}(x)$ is finite. Take any solution $y(x)$ of (5.18) on U. Since the monodromy group is finite, the results of analytic continuations of $\mathcal{Y}(x)$ along all loops with base point x_0 are finite in number. Then it follows that the number of the branches of $y(x)$ is finite, because $y(x)$ is a linear combination of $\mathcal{Y}(x)$. Let

$$F = \{y_1(x), y_2(x), \ldots, y_N(x)\}$$

be the set of all branches of $y(x)$ on U. Take any $\gamma \in \pi_1(D, x_0)$. For each $y_j(x) \in F$, the result $\gamma_* y_j(x)$ of analytic continuation along γ is a branch of $y(x)$, and hence belongs to F. Note that, if $y_j(x) \neq y_k(x)$, we have $\gamma_* y_j(x) \neq \gamma_* y_k(x)$. Then we see that the analytic continuation along γ causes a permutation of F. Hence, if we denote by $s_i(x)$ ($1 \leq i \leq N$) the i-th elementary symmetric polynomial in the elements of F, $s_i(x)$ is invariant under the analytic continuation along γ. Namely $s_i(x)$ is a single valued holomorphic function on D. Moreover the singular points $a_0, a_1, \ldots, a_p$ are at most regular singular, which implies, by Theorem 4.1, the singular points are at most poles. Hence each $s_i(x)$ is a rational function. Now we know that any element in F is a root of the algebraic equation

$$X^N - s_1(x)X^{N-1} + s_2(x)X^{N-2} - \cdots + (-1)^N s_N(x) = 0$$

with coefficients in rational functions, which implies $y(x)$ is an algebraic function.

Conversely, if any solution of (5.18) is an algebraic function, the results of analytic continuations of any fundamental system of solutions is finite in number. Then the monodromy group is finite.   □

We have a similar result for a system of differential equations of the first order.

Theorem 5.12 *The monodromy group of the system of Fuchsian differential equations*

$$\frac{dY}{dx} = A(x)Y$$

of the first order is a finite group if and only if every entry of any solution is an algebraic function.

We can prove this by replacing $y(x)$ in the proof of Theorem 5.11 by any entry $y_i(x)$ of any solution $Y(x)$.

We note that, if any solution of a linear differential equation is an algebraic function, then the differential equation is Fuchsian, because any singular point of an algebraic function is regular singular. However, this fact does not mean that a linear differential equation with a finite monodromy group is Fuchsian.[3]

The following fact can be easily derived from Theorems 5.11 and 5.12, and is useful.

Corollary 5.13 *If any solution of a linear differential equation is an algebraic function, the local monodromy at each singular point is of finite order (as an element of the general linear group). In particular, for such differential equation, every characteristic exponent is a rational number.*

Proof By Theorem 5.11 or 5.12, we see that the monodromy group is finite. Since a local monodromy is a conjugacy class of an element of the monodromy group, which is a finite group, the order of a local monodromy is finite. Then any eigenvalue of a local monodromy is a root of 1. We know, by Theorem 4.13, that an eigenvalue of a local monodromy is given by $e^{2\pi\sqrt{-1}\lambda}$ with a characteristic exponent λ. Hence the characteristic exponent should be a rational number.   □

It was Schwarz who studied the condition for a differential equation whose solutions are algebraic functions and obtained an essential result. He considered the Gauss hypergeometric differential equation

$$x(1-x)y'' + (\gamma - (\alpha + \beta + 1)x)y' - \alpha\beta y = 0, \tag{5.19}$$

---

[3] A simple counter example is $y' - y = 0$.

where $\alpha, \beta, \gamma \in \mathbb{C}$ are parameters. Equation (5.19) is Fuchsian with regular singular points at $x = 0, 1, \infty$. The table of the characteristic exponents at each singular point $0, 1, \infty$ is given by

$$\left\{ \begin{matrix} x = 0 & x = 1 & x = \infty \\ 0 & 0 & \alpha \\ 1 - \gamma & \gamma - \alpha - \beta & \beta \end{matrix} \right\}.$$

This table is called the Riemann scheme, which will be introduced in Sect. 7.2. Here we briefly explain the works of Schwarz and the development caused by the works. A good reference is a book [147] (in Japanese).

If any solution of the hypergeometric differential equation (5.19) is algebraic, we have $\alpha, \beta, \gamma \in \mathbb{Q}$ by Corollary 5.13. Then we assume this condition. Let (y_1, y_2) be a fundamental system of solutions of (5.19). Then the Wronskian $y_1 y_2' - y_2 y_1'$ is a scalar multiple of $x^\gamma (1 - x)^{\alpha + \beta - \gamma + 1}$, and hence is an algebraic function by the assumption $\alpha, \beta, \gamma \in \mathbb{Q}$. Using this fact, we can see that both y_1 and y_2 are algebraic function if and only if the ratio y_1/y_2 is an algebraic function.

Schwarz started to study the map defined by the ratio y_1/y_2 of a fundamental system of solutions of (5.19). This map

$$\begin{aligned} \sigma : \mathbb{P}^1 \setminus \{0, 1, \infty\} &\to \mathbb{P}^1 \\ x &\mapsto y_1(x)/y_2(x) \end{aligned}$$

is called the Schwarz map. The Schwarz map σ depends on the choice of the fundamental system of solutions (y_1, y_2). If we replace it by another fundamental system of solutions $(\tilde{y}_1, \tilde{y}_2)$, we have

$$(\tilde{y}_1, \tilde{y}_2) = (ay_1 + by_2, cy_1 + dy_2), \quad \begin{pmatrix} a & b \\ c & d \end{pmatrix} \in GL(2, \mathbb{C}),$$

and hence

$$\tilde{y}_1(x)/\tilde{y}_2(x) = \frac{ay_1(x) + by_2(x)}{cy_1(x) + dy_2(x)} = \frac{a\sigma(x) + b}{c\sigma(x) + d}.$$

Thus the change of a fundamental system of solutions acts as a fractional linear transformation of σ.

Let us study the image of the map σ. The domain $\mathbb{P}^1 \setminus \{0, 1, \infty\}$ is not simply connected, and then it is hard to see its image. Then we first study the image of its simply connected subdomain $\mathbb{H} = \{x \in \mathbb{C}; \operatorname{Im} x > 0\}$, which is the upper half plane. It is enough to study the image of the three intervals $(-\infty, 0), (0, 1), (1, +\infty)$, which are the intersections of the boundary $\mathbb{R}$ of $\mathbb{H}$ and the domain $\mathbb{P}^1 \setminus \{0, 1, \infty\}$.

In order to study the image of the interval $(0, 1)$, we take solutions $\varphi_1(x)$ and $x^{1-\gamma} \varphi_2(x)$ which are of exponent 0 and $1 - \gamma$, respectively, at $x = 0$, as a fundamental system of solutions, where $\varphi_1(x)$, $\varphi_2(x)$ are convergent power series at $x = 0$ with the radius of convergence 1 and are normalized as $\varphi_1(0) = \varphi_2(0) = 1$. We remark that the solution $\varphi_1(x)$ is nothing but the hypergeometric series $F(\alpha, \beta, \gamma; x)$, however, we do not use this fact in the following argument. We use the following basic fact. Since we have assumed $\alpha, \beta, \gamma \in \mathbb{Q}$, the coefficients of the differential equation (5.19) are real-valued fundtions. Then we can apply the results of ordinary differential equations on $\mathbb{R}$ when we restrict the argument to $x \in \mathbb{R}$. $\varphi_1(x)$ is a solution in a neighborhood of $x = 0$ in $\mathbb{R}$, and the coefficients of the differential equation (5.19) are continuous on the intervals $(-\infty, 0)$ and $(0, 1)$, which implies that $\varphi_1(x)$ is a real-valued function defined on $(-\infty, 1)$. If we apply the gauge transformation $u = x^{\gamma-1}y$ to (5.19), we get a differential equations obtained from (5.19) by replacing the parameters (α, β, γ) by $(\alpha - \gamma + 1, \beta - \gamma + 1, 2 - \gamma)$, and hence also we have a differential equation with coefficients in real-valued functions. Thus by a similar reason, $\varphi_2(x)$ is a real-valued function defined on $(-\infty, 1)$.

By taking $\arg x = 0$ on the interval $(0, 1)$, we have $x^{1-\gamma} > 0$ and

$$\lim_{x \to +0} x^{1-\gamma} = \begin{cases} 0 & (1 - \gamma > 0) \\ +\infty & (1 - \gamma < 0). \end{cases}$$

If $1 - \gamma > 0$, we take $y_1(x) = x^{1-\gamma}\varphi_2(x)$, $y_2(x) = \varphi_1(x)$, and if $1 - \gamma < 0$, we take $y_1(x) = \varphi_1(x)$, $y_2(x) = x^{1-\gamma}\varphi_2(x)$. Then in both cases, by defining $\sigma(x) = y_1(x)/y_2(x)$, we have $\lim_{x \to +0} \sigma(x) = 0$. Since $\varphi_1(0) = \varphi_2(0) = 1 > 0$ and $\sigma'(x) \neq 0$ on $(0, 1)$, we see that $\sigma(x)$ takes a positive real value on $(0, 1)$. Thus the image of $(0, 1)$ by the map σ becomes the interval $(0, \sigma(1))$.

When x moves from $(0, 1)$ to $(-\infty, 0)$ via the upper half plane, we have $\arg x = \pi$, and hence $\arg \sigma(x) = \arg x^{|1-\gamma|} = |1 - \gamma|\pi$. Then the image of $(-\infty, 0)$ by the map σ becomes a line segment from 0 to $\sigma(-\infty)$ with the angle of incline $|1 - \gamma|\pi$.

Similarly, by taking solutions at $x = 1$ of exponents 0 and $\gamma - \alpha - \beta$ as a fundamental system of solutions, we see that the images of $(0, 1)$ and $(1, \infty)$ by σ are two line segments with angle $|\gamma - \alpha - \beta|\pi$. Moreover, by using solutions at $x = \infty$ of exponents α and β, we see that the images of $(1, +\infty)$ and $(-\infty, 0)$ by σ are two line segments with angle $|\alpha - \beta|\pi$.

As we have noted, by the change of fundamental systems of solutions, σ is transformed by a fractional linear transformation, which sends a circle (or a line) to a circle (or a line). Thus for any fundamental system of solutions, the images of $(-\infty, 0)$, $(0, 1)$, $(1, +\infty)$ by σ becomes three arcs. Set

$$\lambda = |1 - \gamma|, \ \mu = |\gamma - \alpha - \beta|, \ \nu = |\alpha - \beta|.$$

Then $\lambda\pi$, $\mu\pi$, $\nu\pi$ are the angles between these three arcs. If we assume $\lambda, \mu, \nu < 2$, these arcs makes an arc triangle with angles $\lambda\pi$, $\mu\pi$, $\nu\pi$. Thus the interior of this arc triangle is the image of the upper half plane $\mathbb{H}$ by the map σ.

There are three ways to go from the upper half plane to the lower half plane: to go through one of three intervals $(-\infty, 0)$, $(0, 1)$, $(1, +\infty)$. If we analytically continue $\sigma(x)$ from the upper half plane to the lower half plane through $(0, 1)$, the image of the lower half plane by the map σ becomes, by the reflection principle of Schwarz, the reflection of the arc triangle with respect to the side that is the image of $(0, 1)$. By repeating such process, we see that the image of σ is the sequence of reflections of arc triangles with respect to each side.

In this way, Schwarz succeeded to describe the behavior of solutions of (5.19) geometrically by using the Schwarz map. In particular, the action of the monodromy group can be realized geometrically. Take a base point b in the upper half plane, and fix a fundamental system of solutions (y_1, y_2) defined in a simply connected neighborhood of b. Let γ be a loop with base point b which encircles $x = 0$ once in the positive direction. We consider the analytic continuation of $\sigma(x) = y_1(x)/y_2(x)$ along γ. The point x on γ first meets with the interval $(-\infty, 0)$, then goes into the lower half plane, and finally goes into the upper half plane via the interval $(0, 1)$ to come back to b. Then, at first $\sigma(x)$ is in the arc triangle Δ_1 which is the image of the upper half plane, then goes into the reflection Δ_2 with respect to the side that is the image of $(-\infty, 0)$, and finally goes into the reflection Δ_3 of Δ_2 with respect to the side that is the image of $(0, 1)$. Thus $\gamma_*\sigma(b)$ is a point in Δ_3. It is similar for any other loops with base point b, and the result of the analytic continuation along such loop is in an arc triangle which is obtained from Δ_1 by reflections in even times (Fig. 5.4).

Schwarz considered the case where the arc triangles do not overlap each other, namely, the case that the inverse function σ^{-1} becomes single valued. For this case, the numbers λ, μ, ν should be inverses of integers or ∞. We set

$$\lambda = \frac{1}{m}, \quad \mu = \frac{1}{n}, \quad \nu = \frac{1}{p}$$

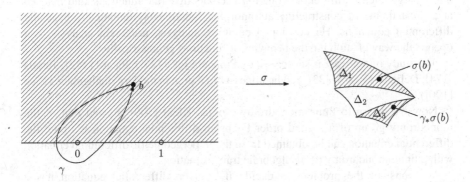

Fig. 5.4 Images of Schwarz map

with $m, n, p \in \mathbb{Z} \cup \{\infty\}$. We divide into three cases:

(1) $\lambda + \mu + \nu > 1$,
(2) $\lambda + \mu + \nu = 1$,
(3) $\lambda + \mu + \nu < 1$.

The triples (m, n, p) satisfying (1) are

$$(2, 2, k), \quad (2, 3, 3), \quad (2, 3, 4), \quad (2, 3, 5),$$

where $k \in \mathbb{Z}_{>0}$. For each triple, $\mathbb{P}^1$, the target space of σ, is completely covered by a finite number of arc triangles of angles $\lambda\pi, \mu\pi, \nu\pi$. This implies that the image of σ is $\mathbb{P}^1$, and that the monodromy group is finite. Thus, for these triples, all solutions of (5.19) are algebraic functions. The monodromy groups are known to be the dihedral group, the tetrahedral group, the octahedral group and the icosahedral group, respectively.

The triples (m, n, p) satisfying (2) are

$$(\infty, 2, 2), \quad (3, 3, 3), \quad (2, 4, 4), \quad (2, 3, 6).$$

In this case, $\mathbb{P}^1$ is covered by an infinite number of arc triangles of angles $\lambda\pi, \mu\pi, \nu\pi$. The inverse function σ^{-1} is written in terms of a trigonometric function for the case $(\infty, 2, 2)$, and is an elliptic function for the other cases.

There exist infinitely many triples (m, n, p) satisfying (3), and for each triple, the image of σ is the interior of a circle C on $\mathbb{P}^1$. We can see that the interior of C is covered by an infinite number of arc triangles given by the arcs perpendicular to the circle C. In this case, the inverse function σ^{-1} becomes an automorphic function. For example, when $(m, n, p) = (\infty, 3, 2)$, we have the elliptic modular function $j(\tau)$ as the inverse of σ, and when $(m, n, p) = (\infty, \infty, \infty)$, we have the lambda function $\lambda(\tau)$.

Thus, Schwarz started his study to determine all the cases where all solutions of the hypergeometric differential equation become algebraic functions, and reached at a great theme of constructing automorphic functions systematically by using differential equations. He developed complex analysis, used real analysis, and opened the way of studying the behavior of functions geometrically.

The study of Schwarz is succeeded by Poincaré (cf. [147]), Picard [142], Terada [174], Deligne-Mostow [33], and becomes a big theme in modern mathematics (See [190]).

Now we return to finite monodromy groups. Klein [109] showed that, if the monodromy group of a second order Fuchsian differential equation is finite, the differential equation can be obtained from the hypergeometric differential equation with a finite monodromy by an algebraic transformation.

We consider the problem to decide if a given differential equation has a finite monodromy. In general, it is very difficult to determine generators of the monodromy group of a given differential equation, and, even if we succeeded to get generators, it is also difficult to decide if the group is finite by looking at the generators. The following fact is useful for such study.

Theorem 5.14 *If a subgroup G of $\mathrm{GL}(n, \mathbb{C})$ is finite, there exists a positive definite G-invariant Hermitian form.*

Proof We set $G = \{g_1, g_2, \ldots, g_k\}$. If we define

$$H = \sum_{i=1}^{k} {}^t\bar{g}_i g_i,$$

we have ${}^t\bar{H} = H$, and hence H is a Hermitian form. For any $g \in G$, it follows directly from the definition of H that

$${}^t\bar{g}Hg = H,$$

and hence H is G-invariant. Since every element of G is a non-singular matrix, $g_i v \neq 0$ for any $v \in \mathbb{C}^n \setminus \{0\}$. Thus we have ${}^t\bar{v}\,{}^t\bar{g}_i \cdot g_i v > 0$. Hence

$${}^t\bar{v}Hv > 0$$

holds for any $v \in \mathbb{C}^n \setminus \{0\}$, which implies that H is positive definite. □

There are several works to determine differential equations which have finite monodromies by using classifications of finite groups.

A matrix $A \in \mathrm{GL}(n, \mathbb{R})$ is called a reflection if it is semi-simple with eigenvalue 1 of multiplicity $n - 1$ and -1 of multiplicity free. Extending this notion, we call a matrix $A \in \mathrm{GL}(n, \mathbb{C})$ a complex reflection if it is semi-simple with eigenvalue 1 of multiplicity $n - 1$. A subgroup of $\mathrm{GL}(n, \mathbb{C})$ generated by complex reflections is called a complex reflection group. Shephard-Todd [159] classified all finite irreducible complex reflection groups. This classification is used to determine differential equations with finite monodromies.

Takano-Bannai [171] determined, for a class of Fuchsian differential equations called Jordan-Pochhammer equations, all cases where the monodromy groups are finite. The monodromy group of a Jordan-Pochhammer equation is generated by complex reflections, and then they can apply the classification of Shephard-Todd.

Beukers-Heckman [18] obtained a similar result for the generalized hypergeometric differential equation, which is another class of Fuchsian differential equations, and satisfied by the generalized hypergeometric series ${}_nF_{n-1}$. In this case, the monodromy group is not generated by complex reflections, however, one of the generators can be a complex reflection. Then they also applied the Shephard-Todd classification.

It is shown in [60] that, for any finite irreducible complex reflection group $G \subset \mathrm{GL}(n, \mathbb{C})$, there exists a completely integrable system $\mathcal{M}$ of differential equations in $n - 1$ variables such that any completely integrable system with G as its monodromy group can be obtained from $\mathcal{M}$ by an algebraic transformation. Note that, for a second order Fuchsian ordinary differential equation, we may transform one of the characteristic exponents to 0 at all but one singular points by a gauge transformation, and hence the monodromy group may be considered as a complex reflection group. Then the result of [60] is an extension of the result of Klein we have mentioned above.

Jordan-Pochhammer equation and the generalized hypergeometric differential equation are rigid Fuchsian differential equations. As we will explain in Sect. 7.5.1, we can obtain generators of the monodromy group for any rigid Fuchsian differential equation. Belkale gave a criterion for the finiteness of the monodromy group for rigid Fuchsian differential equations in terms of Hermitian forms. Here we introduce the result without proof.

As will be shown in Chap. 7, generators of the monodromy group of a rigid Fuchsian differential equation can be expressed rationally by using the eigenvalues of the local monodromies. Since we are interested in the case where the monodromy group is finite, we may assume that the orders of the local monodromies are finite. Hence in this case, the eigenvalues of the local monodromies are roots of 1. Let $K/\mathbb{Q}$ be the field extension obtained from $\mathbb{Q}$ by adjoining the eigenvalues of the local monodromies. Then the monodromy group G becomes a subgroup of $\mathrm{GL}(n, K)$. Note that the Galois group $\Sigma = \mathrm{Gal}(K/\mathbb{Q})$ naturally acts on $\mathrm{GL}(n, K)$. We denote by σG the image of G by the action of $\sigma \in \Sigma$.

Theorem 5.15 (Belkale [16]) *The monodromy group G of a rigid Fuchsian differential equation is finite if and only if there exists a positive definite σG-invariant Hermitian form for any $\sigma \in \Sigma$.*

5.4.4 Differential Galois Theory and Solvable Monodromy

In this section we sketch differential Galois theory and show several useful facts. We will not go into detail, and omit many proofs. For the readers who are interested in differential Galois theory or differential algebra, please refer to Kaplansky [90], van der Put-Singer [178] or Nishioka [128]. The works of Kolchin [113, 114] are also basic.

A field K is called a *differential field* if a map

$$\partial : K \to K$$

satisfying (1) and (2) is equipped:

(1) $\partial(a + b) = \partial(a) + \partial(b)$ $(a, b \in K)$,
(2) $\partial(ab) = \partial(a)b + a\partial(b)$ $(a, b \in K)$.

We call the map ∂ a *differential*. Precisely speaking, a pair (K, ∂) of a field and a differential is a differential field. For a differential field (K, ∂), the subset

$$C_K = \{a \in K \; ; \; \partial(a) = 0\}$$

becomes a subfield of K. We call it the field of constants. In the following we often use ∂a instead of $\partial(a)$.

The field $\mathbb{C}$ of the complex numbers, the field $\mathbb{C}(x)$ of the rational functions, and the field $\mathcal{M}_D$ of the meromorphic functions on a domain $D \subset \mathbb{C}$ are differential fields with the usual differentiation $\dfrac{d}{dx}$ with respect to the coordinate x of $\mathbb{C}$. The fields of constants of these differential fields are $\mathbb{C}$.

For elements $y_1, y_2, \ldots, y_n$ of a differential field K, the Wronskian $W(y_1, y_2, \ldots, y_n)$ is defined by

$$W(y_1, y_2, \ldots, y_n) = \begin{vmatrix} y_1 & y_2 & \cdots & y_n \\ \partial y_1 & \partial y_2 & \cdots & \partial y_n \\ \vdots & \vdots & & \vdots \\ \partial^{n-1} y_1 & \partial^{n-1} y_2 & \cdots & \partial^{n-1} y_n \end{vmatrix},$$

which is same as the definition (3.10) in Chap. 3. The following assertion is obtained by an argument of differential algebra, while is useful in analysis.

Theorem 5.16 *Elements $y_1, y_2, \ldots, y_n$ of a differential field K is linearly dependent over the field C_K of constants if and only if the Wronskian $W(y_1, y_2, \ldots, y_n)$ is equal to $0 \in K$.*

Proof Assume that the elements $y_1, y_2, \ldots, y_n$ are linearly dependent over C_K. Then there exists ${}^t(c_1, c_2, \ldots, c_n) \in (C_K)^n \setminus \{0\}$ such that

$$c_1 y_1 + c_2 y_2 + \cdots + c_n y_n = 0. \tag{5.20}$$

Applying the differentiation up to $n - 1$ times to this equality, we get

$$\begin{pmatrix} y_1 & y_2 & \cdots & y_n \\ \partial y_1 & \partial y_2 & \cdots & \partial y_n \\ \vdots & \vdots & & \vdots \\ \partial^{n-1} y_1 & \partial^{n-1} y_2 & \cdots & \partial^{n-1} y_n \end{pmatrix} \begin{pmatrix} c_1 \\ c_2 \\ \vdots \\ c_n \end{pmatrix} = \begin{pmatrix} 0 \\ 0 \\ \vdots \\ 0 \end{pmatrix}. \tag{5.21}$$

Then the matrix in the left hand side should have an eigenvalue 0. Thus its determinant, which is nothing but the Wronskian, is equal to 0.

Conversely, we assume $W(y_1, y_2, \ldots, y_n) = 0$. Then there exists ${}^t(c_1, c_2, \ldots, c_n) \in K^n \setminus \{0\}$ satisfying (5.21). Without loss of generality, we may assume that $c_1 \neq 0$, and hence may assume $c_1 = 1$. Then applying the differentiation to (5.20), which is the first equality of (5.21), we get

$$\partial y_1 + c_2 \partial y_2 + \cdots + c_n \partial y_n + (\partial c_2 \cdot y_2 + \cdots + \partial c_n \cdot y_n) = 0.$$

Since the second equality of (5.21) holds, we obtain

$$\partial c_2 \cdot y_2 + \cdots + \partial c_n \cdot y_n = 0.$$

Continuing similar arguments, we gt

$$\begin{pmatrix} y_2 & \cdots & y_n \\ \vdots & & \vdots \\ \partial^{n-2} y_2 & \cdots & \partial^{n-2} y_n \end{pmatrix} \begin{pmatrix} \partial c_2 \\ \vdots \\ \partial c_n \end{pmatrix} = \begin{pmatrix} 0 \\ \vdots \\ 0 \end{pmatrix}.$$

Without loss of generality, we may assume $W(y_2, \ldots, y_n) \neq 0$. Then it follows

$$\partial c_2 = \cdots = \partial c_n = 0.$$

Thus we obtain $c_1, c_2, \ldots, c_n \in C_K$, which shows that (5.20) holds for ${}^t(c_1, c_2, \ldots, c_n) \in (C_K)^n \setminus \{0\}$.   □

Let (L, ∂) be a differential field. If a subfield K of L is closed under ∂, (K, ∂) becomes a differential field. In this case, we call the pair L/K a differential field extension. Let L/K be a differential field extension and $a \in L$. We denote by $K\langle a \rangle$ the subfield of L obtained from K by the adjunction of $\partial^j a$ for all $j \in \mathbb{Z}_{\geq 0}$. Namely

$$K\langle a \rangle = K\left(a, \partial a, \partial^2 a, \partial^3 a, \ldots\right).$$

Then $(K\langle a \rangle, \partial)$ becomes a differential field, and $L/K\langle a \rangle$ is a differential field extension. Recursively we define $K\langle a_1, a_2, \ldots, a_l \rangle = K\langle a_1, a_2, \ldots, a_{l-1} \rangle \langle a_l \rangle$.

A homomorphism

$$\sigma : L_1 \to L_2$$

of fields between differential fields (L_1, ∂_1), (L_2, ∂_2) is called a differential homomorphism, if it commutes with the differentiation:

$$\sigma(\partial_1(a)) = \partial_2(\sigma(a)) \qquad (a \in L_1).$$

If moreover σ is an isomorphism, it is called a differential isomorphism.

Let L/K be a differential field extension. A differential isomorphism of L to itself is called a differential isomorphism of L over K if it keeps any element in

K invariant. We call the group of the differential isomorphisms of L over K the differential Galois group of L/K, and denote by $\mathrm{Gal}(L/K)$. Namely we have

$$\mathrm{Gal}(L/K) = \{\sigma : L \to L \mid \sigma\colon \text{differential isomorphism}, \sigma(a) = a \ (\forall a \in K)\}.$$

Differential Galois theory studies the Galois correspondence, for a differential field extension L/K, between intermediate differential fields and subgroups of the differential Galois group. In the case that L is obtained from K by the adjunction of solutions of a differential equation, we can evaluate how transcendental the solutions are by the differential Galois group. Here we explain the classical differential Galois theory, which is called the Picard-Vessiot theory. This theory gives a criterion for the solvability of linear ordinary differential equations by quadrature.

A differential field extension L/K is called a *Picard-Vessiot extension*, if there is a linear ordinary differential equation

$$\partial^n y + a_1 \partial^{n-1} y + \cdots + a_{n-1}\partial y + a_n y = 0 \quad (a_i \in K) \tag{5.22}$$

with coefficients in K, such that L is obtained from K by the adjunction of n solutions $y_1, y_2, \ldots, y_n$ which are linearly independent over the field of constants C_L:

$$L = K\langle y_1, y_2, \ldots, y_n \rangle,$$

and that

$$C_L = C_K$$

holds.

We have another formulation of the definition of a Picard-Vessiot extension. A differential field extension L/K is called a Picard-Vessiot extension if $C_L = C_K$ holds and there exist $y_1, y_2, \ldots, y_n \in L$ satisfying the following conditions (i), (ii), (iii):

(i)

$$L = K\langle y_1, y_2, \ldots, y_n \rangle,$$

(ii)

$$W(y_1, y_2, \ldots, y_n) \neq 0,$$

(iii)

$$\begin{vmatrix} y_1 & y_2 & \cdots & y_n \\ \vdots & \vdots & & \vdots \\ \partial^{i-1}y_1 & \partial^{i-1}y_2 & \cdots & \partial^{i-1}y_n \\ \partial^{i+1}y_1 & \partial^{i+1}y_2 & \cdots & \partial^{i+1}y_n \\ \vdots & \vdots & & \vdots \\ \partial^n y_1 & \partial^n y_2 & \cdots & \partial^n y_n \end{vmatrix} / W(y_1, y_2, \ldots, y_n) \in K \quad (0 \le i \le n-1).$$

We see that the two definitions are equivalent because the differential equation (5.22) is written by using the Wronskian:

$$W(y, y_1, y_2, \ldots, y_n) = 0. \tag{5.23}$$

Noting this fact, we have the following lemma.

Lemma 5.17 *Let K a differential field, and consider a differential equation (5.22) of order n with coefficients in K. Let F/K be a differential field extension, and $y_1, y_2, \ldots, y_n \in F$ solutions of (5.22) which are linearly independent over C_F. Then any solution of (5.22) in F is a linear combination of $y_1, y_2, \ldots, y_n$ with coefficients in C_F.*

Proof Let z be a solution in F. Since the differential equation (5.22) can be given by (5.23), we get

$$W(z, y_1, y_2, \ldots, y_n) = 0.$$

Then, thanks to Theorem 5.16, we see that $z, y_1, y_2, \ldots, y_n$ are linearly dependent over C_F, which implies the assertion. $\square$

Let L/K be a Picard-Vessiot extension obtained by the adjunction of linearly independent solutions $y_1, y_2, \ldots, y_n$ of the differential equation (5.22). We shall show that the differential Galois group $\mathrm{Gal}(L/K)$ is realized as a subgroup of $\mathrm{GL}(n, C_K)$. Take a $\sigma \in \mathrm{Gal}(L/K)$. Since σ is a differential homomorphism over K, we have

$$\partial^n \sigma(y_j) + a_1 \partial^{n-1}\sigma(y_j) + \cdots + a_{n-1}\partial\sigma(y_j) + a_n\sigma(y_j) = 0$$

for $j = 1, 2, \ldots, n$. Thus $\sigma(y_1), \sigma(y_2), \ldots, \sigma(y_n) \in L$ are solutions of (5.22), and linearly independent over $C_L = C_K$ because

$$W(\sigma(y_1), \sigma(y_2), \ldots, \sigma(y_n)) = \sigma(W(y_1, y_2, \ldots, y_n)) \neq 0, \tag{5.24}$$

which holds since σ is a differential isomorphism. Owing to Lemma 5.17, there exist $c_{jk} \in C_L = C_K$ such that

$$\sigma(y_j) = \sum_{k=1}^{n} c_{jk} y_k \quad (1 \le j \le n).$$

We see that the determinant of the matrix (c_{jk}) does not vanish by (5.24). Thus we obtain a group homomorphism

$$\begin{array}{ccc} \mathrm{Gal}(L/K) & \to & \mathrm{GL}(n, C_K) \\ \sigma & \mapsto & (c_{jk}) \end{array},$$

and then $\mathrm{Gal}(L/K)$ is isomorphic to a subgroup of $\mathrm{GL}(n, C_K)$.

Under these preparations, we can state the main results of the Picard-Vessiot theory. The proofs are found in [90], and we omit them.

Theorem 5.18 *The differential Galois group $\mathrm{Gal}(L/K)$ of a Picard-Vessiot extension L/K is isomorphic to an algebraic subgroup of $\mathrm{GL}(n, C_K)$.*

Theorem 5.19 *Let L/K be a Picard-Vessiot extension, and assume that C_K is an algebraically closed field of characteristic 0. Then L/K is a normal extension.*

Here we call an extension L/K normal if all elements of L which are invariant by any element of $\mathrm{Gal}(L/K)$ are elements of K.

Theorem 5.20 *Let L/K be a Picard-Vessiot extension, and assume that C_K is an algebraically closed field of characteristic 0. Then there exists a $1 : 1$ correspondence, called the Galois correspondence, between differential indeterminate fields of L/K and algebraic subgroups of $\mathrm{Gal}(L/K)$.*

The Galois correspondence is obtained as follows. To a differential intermediate field M of L/K, there corresponds the differential Galois group $\mathrm{Gal}(L/M)$. To an algebraic subgroup H of $\mathrm{Gal}(L/K)$, there corresponds a differential intermediate field

$$M = \{a \in L \mid \sigma(a) = a \ (\forall \sigma \in H)\}$$

of L/K. In this correspondence, H is a normal subgroup of $\mathrm{Gal}(L/K)$ if and only if M/K is a normal extension, where M is the corresponding intermediate field. In this case, we have

$$\mathrm{Gal}(M/K) = \mathrm{Gal}(L/K)/H.$$

Now we consider the quadrature, which is a collection of explicit methods of solving differential equations. It is well known that a non-linear ordinary differential equation of the first order is solvable by quadrature if it is of the form of separation of variables. We want to decide if a given differential equation is solvable by quadrature. For the purpose, we should fix the operations used in the quadrature. We may choose several operations according to individual purposes, and one standard choice is given by the generalized Liouvillian extension.

Let K be a differential field, and a a point in K. An element u in a differential field extension of K is called an integral of a if

$$\partial u = a$$

holds. An element u in a differential field extension of K satisfying

$$\partial u - au = 0$$

is called an exponential of an integral of a. If we consider in a category of analytic functions, an integral of a function $a(x)$ is given by the integral $\int a(x)\,dx$, and an exponential of an integral of $a(x)$ is given by $\exp\left(\int a(x)\,dx\right)$.

A differential field extension N/K is called a *generalized Liouvillian extension* if there exists a sequence

$$K = K_0 \subset K_1 \subset K_2 \subset \cdots \subset K_n = N$$

of differential field extensions such that each K_{i+1}/K_i is a finite algebraic extension, or an extension obtained by the adjunction of an integral of an element of K_i, or an extension obtained by the adjunction of an exponential of an integral of an element of K_i. Namely, we choose, for the quadrature, the operations of solving algebraic equations, integrating and solving linear differential equations of the first order as operations. If we remove solving algebraic equations from the definition of the generalized Liouvillian extension, we get a definition of a *Liouvillian extension*.

The following is one of the main results in the Picard-Vessiot theory.

Theorem 5.21 *Let L/K be a Picard-Vessiot extension, and assume that C_K is an algebraically closed field of characteristic 0. L is contained in a generalized Liouvillian extension of K if and only if the connected component of the identity of $\mathrm{Gal}(L/K)$ is solvable.*

We recall the definition of solvability of a group. A group G is said to be solvable if there exists a normal chain

$$G = N_0 \supset N_1 \supset \cdots \supset N_r = 1$$

such that each N_{i-1}/N_1 is commutative. (Here 1 denotes the group consisting of the identity.) Equivalently, the derived chain

$$G \supset D(G) \supset D^2(G) \supset D^3(G) \supset \cdots$$

reaches at 1, where $D(G) = [G, G]$, $D^{i+1}(G) = [D^i(G), D^i(G)]$, and, for subgroups K, H of a group G, we denote by $[K, H]$ the subgroup generated by the commutators $khk^{-1}h^{-1}$ $(k \in K, h \in H)$.

These are the main results of the Picard-Vessiot theory. Kolchin extended the notion of a Picard-Vessiot extension, and obtained the notion of a strongly normal extension. Umemura rediscovered this notion, and used it to the proof the irreducibility of the first Painlevé transcendent [177].

Now we are ready to explain the relation of the monodromy to the differential Galois theory.

It is very difficult to determine a differential Galois group in general. However, for Fuchsian differential equations, we can compute the differential Galois group by using the monodromy group.

Theorem 5.22 *Consider a Fuchsian differential equation*

$$y^{(n)} + a_1(x)y^{(n-1)} + \cdots + a_n(x)y = 0, \tag{5.25}$$

and let $\mathcal{Y} = (y_1, y_2, \ldots, y_n)$ be a fundamental system of solutions. We set

$$L = \mathbb{C}(x)\langle y_1, y_2, \ldots, y_n\rangle.$$

Then the differential Galois group $\mathrm{Gal}(L/\mathbb{C}(x))$ *is the Zariski closure of the monodromy group with respect to $\mathcal{Y}$.*

A Zariski closure of a group G is the minimal algebraic group containing G.

Proof The action of the monodromy group G is analytic continuations along loops, and then a fundamental system of solutions is sent to another fundamental system of solutions and any element of $\mathbb{C}(x)$ is left unchanged. Thus we can regard any element of G as an element of $\mathrm{Gal}(L/\mathbb{C}(x))$. By Theorem 5.18, $\mathrm{Gal}(L/\mathbb{C}(x))$ is an algebraic group, and hence the Zariski closure $\bar{G}$ of G is also contained in $\mathrm{Gal}(L/\mathbb{C}(x))$. Namely we have $\bar{G} \subset \mathrm{Gal}(L/\mathbb{C}(x))$.

Any element of L is a ratio of differential polynomials in $y_1, y_2, \ldots, y_n$ with coefficients in $\mathbb{C}(x)$. Then its singular points are singular points of the differential equation (5.25) or zeros of the denominator. These singular points are at most regular singular. If an element of L is invariant by the action of G, it becomes single valued. Hence the singular points are poles by Theorem 4.1. Thus the element of L becomes a rational function. Thanks to Theorem 5.19, the Picard-Vessiot extension $L/\mathbb{C}(x)$ is normal, which implies $\bar{G} \supset \mathrm{Gal}(L/\mathbb{C}(x))$.

Thus we obtain $\bar{G} = \mathrm{Gal}(L/\mathbb{C}(x))$. $\square$

The hypergeometric differential equation (5.19) has been studied since long ago. There are many works on the description of the solutions where the parameters α, β, γ take particular values. The work of Schwarz we have explained is one of such studies. Gauss gave a list of hypergeometric functions which represent

elementary functions. Here the hypergeometric function is a function defined by
a power series

$$F(\alpha, \beta, \gamma; x) = \sum_{n=0}^{\infty} \frac{(\alpha, n)(\beta, n)}{(\gamma, n)n!} x^n, \quad (\alpha, n) = \frac{\Gamma(\alpha + n)}{\Gamma(\alpha)},$$

which is the solution of the hypergeometric differential equation at $x = 0$ of
characteristic exponent 0. We find in the Works of Gauss the following list:

$$(t + u)^n = t^n F\left(-n, \beta, \beta; -\frac{u}{t}\right),$$

$$\log(1 + t) = tF(1, 1, 2; -t),$$

$$e^t = 2 \lim_{k \to \infty} F\left(1, k, 1; \frac{t}{k}\right),$$

$$\sin t = t \lim_{k,k' \to \infty} F\left(k, k', \frac{3}{2}; -\frac{t^2}{4kk'}\right),$$

$$\cos t = \lim_{k,k' \to \infty} F\left(k, k', \frac{1}{2}; -\frac{t^2}{4kk'}\right),$$

$$t = \sin t\, F\left(\frac{1}{2}, \frac{1}{2}, \frac{3}{2}; \sin^2 t\right),$$

$$t = \tan t\, F\left(\frac{1}{2}, 1, \frac{3}{2}; -\tan^2 t\right).$$

The last two equalities can be understood as representations of $\sin^{-1} t, \tan^{-1} t$,
respectively. The elementary functions can be regarded as elements of a differential
field generated over $\mathbb{C}(x)$ by integrals and exponentials of integrals. Then it may
be a natural problem to determine all sets of parameters of the hypergeometric
differential equation which define a generalized Liouvillian extension. This problem
is reduced, owing to Theorems 5.22 and 5.21, to the analysis of the monodromy
group, and solved by Kimura [104].

For differential equations containing irregular singular points, we can define
the generalized monodromy group which is generated by the monodromy group
together with the transformations given by Stokes matrices. Stokes matrices will
be explained in Chap. 10. It is shown that the Zariski closure of the generalized
monodromy group coincides with the differential Galois group $\mathrm{Gal}(L/\mathbb{C}(x))$
(Martinet-Ramis [121]).

Chapter 6
Connection Problem

6.1 Physics and Connection Problem

We start with considering a vibration of a string. Let a string of length ℓ be set on the interval $[0, \ell]$ of x-axis in xy-plane. Then we can regard the vibration of the string as a motion of each point on $x \in [0, \ell]$ in y-direction. We denote the deviation in y-direction of a point at x at the time t by $y(t, x)$ (Fig. 6.1). Then the function $y(t, x)$ in two variables represents the vibration. Applying Newton's law of motion, we get the wave equation for $y(t, x)$

$$\frac{\partial^2 y}{\partial t^2} = c^2 \frac{\partial^2 y}{\partial x^2}, \tag{6.1}$$

where c is a constant determined by the tension and the density of the string.

We are going to solve the wave equation (6.1) by the method of separation of variables. We assume

$$y(t, x) = v(t)z(x). \tag{6.2}$$

Namely, $y(t, x)$ is a product of a function in t and a function in x. We denote by the same symbol $'$ the differentiation with respect to each variable. Putting (6.2) into (6.1), we obtain

$$v''z = c^2 vz''.$$

© The Editor(s) (if applicable) and The Author(s), under exclusive license to Springer Nature Switzerland AG 2020
Y. Haraoka, *Linear Differential Equations in the Complex Domain*, Lecture Notes in Mathematics 2271, https://doi.org/10.1007/978-3-030-54663-2_6

Fig. 6.1 The shape of a
string at the time t

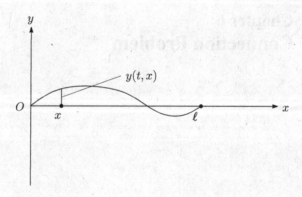

We rewrite this as

$$\frac{1}{c^2}\frac{v''}{v} = \frac{z''}{z}. \tag{6.3}$$

Then we see that both sides are constant, because the left hand side does not depend on x and the right hand side does not depend on t. We denote this constant by $-\lambda$, and then have the following two differential equations

$$v'' + c^2\lambda v = 0, \tag{6.4}$$

$$z'' + \lambda z = 0. \tag{6.5}$$

Any solution $v(t)$ of (6.4) is a linear combination of $\sin c\sqrt{\lambda}t$ and $\cos c\sqrt{\lambda}t$, and hence can be written as

$$v(t) = v_0 \cos(c\sqrt{\lambda}t + \omega) \tag{6.6}$$

with constants v_0, ω. The expression (6.2) means that $z(x)$ represents the shape of the string and $v(t)$ represents the scale and the direction of the shape. Then, by the explicit form (6.6) of $v(t)$, we see that $v(t)$ is a periodic function with period $\frac{2\pi}{c\sqrt{\lambda}}$, which is the period of the vibration. Therefore the frequency is given by the reciprocal $\frac{c\sqrt{\lambda}}{2\pi}$. Thus the frequency of this vibration is determined by λ. We call λ the spectral parameter. At this moment we know only that λ is a constant.

Since the string is fixed at both ends (boundaries) $x = 0$, $x = \ell$, we have

$$y(t, 0) = y(t, \ell) = 0.$$

Then we get

$$z(0) = z(\ell) = 0. \tag{6.7}$$

Therefore the solution $z(x)$ of the differential equation (6.5) should satisfy the boundary condition (6.7), which will specify particular values of λ. We shall explain the mechanism in some general way.

The linear differential equation (6.5) has a singular point only at $x = \infty$, and the points $x = 0, x = \ell$ are both regular. We denote the solution satisfying the initial condition

$$z(0) = 0, \ z'(0) = 1$$

at $x = 0$ by $s_0(x)$, and the solution satisfying the initial condition

$$z(0) = 1, \ z'(0) = 0$$

by $c_0(x)$. Explicitly, we have

$$s_0(x) = \frac{1}{\sqrt{\lambda}} \sin \sqrt{\lambda} x, \ c_0(x) = \cos \sqrt{\lambda} x.$$

Moreover, we denote the solution satisfying the initial condition

$$z(\ell) = 0, \ z'(\ell) = 1$$

at $x = \ell$ by $s_\ell(x)$, and the solution satisfying the initial condition

$$z(\ell) = 1, \ z'(\ell) = 0$$

by $c_\ell(x)$. Explicitly, we have

$$s_\ell(x) = \frac{1}{\sqrt{\lambda}} \sin \sqrt{\lambda}(x - \ell), \ c_\ell(x) = \cos \sqrt{\lambda}(x - \ell).$$

The set $(s_0(x), c_0(x))$ makes a fundamental system of solutions at $x = 0$, and $(s_\ell(x), c_\ell(x))$ a fundamental system of solutions at $x = \ell$.

Now by the boundary condition (6.7), we have

$$z(x) = a s_0(x)$$

in a neighborhood of $x = 0$ with some constant a. Continue $s_0(x)$ analytically to $x = \ell$. Then the result can be written as a linear combination of the fundamental system of solutions $(s_\ell(x), c_\ell(x))$, and then there exists constants A, B such that

$$s_0(x) = A s_\ell(x) + B c_\ell(x).$$

We call these constants A, B connection coefficients. Then the boundary condition (6.7) holds if and only if

$$B = 0.$$

By using properties of trigonometric functions, we have

$$s_0(x) = \frac{1}{\sqrt{\lambda}} \sin \sqrt{\lambda}(x - \ell + \ell)$$

$$= \frac{1}{\sqrt{\lambda}} \left(\sin \sqrt{\lambda}(x - \ell) \cos \sqrt{\lambda}\ell + \cos \sqrt{\lambda}(x - \ell) \sin \sqrt{\lambda}\ell \right)$$

$$= \cos \sqrt{\lambda}\ell \, s_\ell(x) + \frac{1}{\sqrt{\lambda}} \sin \sqrt{\lambda}\ell \, c_\ell(x),$$

which shows

$$A = \cos \sqrt{\lambda}\ell, \quad B = \frac{1}{\sqrt{\lambda}} \sin \sqrt{\lambda}\ell.$$

Then from $B = 0$ we obtain

$$\sin \sqrt{\lambda}\ell = 0,$$

which implies $\sqrt{\lambda}\ell = n\pi \ (n \in \mathbb{Z})$. Hence we have

$$\lambda = \left(\frac{n\pi}{\ell} \right)^2.$$

Thus we can specify the value of the spectral parameter λ by the vanishing of the connection coefficient B.

Next, we consider a potential of electrostatic field. If we put a particle with electric charge, there appear a vector field that describes Coulomb forces acting on other particles (Fig. 6.2). The vector field is called the electrostatic field. It is known that the electrostatic field has a potential. Namely, there exists a function $\phi(x, y, z)$ in three variables called a potential such that the vector at a point (x, y, z) is given by

$$\mathrm{grad}\, \phi(x, y, z) = \left(\frac{\partial \phi}{\partial x}(x, y, z), \frac{\partial \phi}{\partial y}(x, y, z), \frac{\partial \phi}{\partial z}(x, y, z) \right).$$

The potential ϕ satisfies the Laplace equation

$$\Delta \phi = \frac{\partial^2 \phi}{\partial x^2} + \frac{\partial^2 \phi}{\partial y^2} + \frac{\partial^2 \phi}{\partial z^2} = 0 \tag{6.8}$$

Fig. 6.2 Coulomb forces

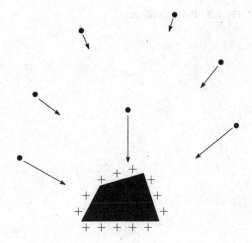

in the complement domain of the charged particle. We shall solve this Laplace equation to obtain the potential ϕ.

We apply the method of separation of variables in the polar coordinate. The polar coordinate of the three dimensional space is given by

$$\begin{cases} x = r \sin \theta \cos \varphi, \\ y = r \sin \theta \sin \varphi, \\ z = r \cos \theta. \end{cases}$$

The geometric meaning of the coordinate (r, θ, φ) is illustrated in Fig. 6.3.

We rewrite the Laplace equation (6.8) in the polar coordinate (r, θ, φ), and then assume that the solution ϕ is a product of one variable functions in r, θ and φ:

$$\phi(x, y, z) = R(r)\Theta(\theta)\Phi(\varphi),$$

We can decompose the Laplace equation into the following three ordinary differential equations

$$r^2 R'' + 2r R' - \lambda R = 0, \tag{6.9}$$

$$\Theta'' + \frac{\cos \theta}{\sin \theta} \Theta' - \left(\lambda - \frac{\mu}{\sin^2 \theta} \right) \Theta = 0, \tag{6.10}$$

$$\Phi'' + \mu \Phi = 0, \tag{6.11}$$

Fig. 6.3 Polar coordinate
in $\mathbb{R}^3$

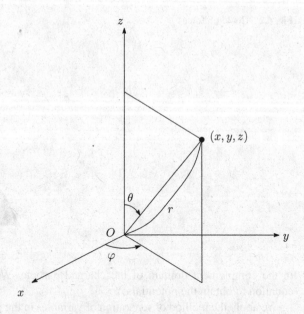

where λ and μ are spectral parameters. Since ϕ is single valued on $\mathbb{R}^3$, we should
have $\Phi(\varphi + 2\pi) = \Phi(\varphi)$. Equation (6.11) has such solution if and only if $\mu = m^2$ ($m \in \mathbb{Z}$). For simplicity, we assume $\mu = 0$ in the following. Then Eq. (6.10)
becomes

$$\Theta'' + \frac{\cos\theta}{\sin\theta}\Theta' - \lambda\Theta = 0.$$

We change the independent variable by

$$\cos\theta = t,$$

and then get the differential equation

$$(1 - t^2)\frac{d^2\Theta}{dt^2} - 2t\frac{d\Theta}{dt} - \lambda\Theta = 0. \tag{6.12}$$

This differential equation is called Legendre's differential equation. This equation
has regular singular points at $t = \pm 1, \infty$, and the characteristic exponents are $0, 0$
at both $t = \pm 1$, and ρ_1, ρ_2 at $t = \infty$, where ρ_1, ρ_2 are solutions of the equation

$$\rho^2 - \rho - \lambda = 0.$$

We collect these data into the table

$$\left\{\begin{matrix} t = 1 & t = -1 & t = \infty \\ 0 & 0 & \rho_1 \\ 0 & 0 & \rho_2 \end{matrix}\right\}, \tag{6.13}$$

which we call the Riemann scheme (cf. Sect. 7.2).

The regular singular points $t = \pm 1$ corresponds to $\theta = 0, \pi$, which represents the z-axis in xyz-space. However, there is no reason that the potential ϕ has singularity at the z-axis, because the xyz-coordinate is introduced artificially for the analysis and then z-axis has no special meaning for the potential. Therefore, $\Theta(t)$ should be continuous (regular) at $t = \pm 1$.

Then we analyze Legendre's differential equation (6.12) at $t = \pm 1$. We have a fundamental system of solutions

$$\Theta_1^+(t) = 1 + \sum_{m=1}^{\infty} a_m (t - 1)^m,$$

$$\Theta_2^+(t) = \Theta_1^+(t) \log(t - 1) + \sum_{m=0}^{\infty} b_m (t - 1)^m$$

at $t = 1$, and similarly

$$\Theta_1^-(t) = 1 + \sum_{m=1}^{\infty} c_m (t + 1)^m,$$

$$\Theta_2^-(t) = \Theta_1^-(t) \log(t + 1) + \sum_{m=0}^{\infty} d_m (t + 1)^m$$

at $t = -1$. Since $\Theta(t)$ is regular at $t = 1$, $\Theta(t)$ should be a constant multiple of $\Theta_1^+(t)$. Continuing $\Theta_1^+(t)$ to $t = 1$, we get

$$\Theta_1^+(t) = A \Theta_1^-(t) + B \Theta_2^-(t)$$

with some constants A, B. Then in order that $\Theta(t)$ is also regular at $t = -1$, we should have $B = 0$.

We can compute these constants A, B, which are called the connection coefficients. After the argument in Sect. 9.10, we derive

$$A = e^{\pi i (1 - \alpha)}, \quad B = e^{\pi i (1 - \alpha)} \frac{1 - e^{-2\pi i \alpha}}{2\pi i},$$

where α is determined by

$$\lambda = \alpha(\alpha - 1).$$

In deriving these constants A, B, we assumed $\alpha \notin \mathbb{Z}$. By the explicit form of B, we see that $B \neq 0$ if $\alpha \notin \mathbb{Z}$. Thus the condition $\alpha \in \mathbb{Z}$ is necessary for $B = 0$. On the other hand, if $\alpha \in \mathbb{Z}$, Legendre's equation (6.12) has a polynomial solution. Namely, if we define the Legendre polynomial $P_n(t)$ for each $n \in \mathbb{Z}_{\geq 0}$ by

$$P_n(t) = \frac{1}{2^n n!} \frac{d^n}{dt^n} (t^2 - 1)^n,$$

$P_{\alpha-1}(t)$ is a solution of (6.12) if $\alpha \in \mathbb{Z}_{\geq 1}$, and $P_{-\alpha}(t)$ is a solution of (6.12) if $\alpha \in \mathbb{Z}_{\leq 0}$. Thus in both cases, we have a solution which is regular at both $t = \pm 1$. This means that $B = 0$ if and only if $\alpha \in \mathbb{Z}$.

In this manner, we see that the value of the spectral parameter λ is given by

$$\lambda = \alpha(\alpha - 1) \quad (\alpha \in \mathbb{Z}).$$

By these two examples, we have seen how the values of spectral parameters are determined by the vanishing of the connection coefficients.

6.2 Formulation of Connection Problem

Generally speaking, the connection problem for linear differential equations is a problem of obtaining a linear relation between two fundamental systems of solutions arising from the analytic continuation. For example the monodromy is a kind of connection problems because it describes the relation between a fundamental system of solutions and the result of its analytic continuation along a loop. Connection problems appear in various problems in mathematics and physics. We have seen such examples in the previous section. We note that connection problems are problems concerning analytic continuations, and then very difficult in nature.

For linear differential equations, we can easily construct local solutions in a neighborhood of a singular point, and then we usually consider a connection problem between fundamental systems of solutions consisting of such local solutions. Now we formulate such connection problem.

Take $a_0, a_1, \ldots, a_p \in \mathbb{P}^1$, and set

$$D = \mathbb{P}^1 \setminus \{a_0, a_1, \ldots, a_p\}.$$

Consider a linear differential equation (E) whose coefficients are holomorphic and single-valued on D with singular points at $a_0, a_1, \ldots, a_p$. For each a_j, take an open disc B_j with center a_j and of radius $r_j > 0$. We take r_j so small that each B_j

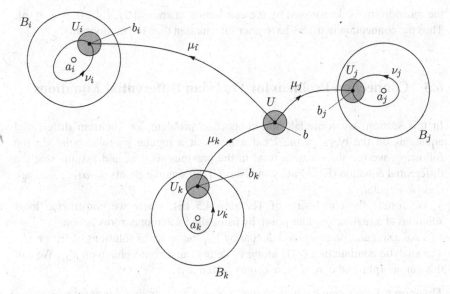

Fig. 6.4 Paths $\mu_\bullet$ and loops $\nu_\bullet$

contains no singular point other than a_j. Take a simply connected domain U_j in $D \cap B_j$, and a point $b_j \in U_j$. Take a loop ν_j in $D \cap B_j$ with base point b_j that encircles a_j once in the positive direction. Moreover, take a point $b \in D$ and a simply connected neighborhood U of b in D, and, for each j, a curve μ_j in D with initial point b and end point b_j (Fig. 6.4).

We consider a fundamental system of solutions $\mathcal{Y}$ on U, and for each j, a fundamental system of solutions $\mathcal{Y}_j$ on U_j. Usually we take $\mathcal{Y}_j$ as a set of local solutions specified by the local behaviors at $x = a_j$. Continuing analytically these fundamental systems of solutions along the curves μ_j, ν_j, we have matrices C_j, L_j such that

$$(\mu_j)_* \mathcal{Y} = \mathcal{Y}_j C_j, \quad (\nu_j)_* \mathcal{Y}_j = \mathcal{Y}_j L_j. \quad (0 \le j \le p)$$

We call these matrices the *connection matrices*. As is seen by Definitions 4.6 and 4.7, the conjugacy class of L_j is the local monodromy at $x = a_j$. Then we often call C_j the connection matrices, while L_j the local monodromies.

For each j, we set

$$\gamma_j = \mu_j \nu_j \mu_j^{-1}.$$

Then $\gamma_0, \gamma_1, \ldots, \gamma_p$ are generators of $\pi_1(D, b)$. Since we have

$$(\gamma_j)_* \mathcal{Y} = \mathcal{Y} C_j^{-1} L_j C_j,$$

the monodromy is determined by the connection matrices $\{C_j, L_j \mid 0 \le j \le p\}$. Thus the connection matrices have finer information than the monodromy.

6.3   Connection Problem for Fuchsian Differential Equations

In this section, we formulate the connection problem for Fuchsian differential equations on the basis of the local analysis at a regular singular point. In the following, we use the notation used in the previous section, and assume that the differential equation (E) is Fuchsian. Thus all the singular points $a_0, a_1, \ldots, a_p$ are regular singular.

We restate the conclusion of Theorem 4.5 (ii), where we constructed local solutions at a regular singular point, by using the local monodromy action.

Take a singular point a_j, and denote by V_{a_j} the space of solutions of (E) on U_j. The analytic continuation $(v_j)_*$ along v_j defines an automorphism on V_{a_j}. We call this automorphism the local monodromy action at a_j.

Theorem 6.1 *Notation being as above. For each regular singular point $x = a_j$, we have a direct sum decomposition*

$$V_{a_j} = \bigoplus_\alpha V_{a_j}^\alpha, \tag{6.14}$$

of the local solution space V_{a_j} into generalized eigenspaces with respect to the local monodromy action $(v_j)_*$, where $V_{a_j}^\alpha$ denotes the generalized eigenspace for an eigenvalue α. For each generalized eigenspace $V_{a_j}^\alpha$, we have a filtration

$$V_{a_j}^{\alpha,0} \subset V_{a_j}^{\alpha,1} \subset \cdots \subset V_{a_j}^{\alpha,k_j} = V_{a_j}^\alpha$$

such that each $V_{a_j}^{\alpha,i}$ is invariant under the action of $(v_j)_*$. Explicitly, each $V_{a_j}^{\alpha,i}$ consists of the solutions of the form

$$(x - a_j)^\rho \sum_{m=0}^{i} \varphi_m(x)(\log(x - a_j))^m,$$

where $\varphi_m(x)$'s are convergent power series in $(x - a_j)$ and ρ is a branch of

$$\frac{1}{2\pi\sqrt{-1}} \log \alpha.$$

Based on the direct sum decomposition (6.14), we are going to formulate the connection problem between a_i and a_j. Take any solution $y \in V_{a_i}^\alpha$, and a path γ_{ij} in D which starts from b_i and ends at b_j. For example, we may take $\mu_i^{-1}\mu_j$ as γ_{ij}.

The result of the analytic continuation of y along γ_{ij} is contained in V_{a_j}. According to the direct sum decomposition (6.14), it can be expressed by the sum

$$(\gamma_{ij})_* y = y_1 + y_2 + \cdots + y_l,$$

where each y_m is a solution in $V_{a_j}^{\beta_m}$, $\beta_1, \beta_2, \ldots, \beta_l$ being the eigenvalues of $(v_j)_*$ on V_{a_j}. Thus, for each m, we have a map from $V_{a_i}^\alpha$ to $V_{a_j}^{\beta_m}$, which sends y to y_m. The objective of the connection problem is this map.

In order to formulate the problem, we introduce the following maps. For each generalized eigenspace $V_{a_j}^\alpha$, we define

$$\pi_{a_j}^\alpha : V_{a_j} \to V_{a_j}^\alpha$$

to be the projection with respect to the direct sum decomposition (6.14), and

$$\iota_{a_j}^\alpha : V_{a_j}^\alpha \to V_{a_j}$$

to be the inclusion. Then the map representing the connection problem is given by

$$c_{a_i,a_j}^{\alpha,\beta} = \pi_{a_j}^\beta \circ (\gamma_{ij})_* \circ \iota_{a_i}^\alpha : V_{a_i}^\alpha \to V_{a_j}^\beta.$$

If $\dim V_{a_i}^\alpha = \dim V_{a_i}^\beta = 1$, by specifying bases of both spaces, the map $c_{a_i,a_j}^{\alpha,\beta}$ is represented by a constant, which is the connection coefficient. In general case, by taking bases of the linear spaces $V_{a_i}^\alpha$ and $V_{a_j}^\beta$, the map $c_{a_i,a_j}^{\alpha,\beta}$ is represented by a matrix, which is called the connection matrix and can be regarded as a correspondent of the connection coefficient. To describe this matrix is a solution of the connection problem. We shall study the connection problem in general case in Sect. 6.3.3.

6.3.1   Connection Problem for Hypergeometric Differential Equation

Gauss solved the connection problem for hypergeometric differential equation. We look at his method.

The characteristic exponents of the hypergeometric differential equation (5.19) at regular singular points $x = 0, 1$ are $\{0, 1 - \gamma\}$ and $\{0, \gamma - \alpha - \beta\}$, respectively. We assume

$$\gamma \notin \mathbb{Z}, \ \gamma - \alpha - \beta \notin \mathbb{Z}.$$

By this assumption, both of the direct sum decompositions (6.14) at $x = 0, 1$ are direct sums of subspaces of dimension 1. Analytically, this implies that there appear no logarithmic solution, and hence we have fundamental systems of solutions $\mathcal{Y}_0(x) = (y_{01}(x), y_{02}(x))$ at $x = 0$ and $\mathcal{Y}_1(x) = (y_{11}(x), y_{12}(x))$ at $x = 1$ of the form

$$y_{01}(x) = \varphi_1(x),$$

$$y_{02}(x) = x^{1-\gamma}\varphi_2(x);$$

$$y_{11}(x) = \psi_1(x-1),$$

$$y_{12}(x) = (1-x)^{\gamma-\alpha-\beta}\psi_2(x-1),$$

where $\varphi_1(x), \varphi_2(x), \psi_1(x), \psi_2(x)$ are Taylor series at $x = 0$ with the initial term 1. By Theorem 4.5, we see that the radii of convergence of these 4 Taylor series are at least 1, and then the above 4 solutions are defined on the simply connected domain

$$U = \{|x| < 1\} \cap \{|x-1| < 1\}.$$

Hence we have a linear relation

$$\mathcal{Y}_0(x) = \mathcal{Y}_1(x)C \tag{6.15}$$

between $\mathcal{Y}_0(x)$ and $\mathcal{Y}_1(x)$ on U. We consider the problem of determining the connection matrix $C = (c_{ij}) \in \mathrm{GL}(2, \mathbb{C})$. We fix the branches of $y_{02}(x)$, $y_{12}(x)$ on U by

$$\arg x = 0, \ \arg(1-x) = 0.$$

We have the relation

$$y_{01}(x) = c_{11}y_{11}(x) + c_{21}y_{12}(x) \tag{6.16}$$

by (6.15). Assume the inequality

$$\mathrm{Re}(\gamma - \alpha - \beta) > 0. \tag{6.17}$$

Then, by taking a limit $x \to 1$ in U, we have

$$y_{11}(x) \to \psi_1(0) = 1,$$

$$y_{12}(x) \to 0$$

in the right hand side of (6.16). Then we obtain

$$c_{11} = \lim_{\substack{x \to 1 \\ x \in U}} y_{01}(x) = \lim_{\substack{x \to 1 \\ x \in U}} \varphi_1(x).$$

Thus the connection coefficient c_{11} is obtained as a special value $\varphi_1(1)$ of the power series $\varphi_1(x)$ at a point $x = 1$ which is on the circle of convergence.

Gauss calculated the value $\varphi_1(1)$ under the assumption (6.17). The power series $\varphi_1(x)$ is nothing but the hypergeometric series $F(\alpha, \beta, \gamma; x)$, and Gauss's result is the evaluation of the special value

$$F(\alpha, \beta, \gamma; 1) = \frac{\Gamma(\gamma)\Gamma(\gamma - \alpha - \beta)}{\Gamma(\gamma - \alpha)\Gamma(\gamma - \beta)} \tag{6.18}$$

of the hypergeometric series. The connection coefficient c_{11} is given by this value. The result (6.18) is called the Gauss-Kummer formula. The Gauss-Kummer formula can be derived in several ways. For example, we can use the integral representation which will be explained in Chap. 9.

We can express $\varphi_2(x)$, $\psi_1(x)$, $\psi_2(x)$ also by the hypergeometric series by using its symmetry. In fact, by combining transformation formulas of the Riemann scheme which will be explained in Sect. 7.2 and the rigidity of the hypergeometric differential equation, we obtain

$$y_{01}(x) = F(\alpha, \beta, \gamma; x),$$

$$y_{02}(x) = x^{1-\gamma} F(\alpha - \gamma + 1, \beta - \gamma + 1, 2 - \gamma; x),$$

$$y_{11}(x) = F(\alpha, \beta, \alpha + \beta - \gamma + 1; 1 - x),$$

$$y_{12}(x) = (1 - x)^{\gamma - \alpha - \beta} F(\gamma - \alpha, \gamma - \beta, \gamma - \alpha - \beta + 1; 1 - x).$$

By using the above value of c_{11}, we have

$$y_{01}(x) = \frac{\Gamma(\gamma)\Gamma(\gamma - \alpha - \beta)}{\Gamma(\gamma - \alpha)\Gamma(\gamma - \beta)} y_{11}(x) + c_{21} y_{12}(x).$$

Taking a limit $x \to 0$ with $x \in U$, we get

$$1 = \frac{\Gamma(\gamma)\Gamma(\gamma - \alpha - \beta)}{\Gamma(\gamma - \alpha)\Gamma(\gamma - \beta)} F(\alpha, \beta, \alpha + \beta - \gamma + 1; 1)$$

$$+ c_{21} F(\gamma - \alpha, \gamma - \beta, \gamma - \alpha - \beta + 1; 1).$$

The special values of the hypergeometric series in the right hand side can be evaluated by the Gauss-Kummer formula (6.18), and hence we have

$$c_{21} = \frac{\Gamma(\gamma)\Gamma(\alpha + \beta - \gamma)}{\Gamma(\alpha)\Gamma(\beta)}.$$

In a similar way, we can determine the connection coefficients c_{12}, c_{22} in the relation

$$y_{02}(x) = c_{12}y_{11}(x) + c_{22}y_{12}(x)$$

by using Gauss-Kummer formula (6.18).

Consequently, we obtain the following connection relation

$$y_{01}(x) = \frac{\Gamma(\gamma)\Gamma(\gamma - \alpha - \beta)}{\Gamma(\gamma - \alpha)\Gamma(\gamma - \beta)} y_{11}(x) + \frac{\Gamma(\gamma)\Gamma(\alpha + \beta - \gamma)}{\Gamma(\alpha)\Gamma(\beta)} y_{12}(x),$$

$$y_{02}(x) = \frac{\Gamma(2 - \gamma)\Gamma(\gamma - \alpha - \beta)}{\Gamma(1 - \alpha)\Gamma(1 - \beta)} y_{11}(x) + \frac{\Gamma(2 - \gamma)\Gamma(\alpha + \beta - \gamma)}{\Gamma(\alpha - \gamma + 1)\Gamma(\beta - \gamma + 1)} y_{12}(x).$$

$$(6.19)$$

We find that each connection coefficient c_{ij} is expressed as a ratio of products of Gamma functions. This is an important fact, from which we can easily derive special values of the parameter (α, β, γ) for $c_{ij} = 0$.

We restate this result (6.19) in the formulation proposed in this section. In the notion given in Fig. 6.4, we take $a_i = 0$ and $a_j = 1$, and set $U_i = U_j = U = \{|x| < 1\} \cap \{|x - 1| < 1\}$. The solutions y_{01}, y_{02}, y_{11}, y_{12} are defined on U, and we regard y_{01}, y_{02} as a basis of the space V_0 of local solutions at $x = 0$, and y_{11}, y_{12} as a basis of the space V_1 of local solutions at $x = 1$. Note that $V_0 = V_1$ as sets. By the local behavior, we have

$$V_0 = V_0^0 \oplus V_0^{1-\gamma}, \quad V_0^0 = \langle y_{01} \rangle, \quad V_0^{1-\gamma} = \langle y_{02} \rangle,$$

and

$$V_1 = V_1^0 \oplus V_1^{\gamma - \alpha - \beta}, \quad V_1^0 = \langle y_{11} \rangle, \quad V_1^{\gamma - \alpha - \beta} = \langle y_{12} \rangle.$$

Then the connection relation (6.19) can be expressed by the maps

$$c_{0,1}^{\lambda, \mu} : V_0^\lambda \to V_1^\mu$$

with $\lambda = 0, 1 - \gamma$ and $\mu = 0, \gamma - \alpha - \beta$. Namely we have

$$c_{0,1}^{0,0}(y_{01}) = \frac{\Gamma(\gamma)\Gamma(\gamma - \alpha - \beta)}{\Gamma(\gamma - \alpha)\Gamma(\gamma - \beta)} y_{11},$$

$$c_{0,1}^{0,\gamma - \alpha - \beta}(y_{01}) = \frac{\Gamma(\gamma)\Gamma(\alpha + \beta - \gamma)}{\Gamma(\alpha)\Gamma(\beta)} y_{12},$$

$$c_{0,1}^{1-\gamma,0}(y_{02}) = \frac{\Gamma(2 - \gamma)\Gamma(\gamma - \alpha - \beta)}{\Gamma(1 - \alpha)\Gamma(1 - \beta)} y_{11},$$

$$c_{0,1}^{1-\gamma,\gamma - \alpha - \beta}(y_{02}) = \frac{\Gamma(2 - \gamma)\Gamma(\alpha + \beta - \gamma)}{\Gamma(\alpha - \gamma + 1)\Gamma(\beta - \gamma + 1)} y_{12}.$$

$$(6.20)$$

6.3.2 Oshima's Result

Oshima obtained a decisive result for the connection problem of Fuchsian ordinary differential equations. We overview his result. Please refer to [137–139] for details.

In Chap. 7, we will define two operations—addition and middle convolution. These operations are invertible, and send a Fuchsian equation to another Fuchsian equation of, in general, different rank. Thus we may connect a Fuchsian equation to another Fuchsian equation of lower rank by a finite iteration of these operations, and then we can see how several quantities of a Fuchsian differential equation are related to those of a lower rank equation. Oshima realized such relation for the connection coefficients. His result can be stated as follows. We call a Fuchsian differential equation *basic* if it has the minimal rank among the Fuchsian differential equations connected by additions and middle convolutions. In the following, we use V_a^ρ instead of $V_a^{e^{2\pi\sqrt{-1}\rho}}$.

Theorem 6.2 *Let $x = a_i, a_j$ be two singular points of a Fuchsian differential equation (E), and ρ, ρ' characteristic exponents at respective singular points. We assume*

$$\dim V_{a_i}^\rho = \dim V_{a_j}^{\rho'} = 1.$$

Take a basis $y(x)$ (resp. $z(x)$) of $V_{a_i}^\rho$ (resp. $V_{a_j}^{\rho'}$). Then the coefficient of $z(x)$ in the direct sum decomposition (6.14) of the result of analytic continuation of $y(x)$ to $x = a_j$ is explicitly expressed in terms of a connection coefficient of the basic equation for (E).

Precisely speaking, the connection coefficient of (E) can be given by a product of a connection coefficient of the basic equation and a ratio of products of Gamma functions. In particular, if (E) is rigid, which will be defined in Chap. 7, the connection coefficient coincides with the ratio of products of Gamma functions, because the basic equation for a rigid equation is of rank one and then has trivial connection coefficients. As a simplest example, the result (6.20) for the hypergeometric differential equation (5.19) can be derived by applying Theorem 6.2, since Eq. (5.19) is rigid and the eigenspaces V_a^ρ are one dimensional.

We give another illustrative example.

Example 6.1 We use notions that will be explained in Chap. 7.

We consider the third order Fuchsian differential equation

$$x^2(x-1)y''' + xp_1(x)y'' + p_2(x)y' + p_3(x)y = 0 \tag{6.21}$$

satisfied by the generalized hypergeometric series

$$_3F_2\left(\begin{matrix}\alpha_1, \alpha_2, \alpha_3 \\ \beta_1, \beta_2\end{matrix}; x\right) = \sum_{n=0}^{\infty} \frac{(\alpha_1, n)(\alpha_2, n)(\alpha_3, n)}{(\beta_1, n)(\beta_2, n)n!} x^n.$$

The Riemann scheme is given by

$$\left\{\begin{matrix} x = 0 & x = 1 & x = \infty \\ 0 & [0]_{(2)} & \alpha_1 \\ 1 - \beta_1 & & \alpha_2 \\ 1 - \beta_2 & -\beta_3 & \alpha_3 \end{matrix}\right\},$$

where

$$\alpha_1 + \alpha_2 + \alpha_3 = \beta_1 + \beta_2 + \beta_3.$$

Equation (6.21) is connected to the Gauss hypergeometric differential equation (5.19) by an addition and a middle convolution. We take the parameters of α, β, γ of (5.19) by

$$\alpha = \alpha_2 - \beta_1 + 1,$$

$$\beta = \alpha_3 - \beta_1 + 1,$$

$$\gamma = \beta_2 - \beta_1 + 1,$$

and operate the gauge transformation $y \mapsto x^\lambda y$ with $\lambda = \alpha_1 - \beta_1$ and the middle convolution with parameter $\mu = 1 - \alpha_1$ successively. Then we obtain Eq. (6.21).

Oshima gave a formula that relates the connection coefficients of equations before middle convolution and after middle convolution by multiplying gamma functions [137, Lemma12.2]. By applying the formula, we can obtain connection coefficients for (6.21) from ones for (5.19). For example, we take the connection coefficient c_{21} representing the map $c_{0,1}^{0,\gamma-\alpha-\beta}$ for Eq. (5.19). By Oshima's formula, we get

$$\tilde{c} = c_{21} \cdot \frac{\Gamma(\lambda + \mu + 1)\Gamma(\alpha + \beta - \gamma - \mu)}{\Gamma(\lambda + 1)\Gamma(\alpha + \beta - \gamma)}$$

$$= \frac{\Gamma(\gamma)\Gamma(\alpha + \beta - \gamma)}{\Gamma(\alpha)\Gamma(\beta)} \cdot \frac{\Gamma(\lambda + \mu + 1)\Gamma(\alpha + \beta - \gamma - \mu)}{\Gamma(\lambda + 1)\Gamma(\alpha + \beta - \gamma)}$$

$$= \frac{\Gamma(\gamma)\Gamma(\lambda + \mu + 1)\Gamma(\alpha + \beta - \gamma - \mu)}{\Gamma(\alpha)\Gamma(\beta)\Gamma(\lambda + 1)}$$

$$= \frac{\Gamma(\beta_2 - \beta_1 + 1)\Gamma(2 - \beta_1)\Gamma(\beta_3)}{\Gamma(\alpha_1 - \beta_1 + 1)\Gamma(\alpha_2 - \beta_1 + 1)\Gamma(\alpha_3 - \beta_1 + 1)},$$

which is the connection coefficient representing the map

$$c_{0,1}^{1-\beta_1,-\beta_3} : V_0^{1-\beta_1} \to V_1^{-\beta_3}$$

for Eq. (6.21).

6.3.3   Connection Problem for Higher Dimensional Eigenspaces

The result of Oshima (Theorem 6.2) seems to be decisive, however, it is restricted to the case that the dimensions of the eigenspaces $V_{a_j}^{\rho}$ of the local monodromies are one. In this section, we discuss how we would study the connection problem for eigenspaces of dimension greater than one. In this subsection, we use some notions introduced in Chap. 7 and an example from Sect. 9.8.

As we have considered, the connection problem is to describe the linear map $c_{a_i,a_j}^{\alpha,\beta}$. This map is represented by a matrix, which we call the connection matrix, if we specify bases of $V_{a_i}^{\alpha}$ and $V_{a_j}^{\beta}$.

For example, we consider a scalar Fuchsian differential equation with a regular singular point at $x = 0$ and assume $\dim V_0^{\rho} = 2$ for some characteristic exponent ρ. Moreover we assume that V_0^{ρ} does not contain any logarithmic solution. Then V_0^{ρ} is spanned by solutions of the form

$$x^{\rho}(a_0 + a_1 x + a_2 x^2 + \cdots).$$

As a basis of V_0^{ρ}, we may take the solutions determined by $(a_0, a_1) = (1, 0)$ and $(a_0, a_1) = (0, 1)$, since $\{(1, 0), (0, 1)\}$ is a basis of $\mathbb{C}^2$. However, there is no canonical basis for $\mathbb{C}^2$. Hence the corresponding basis of V_0^{ρ} has no intrinsic meaning, so that we may not expect the entries of the connection matrix to be simple.

To resolve this difficulty, we will consider an expression which is independent of the choice of the basis. The Wronskian of a basis of V_0^{ρ} changes only by a scalar multiple if we change the basis. Then the connection matrix for the Wronskian may become simple. Usually we consider a Wronskian of a full basis of solutions of the differential equation. Then we call the Wronskian of a basis of a subspace of the full solution space a partial Wronskian. We can derive a differential equation satisfied by a partial Wronskian. In the derived differential equation, the eigenspace V_0^{ρ} may be sent to an eigenspace of dimension one. As we explained after Theorem 6.2, connection coefficients among eigenspaces of dimension one for a rigid Fuchsian differential equation can be expressed in a simple form. Then the connection coefficients for a partial Wronskian is expected to be simple. Note that the operation of deriving a differential equation for a partial Wronskian is independent of the rigidity. Then a differential equation for a partial Wronskian of a rigid differential equation may become non-rigid, and the converse may occur.

In the following, we consider another way. We noted that there is no canonical basis for $V_{a_j}^\rho$ as a linear space. Nevertheless, there may exist some intrinsic basis for $V_{a_j}^\rho$ as a space of solutions of the differential equation. We study two examples.

The first example is the following integral

$$y(x) = \int_\Delta s^{\lambda_1}(s-1)^{\lambda_2}t^{\lambda_3}(t-x)^{\lambda_4}(s-t)^{\lambda_5}\,ds\,dt, \tag{6.22}$$

which we will study intensively in the Chap. 9 (Sect. 9.8). It is shown that the integral (6.22) satisfies a Fuchsian differential equation $(_3E_2)$ of third order with regular singular points $x = 0, 1, \infty$ for an appropriate domain Δ of integration. The differential equation $(_3E_2)$ is the differential equation satisfied by the generalized hypergeometric function $_3F_2\left(\begin{matrix}\alpha_1, \alpha_2, \alpha_3 \\ \beta_1, \beta_2\end{matrix}; x\right)$. The characteristic exponents at $x = 0$ are $0, \lambda_1+\lambda_3+\lambda_4+\lambda_5+2, \lambda_3+\lambda_4+1$, and those at $x = 1$ are $0, 1, \lambda_2+\lambda_4+\lambda_5+2$. Hence we have $\dim V_1^0 = 2$. In the following, we use the notation

$$\lambda_{ij\cdots k} = \lambda_i + \lambda_j + \cdots + \lambda_k. \tag{6.23}$$

In Sect. 9.8, we assume $0 < x < 1$ and give appropriate domains of integration Δ_j $(1 \le j \le 13)$. Let $y_j(x)$ be the integral (6.22) with $\Delta = \Delta_j$. By using these integrals, we obtain bases of eigenspaces of the local monodromies at $x = 0$ and $x = 1$. We have

$$V_0^0 = \langle y_{12} \rangle,$$

$$V_0^{\lambda_{1345}+2} = \langle y_6 \rangle,$$

$$V_0^{\lambda_{34}+1} = \langle y_8 \rangle,$$

$$V_1^{\lambda_{245}+2} = \langle y_{11} \rangle,$$

$$V_1^0 = \langle y_1, y_2, y_3, y_4 \rangle.$$

If we choose two integrals among y_1, y_2, y_3, y_4 spanning V_1^0, we get a basis of V_1^0. The problem is whether there is a good choice of the basis.

As is explained in Sect. 6.1, in applying the connection problem, it is important to see when the connection coefficient vanishes. Then, taking our formulation of the connection problem into account, it will be important to see when a direct sum component in the direct sum decomposition of the analytic continuation vanishes. This problem is equivalent to the vanishing of the connection coefficient if the dimension of the eigenspace is one.

Thus we want to choose a basis of V_1^0 so that we can directly see whether the analytic continuation of each eigenfunction in V_0 to $x = 1$ contains an element in V_1^0. In Sect. 9.8, we will get linear relations (9.36), (9.37) among $y_j(x)$, where I_{Δ_j}

means $y_j(x)$. The connection coefficients can be obtained by solving these relations. For example, we choose y_1, y_2 as a basis of V_1^0, and express the eigenfunctions y_{12}, y_6, y_8 in V_0 as linear combinations of the basis y_{11}, y_1, y_2 of V_1. Then, for y_8, we have

$$a_{(8:1)} = \frac{A \cdot \Gamma(\lambda_2)\Gamma(1-\lambda_2)\Gamma(\lambda_4)\Gamma(1-\lambda_4)\Gamma(\lambda_{245})\Gamma(1-\lambda_{245})}{e_1 e_2 e_3 e_4{}^2 e_5{}^2},$$

$$a_{(8:2)} = -\frac{\Gamma(\lambda_2)\Gamma(1-\lambda_2)\Gamma(\lambda_4)\Gamma(1-\lambda_4)\Gamma(\lambda_{245})\Gamma(1-\lambda_{245})}{\Gamma(\lambda_1)\Gamma(1-\lambda_1)\Gamma(\lambda_{24})\Gamma(1-\lambda_{24})\Gamma(\lambda_{345})\Gamma(1-\lambda_{345})},$$

$$a_{(8:11)} = \frac{\Gamma(\lambda_{245})\Gamma(1-\lambda_{245})}{\Gamma(\lambda_5)\Gamma(1-\lambda_5)},$$

where we set

$$y_8 = a_{(8:1)}y_1 + a_{(8:2)}y_2 + a_{(8:11)}y_{11}$$

and

$$e_j = e^{\pi\sqrt{-1}\lambda_j} \quad (1 \le j \le 5).$$

The factor A in the numerator of $a_{(8:1)}$ is given by

$$
\begin{aligned}
A ={}& 1 - e_4{}^2 - e_1{}^2 e_5{}^2 + e_4{}^2 e_5{}^2 - e_2{}^2 e_4{}^2 e_5{}^2 + e_1{}^2 e_2{}^2 e_4{}^2 e_5{}^2 - e_3{}^2 e_4{}^2 e_5{}^2 \\
&+ e_1{}^2 e_3{}^2 e_4{}^2 e_5{}^2 - e_1{}^2 e_2{}^2 e_3{}^2 e_4{}^2 e_5{}^2 + e_2{}^2 e_3{}^2 e_4{}^4 e_5{}^2 + e_1{}^2 e_2{}^2 e_3{}^2 e_4{}^2 e_5{}^4 \\
&- e_1{}^2 e_2{}^2 e_3{}^2 e_4{}^4 e_5{}^4,
\end{aligned}
$$

which cannot be factorized. Therefore it is hard to see whether $a_{(8:1)}$ vanishes. The Gamma factors in $a_{(8:2)}$ and $a_{(8:11)}$ appear because the numerators and the denominators can be factorized into factors of degree one and of two terms. Then we apply the formula

$$e^{\pi\sqrt{-1}\lambda} - e^{-\pi\sqrt{-1}\lambda} = \frac{2\pi\sqrt{-1}}{\Gamma(\lambda)\Gamma(1-\lambda)}$$

to get the Gamma factors.

There are six choices of two among y_1, y_2, y_3, y_4 as a basis of V_1^0. It turns out that, for five bases among these six, we have factors like A. There remains only one basis y_1, y_3 of V_1^0 such that every connection coefficients can be expressed by

Gamma factors. Explicitly, we get

$$c_{(12:1)} = \frac{\Gamma(\lambda_{12})\Gamma(1-\lambda_{12})\Gamma(\lambda_{245})\Gamma(1-\lambda_{245})}{\Gamma(\lambda_{13})\Gamma(1-\lambda_{13})\Gamma(\lambda_{125})\Gamma(1-\lambda_{125})},$$

$$c_{(12:3)} = \frac{\Gamma(\lambda_{12})\Gamma(1-\lambda_{12})\Gamma(\lambda_{245})\Gamma(1-\lambda_{245})}{\Gamma(\lambda_1)\Gamma(1-\lambda_1)\Gamma(\lambda_2-\lambda_3)\Gamma(1-\lambda_2+\lambda_3)},$$

$$c_{(12:11)} = \frac{\Gamma(\lambda_{245})\Gamma(1-\lambda_{245})}{\Gamma(\lambda_4)\Gamma(1-\lambda_4)},$$

$$c_{(6:1)} = -\frac{\Gamma(\lambda_{12})\Gamma(1-\lambda_{12})\Gamma(\lambda_4)\Gamma(1-\lambda_4)\Gamma(\lambda_{245})\Gamma(1-\lambda_{245})}{\Gamma(\lambda_{125})\Gamma(1-\lambda_{125})\Gamma(\lambda_{45})\Gamma(1-\lambda_{45})\Gamma(\lambda_{12345})\Gamma(1-\lambda_{12345})},$$

$$c_{(6:3)} = -\frac{\Gamma(\lambda_{12})\Gamma(1-\lambda_{12})\Gamma(\lambda_4)\Gamma(1-\lambda_4)\Gamma(\lambda_{245})\Gamma(1-\lambda_{245})}{\Gamma(\lambda_2)\Gamma(1-\lambda_2)\Gamma(\lambda_{1245})\Gamma(1-\lambda_{1245})\Gamma(\lambda_{345})\Gamma(1-\lambda_{345})},$$

$$c_{(6:11)} = \frac{\Gamma(\lambda_{245})\Gamma(1-\lambda_{245})}{\Gamma(\lambda_2)\Gamma(1-\lambda_2)},$$

$$c_{(8:1)} = -\frac{\Gamma(\lambda_{12})\Gamma(1-\lambda_{12})\Gamma(\lambda_4)\Gamma(1-\lambda_4)\Gamma(\lambda_{245})\Gamma(1-\lambda_{245})}{\Gamma(\lambda_4-\lambda_1)\Gamma(1-\lambda_4+\lambda_1)\Gamma(\lambda_5)\Gamma(1-\lambda_5)\Gamma(\lambda_{12345})\Gamma(1-\lambda_{12345})},$$

$$c_{(8:3)} = \frac{\Gamma(\lambda_{12})\Gamma(1-\lambda_{12})\Gamma(\lambda_4)\Gamma(1-\lambda_4)\Gamma(\lambda_{245})\Gamma(1-\lambda_{245})}{\Gamma(\lambda_1)\Gamma(1-\lambda_1)\Gamma(\lambda_{24})\Gamma(1-\lambda_{24})\Gamma(\lambda_{345})\Gamma(1-\lambda_{345})},$$

$$c_{(8:11)} = \frac{\Gamma(\lambda_{245})\Gamma(1-\lambda_{245})}{\Gamma(\lambda_5)\Gamma(1-\lambda_5)},$$

where we set

$$y_{12} = c_{(12:1)}y_1 + c_{(12:3)}y_3 + c_{(12:11)}y_{11},$$

$$y_6 = c_{(6:1)}y_1 + c_{(6:3)}y_3 + c_{(6:11)}y_{11},$$

$$y_8 = c_{(8:1)}y_1 + c_{(8:3)}y_3 + c_{(8:11)}y_{11}.$$

Then it is easy to see for which special values of the parameters a connection coefficient vanishes. Therefore, as a basis of V_1^0, the basis y_1, y_3 is useful.

The second example is the fourth order differential equation (E$_4$) in the series of rigid Fuchsian ordinary differential equations called the even family of Simpson. The differential equation (E$_4$) has regular singular points at $x = 0, 1, \infty$, and the multiplicities of the characteristic exponents (the spectral types) are

$$((112), (22), (1111)).$$

For the spectral type, we will explain in Chap. 7, Sect. 7.3.1. It is known that the equation (E$_4$) has the following integral representation of solutions

$$y(x) = \int_\Delta t_1{}^{\lambda_1}(t_1-1)^{\lambda_2}(t_1-t_2)^{\lambda_3}t_2{}^{\lambda_4}(t_2-t_3)^{\lambda_5}(t_3-1)^{\lambda_6}(t_3-x)^{\lambda_7}\, dt_1\, dt_2\, dt_3,$$

$$(6.24)$$

where $\lambda_1, \lambda_2, \ldots, \lambda_7$ are parameters satisfying

$$\lambda_2 + \lambda_3 + \lambda_5 + \lambda_6 + 2 = 0.$$

By using this integral representation, we can derive the connection coefficients.

We consider the connection problem between $x = 0$ and $x = \infty$. The spectral type at $x = \infty$ is $(1, 1, 1, 1)$, which means that the dimension of each eigenspace is one. On the other hand, at $x = 0$, there are two eigenspaces of dimension one and one eigenspace of dimension two. The connection coefficients between the eigenspaces of dimension one at $x = \infty$ and $x = 0$ can be expressed as a ratio of products of Gamma functions. This is a consequence of the result of Oshima (Theorem 6.2), and also can be derived by using the integral representation [63]. Here we are interested in the connection problem between the eigenspaces at $x = \infty$ and the eigenspace at $x = 0$ of dimension two.

Bases of these eigenspaces can be given by choosing domains of integration for (6.24). We assume $-\infty < x < 0$. The characteristic exponents at $x = \infty$ are

$$-\lambda_{1234567} - 3, \quad -\lambda_{34567} - 2, \quad -\lambda_{567} - 1, \quad -\lambda_7,$$

where we use the notation (6.23). We can find the domains of integration which give the solutions of these characteristic exponents. We denote them by

$$\Delta_{\infty,1}, \quad \Delta_{\infty,2}, \quad \Delta_{\infty,3}, \quad \Delta_{\infty,4},$$

respectively. At $x = 0$, there are one dimensional eigenspaces for the exponents

$$\lambda_{13457} + 3, \quad \lambda_{457} + 2$$

and two dimensional eigenspace for holomorphic solutions. We denote the domains of integration corresponding to the above two exponents by

$$\Delta_{0,1}, \quad \Delta_{0,2},$$

respectively. We can find six domains of integration corresponding to holomorphic solutions at $x = 0$. We choose two among them, and denote these two by

$$\Delta_{0,h1}, \quad \Delta_{0,h2}.$$

We denote the solution of (6.24) obtained by the integration on the domain $\Delta_{a,k}$ by $y_{a,k}(x)$.

The left half plane $\{\operatorname{Re} x < 0\}$ is a common domain of definition for the above local solutions at $x = \infty$ and for ones at $x = 0$. Then there exists a linear relation between the two fundamental systems

$$(y_{\infty,1}(x), y_{\infty,2}(x), y_{\infty,3}(x), y_{\infty,4}(x)), \quad (y_{0,1}(x), y_{0,2}(x), y_{0,h1}(x), y_{0,h2}(x))$$

of solutions on the domain. We write it as

$$y_{\infty,1}(x) = c_{11}y_{0,1}(x) + c_{21}y_{0,2}(x) + c_{31}y_{0,h1}(x) + c_{41}y_{0,h2}(x),$$

$$y_{\infty,2}(x) = c_{12}y_{0,1}(x) + c_{22}y_{0,2}(x) + c_{32}y_{0,h1}(x) + c_{42}y_{0,h2}(x),$$

$$y_{\infty,3}(x) = c_{13}y_{0,1}(x) + c_{23}y_{0,2}(x) + c_{33}y_{0,h1}(x) + c_{43}y_{0,h2}(x),$$

$$y_{\infty,4}(x) = c_{14}y_{0,1}(x) + c_{24}y_{0,2}(x) + c_{34}y_{0,h1}(x) + c_{44}y_{0,h2}(x).$$

It turns out that c_{1j}, c_{2j} $(1 \leq j \leq 4)$ can be written as ratios of products of Gamma functions, while, for any choice of two domains among six corresponding to holomorphic solutions at $x = 0$, the remaining coefficients c_{3j}, c_{4j} $(1 \leq j \leq 4)$ cannot be of such simple form.

We find that there are good bases of the eigenspace V_0^0 of holomorphic solutions at $x = 0$. Choose any two among $y_{\infty,j}(x)$ $(1 \leq j \leq 4)$. For example, we choose $y_{\infty,1}(x)$ and $y_{\infty,2}(x)$. Analytic continuation of these to $x = 0$ yields the decompositions

$$y_{\infty,1}(x) = c_{11}y_{0,1}(x) + c_{21}y_{0,2}(x) + h_1(x),$$

$$y_{\infty,2}(x) = c_{12}y_{0,1}(x) + c_{22}y_{0,2}(x) + h_2(x)$$

according to the direct sum decomposition of V_0. Here we have $h_1(x), h_2(x) \in V_0^0$. We use $h_1(x), h_2(x)$ as a basis of V_0^0, and compute the analytic continuation of the remaining $y_{\infty,3}(x), y_{\infty,4}(x)$. Then we have

$$y_{\infty,3}(x) = c_{13}y_{0,1}(x) + c_{23}y_{0,2}(x) + b_{33}h_1(x) + b_{43}h_2(x),$$

$$y_{\infty,4}(x) = c_{14}y_{0,1}(x) + c_{24}y_{0,2}(x) + b_{34}h_1(x) + b_{44}h_2(x),$$

where the coefficients b_{3j}, b_{4j} $(j = 3, 4)$ of $h_1(x), h_2(x)$ are ratios of products of Gamma functions. Therefore we can easily see whether $y_{\infty,3}(x)$ or $y_{\infty,4}(x)$ contains components in V_0^0 by using this basis.

These good bases are obtained by the analytic continuation of eigenfunctions at another singular point. This idea comes from Okubo [134], and will be applied in various aspects.

Remark 6.1 So far we studied the case where the connection coefficients are described by exact formulas. However, even for Fuchsian ordinary differential equations, the connection coefficients are believed to be highly transcendental in general, and we can seldom describe them in exact formulas. Then it will be an important problem to give some approximation formula for connection coefficients. Schäfke and Schmidt have derived convergent sequences and series whose limits yield connection coefficients [155, 156]. See also [119].

Chapter 7
Fuchsian Differential Equations

As we defined in Chap. 5 (Definition 5.2), a differential equation with coefficient in rational functions is called Fuchsian if all singular points are regular singular points. Note that $x = \infty$ is a regular singular point if, for a local coordinate t at $\infty \in \mathbb{P}^1$, $t = 0$ is a regular singular point.

Fuchsian ordinary differential equations have been studied since Euler and Gauss, and the basis of the theory was constructed by Riemann, Fuchs, Frobenius, and so on, in nineteenth century. Recently, a big progress has been brought by the work of Katz [100], and then we should rewrite the basic part of the theory. In this chapter, we give the new basic part of the theory of Fuchsian differential equations in explaining the Katz theory.

7.1 Fuchs Relation

We consider a scalar Fuchsian differential equation

$$y^{(n)} + p_1(x)y^{(n-1)} + \cdots + p_n(x)y = 0. \tag{7.1}$$

We denote by S the set of regular singular points of (7.1). For each regular singular point $a \in S$, the characteristic exponents are roots of the characteristic equation, which is an algebraic equation of degree n. Then we have n characteristic exponents with possible multiplicities, which we denote by

$$\lambda_{a,1}, \lambda_{a,2}, \ldots, \lambda_{a,n}. \tag{7.2}$$

The *Fuchs relation* is an equation for the sum of these quantities for all $a \in S$.

© The Editor(s) (if applicable) and The Author(s), under exclusive license to Springer Nature Switzerland AG 2020
Y. Haraoka, *Linear Differential Equations in the Complex Domain*, Lecture Notes in Mathematics 2271, https://doi.org/10.1007/978-3-030-54663-2_7

Theorem 7.1 (Fuchs Relation) *For a scalar Fuchsian differential equation (7.1), we denote by S the set of regular singular points, and by (7.2) the set of characteristic exponents at $a \in S$. Then we have*

$$\sum_{a \in S} \sum_{k=1}^{n} \lambda_{a,k} = \frac{n(n-1)(\#S - 2)}{2}, \tag{7.3}$$

where #S is the number of the elements of S.

Proof Let $a_1, a_2, \ldots, a_m$ be the regular singular points of (7.1) in $\mathbb{C}$. By Fuchs theorem (Theorem 4.3), the points $x = a_1, a_2, \ldots, a_m$ are at most poles of order 1 of $p_1(x)$. Then we can write

$$p_1(x) = \frac{q(x)}{\prod_{i=1}^{m}(x - a_i)},$$

where $q(x)$ is a polynomial. Let l be the degree of $q(x)$, and set

$$q(x) = q_0 x^l + \cdots .$$

The point $x = \infty$ is a regular point or a regular singular point of (7.1). We rewrite Eq. (7.1) in the new variable $t = \dfrac{1}{x}$. Note that, for $k = 1, 2, \ldots$, we have

$$\frac{d^k}{dx^k} = (-1)^k \left(t^{2k} \frac{d^k}{dt^k} + k(k-1)t^{2k-1} \frac{d^{k-1}}{dt^{k-1}} + \cdots \right).$$

Then we obtain

$$\frac{d^n y}{dt^n} + \left(\frac{n(n-1)}{t} - \frac{1}{t^2} p_1 \left(\frac{1}{t} \right) \right) \frac{d^{n-1} y}{dt^{n-1}} + \cdots = 0. \tag{7.4}$$

Since $t = 0$ is a regular point or a regular singular point, again by Fuchs theorem, the coefficient of $d^{n-1}y/dt^{n-1}$ has at most a pole of order 1 at $t = 0$. On the other hand, we have

$$\frac{n(n-1)}{t} - \frac{1}{t^2} p_1 \left(\frac{1}{t} \right) = \frac{n(n-1)}{t} - t^{m-2-l}(q_0 + O(t)),$$

and then we get $m - 2 - l \geq -1$. Thus we obtain $l \leq m - 1$, which induces the partial fractional expansion of $p_1(x)$

$$p_1(x) = \sum_{i=1}^{m} \frac{r_i}{x - a_i}.$$

Moreover we get

$$p_1(x) = \frac{\left(\sum_{i=1}^m r_i\right) x^{m-1} + \cdots}{\prod_{i=1}^m (x - a_i)}.$$

The characteristic polynomial at a regular singular point $x = a_i$ is

$$\rho(\rho - 1) \cdots (\rho - n + 1) + r_i \rho(\rho - 1) \cdots (\rho - n + 2) + \cdots$$

$$= \rho^n - \left(\frac{n(n-1)}{2} - r_i\right) \rho^{n-1} + \cdots,$$

and then the sum of the characteristic exponents at $x = a_i$ is equal to

$$\frac{n(n-1)}{2} - r_i.$$

On the other hand, the residue of the coefficient of $d^{n-1}y/dt^{n-1}$ in (7.4) at $t = 0$ is

$$n(n-1) - \sum_{i=1}^m r_i. \tag{7.5}$$

If this value is different from 0, $x = \infty$ is a regular singular point, and the sum of the characteristic exponents at $x = \infty$ is given by

$$\frac{n(n-1)}{2} - \left(n(n-1) - \sum_{i=1}^m r_i\right) = -\frac{n(n-1)}{2} + \sum_{i=1}^m r_i.$$

Then in this case, the sum of all characteristic exponents at all regular singular points is

$$\sum_{i=1}^m \left(\frac{n(n-1)}{2} - r_i\right) + \left(-\frac{n(n-1)}{2} + \sum_{i=1}^m r_i\right) = \frac{n(n-1)(m-1)}{2},$$

which coincides with the right hand side of (7.3) because $\#S = m + 1$. If the value (7.5) is equal to 0, $x = \infty$ is a regular point, and then the regular singular points are $a_1, a_2, \ldots, a_m$. In this case, the sum of the characteristic exponents at these points is

$$\sum_{i=1}^m \left(\frac{n(n-1)}{2} - r_i\right) = \frac{n(n-1)}{2} m - \sum_{i=1}^m r_i$$

$$= \frac{n(n-1)m}{2} - n(n-1)$$

$$= \frac{n(n-1)(m-2)}{2},$$

where the second equality is shown by the fact that (7.5) is 0. Then we also have (7.3) because $\#S = m$.   □

The Fuchs relation is a necessary condition for the existence of Fuchsian differential equations, however, not a sufficient condition. A sufficient condition will be discussed in Sect. 7.6.

7.2 Riemann Scheme

Consider a scalar Fuchsian differential equation (7.1) at a regular singular point $x = a$. If there is no integral difference between any two characteristic exponents, logarithmic solutions do not appear in the local solutions. A logarithmic solution may appear if there is an integral difference among characteristic exponents. In particular, if there is a multiplicity in the characteristic exponents, a logarithmic solution necessarily appears. We shall examine how logarithmic solutions appear.

We consider the case where integral differences exists among characteristic exponents. Let λ be a characteristic exponent, and

$$\lambda,\ \lambda + k_1,\ \lambda + k_2, \ldots,\ \lambda + k_{m-1}$$

the set of characteristic exponents which have integral differences with λ, where $k_1, k_2, \ldots, k_{m-1}$ are integers satisfying $0 \le k_1 \le k_2 \le \cdots \le k_{m-1}$. We shall show that, in the case

$$(k_1, k_2, \ldots, k_{m-1}) = (1, 2, \ldots, m - 1), \tag{7.6}$$

something special happens. By operating the gauge transformation

$$y = (x - a)^{\lambda} z$$

to the differential equation (7.1), we can reduce λ to 0.

Proposition 7.2 *We consider a scalar differential equation*

$$z^{(n)} + q_1(x)z^{(n-1)} + \cdots + q_n(x)z = 0. \tag{7.7}$$

Let $x = a$ be a regular singular point of the equation, and assume that $0, 1, \ldots, m-1$ are characteristic exponents at $x = a$ and that no other integer is a characteristic exponent at $x = a$. Then the solutions of characteristic exponents $0, 1, \ldots, m - 1$ contain no logarithmic functions if and only if

$$b_i \le n - m \quad (n - m + 1 \le i \le n)$$

holds, where b_i is the order of pole of $q_i(x)$ at $x = a$.

Proof Without loss of generality, we may assume $a = 0$. Since $x = 0$ is a regular singular point, the order of pole of $q_i(x)$ at $x = 0$ is at most i. Set

$$q_i(x) = \frac{1}{x^i} \sum_{k=0}^{\infty} q_{ik} x^k \ (1 \le i \le n), \quad z(x) = x^\rho \sum_{k=0}^{\infty} z_k x^k,$$

and put them into the differential equation (7.7). By setting

$$f_0(\rho) = \rho(\rho - 1) \cdots (\rho - n + 1) + \sum_{i=1}^{n} q_{i0} \rho(\rho - 1) \cdots (\rho - n + i + 1),$$

$$f_k(\rho) = \sum_{i=1}^{n} q_{ik} \rho(\rho - 1) \cdots (\rho - n + i + 1) \ (k \ge 1),$$

we have

$$f_0(\rho) z_0 = 0 \tag{7.8}$$

by the vanishing of the coefficient of $x^{\rho - n}$ in the left hand side of (7.7), and

$$\sum_{i=0}^{k} f_{k-i}(\rho + i) z_i = 0 \tag{7.9}$$

by the vanishing of the coefficient of $x^{\rho - n + k}$ ($k \ge 1$). The polynomial $f_0(\rho)$ is the characteristic polynomial, and $\rho = 0, 1, \ldots, m - 1$ are characteristic exponents, so that we get

$$f_0(0) = f_0(1) = \cdots = f_0(m - 1) = 0.$$

Also, by the assumption, we have $f_0(k) \ne 0$ for $k \ge m$. Then, if $z_0, z_1, \ldots, z_{m-1}$ are determined, the remaining coefficients z_k ($k \ge m$) are uniquely determined by (7.9).

The solutions of exponents $0, 1, \ldots, m - 1$ contain no logarithmic functions if and only if z_k ($k \ge 0$) are determined for each $\rho = 0, 1, \ldots, m - 1$, and this is equivalent to the relation (7.9) for any $z_0, z_1, \ldots, z_{m-1}$. The last condition is given by

$$f_k(\rho) = 0 \ (1 \le k \le m - 1, \ 0 \le \rho \le m - k - 1),$$

and, by the definition of $f_k(\rho)$, this is equivalent to

$$q_{ik} = 0 \ (n - m + 1 \le i \le n, \ 0 \le k \le i - (n - m + 1)).$$

This condition means $b_i \leq n - m$ $(n - m + 1 \leq i \leq n)$.   □

Example 7.1 We consider a third order scalar differential equation

$$z''' + q_1(x)z'' + q_2(x)z' + q_3(x)z = 0$$

which has a regular singular point at $x = 0$ of characteristic exponents $0, 1, \rho$, where $\rho \notin \mathbb{Z}$. By Proposition 7.2, there is no logarithmic solution at $x = 0$ if and only if the differential equation can be written in the form

$$z''' + \frac{Q_1(x)}{x} z'' + \frac{Q_2(x)}{x} z' + \frac{Q_3(x)}{x} z = 0,$$

where $Q_1(x)$, $Q_2(x)$, $Q_3(x)$ are holomorphic at $x = 0$. On the other hand, there is a logarithmic solution at $x = 0$ if and only if the differential equation can be written in the form

$$z''' + \frac{Q_1(x)}{x} z'' + \frac{Q_2(x)}{x} z' + \frac{R_3(x)}{x^2} z = 0,$$

where $R_3(x)$ is holomorphic at $x = 0$ and $R_3(0) \neq 0$.

As we can see in the proof of Proposition 7.2, if (7.6) holds with $\lambda = 0$ and if there is no logarithmic solution, there exists a solution at $x = a$ of the differential equation (7.1) of the form

$$y(x) = g(x - a) + \sum_{k=m}^{\infty} y_m(x - a)^m,$$

where $g(X)$ is any polynomial of degree at most $m-1$. Then these solutions make an m dimensional subspace of solutions. Moreover, if $m = n$, $x = a$ is a regular point of the differential equation (7.1). Also in the case $\lambda \neq 0$ and $m = n$, any solution at $x = a$ can be obtained from a holomorphic function multiplied by $(x - a)^\lambda$, and then we can regard $x = a$ as an almost regular point. Hence, for the case where (7.6) holds, it may be natural that there is no logarithmic solution. We denote this case by the symbol

$$[\lambda]_{(m)}.$$

Namely, this symbol represents a set $\lambda, \lambda + 1, \ldots, \lambda + m - 1$ of characteristic exponents together with the property that there is no logarithmic solution of these exponents. In other words, this symbol implies that there exist solutions of the form

$$y(x) = (x - a)^\lambda [g(x - a) + O((x - a)^m)],$$

where $g(X)$ is any polynomial of degree at most $m - 1$. We call $[\lambda]_{(m)}$ an extended characteristic exponent.

At a regular singular point, the local behaviors represented by the extended characteristic exponents

$$[\lambda_1]_{(m_1)}, [\lambda_2]_{(m_2)}, \ldots, [\lambda_l]_{(m_l)}$$

are basic. Here $\lambda_1, \lambda_2, \ldots, \lambda_l$ are complex numbers mutually distinct modulo integers, and $(m_1, m_2, \ldots, m_l)$ is a partition of n. In other words, this is the case where a fundamental system of solutions consisting of the solutions of the form

$$(x - a)^{\lambda_j}[g_j(x - a) + O((x - a)^{m_j})] \quad (j = 1, 2, \ldots, l)$$

exists, where each $g_j(X)$ is a polynomial of degree at most $m_j - 1$. A logarithmic solution may appear when there is an integral difference among λ_j.

Let $a_0, a_1, \ldots, a_p$ be the regular singular points of the Fuchsian differential equation (7.1). Assume that, for each $x = a_j$, extended characteristic exponents are given by

$$[\lambda_{j,1}]_{(m_{j,1})}, [\lambda_{j,2}]_{(m_{j,2})}, \ldots, [\lambda_{j,l_j}]_{(m_{j,l_j})}.$$

We call the table of these data

$$\left\{ \begin{array}{cccc} x = a_0 & x = a_1 & \cdots & x = a_p \\ [\lambda_{0,1}]_{(m_{0,1})} & [\lambda_{1,1}]_{(m_{1,1})} & \cdots & [\lambda_{p,1}]_{(m_{p,1})} \\ \vdots & \vdots & & \vdots \\ [\lambda_{0,l_0}]_{(m_{0,l_0})} & [\lambda_{1,l_1}]_{(m_{1,l_1})} & \cdots & [\lambda_{p,l_p}]_{(m_{p,l_p})} \end{array} \right\} \tag{7.10}$$

the *Riemann scheme* of the differential equation (7.1). When $m_{j,k} = 1$, we write $\lambda_{j,k}$ in place of $[\lambda_{j,k}]_{(1)}$. The Riemann scheme represents all local behaviors of a Fuchsian differential equation, and is useful. However, if there is an integral difference among $\lambda_{j,k}$ ($k = 1, 2, \ldots, l_j$), the Riemann scheme does not tell whether there is a logarithmic solution.

The Fuchs relation (7.3) can be written in terms of the data of the Riemann scheme (7.10) as

$$\sum_{j=0}^{p} \sum_{k=1}^{l_j} \sum_{i=0}^{m_{j,k}-1} (\lambda_{j,k} + i) = \frac{n(n-1)(p-1)}{2}.$$

For a system of differential equation of the first order, we can also define a table which represents all local behaviors, however, there may be no standard notation. In the next section we shall study systems of non-resonant Fuchsian differential equations of the first order. For such systems we can define the spectral data, which corresponds the Riemann scheme.

Example 7.2 The Riemann scheme of the Gauss hypergeometric differential equation

$$x(1 - x)y'' + (\gamma - (\alpha + \beta + 1)x)y' - \alpha\beta y = 0$$

is given by

$$\left\{ \begin{matrix} x = 0 & x = 1 & x = \infty \\ 0 & 0 & \alpha \\ 1 - \gamma & \gamma - \alpha - \beta & \beta \end{matrix} \right\}. \tag{7.11}$$

The generalized hypergeometric differential equation ($_3E_2$) of the third order can be written, by using the Euler operator $\delta = x\dfrac{d}{dx}$, as

$$\delta(\delta + \beta_1 - 1)(\delta + \beta_2 - 1)y = x(\delta + \alpha_1)(\delta + \alpha_2)(\delta + \alpha_3)y. \tag{7.12}$$

The Riemann scheme of this differential equation is given by

$$\left\{ \begin{matrix} x = 0 & x = 1 & x = \infty \\ 0 & [0]_{(2)} & \alpha_1 \\ 1 - \beta_1 & & \alpha_2 \\ 1 - \beta_2 & -\beta_3 & \alpha_3 \end{matrix} \right\},$$

where β_3 is defined by $\alpha_1 + \alpha_2 + \alpha_3 = \beta_1 + \beta_2 + \beta_3$.

Now we study the changes of Riemann schemes by transformations of the independent variable or of the unknown functions. Let (7.10) be the Riemann scheme of the Fuchsian differential equation (7.1).

First we consider a fractional linear transformation

$$\xi = \frac{ax + b}{cx + d}$$

of the independent variable x. Each singular point a_j is sent to

$$b_j = \frac{aa_j + b}{ca_j + d}.$$

By noting

$$x - a_j = (\xi - b_j)(C_j + O(\xi - b_j)) \quad (C_j \neq 0),$$

we see that the local behaviors including characteristic exponents do not change by this transformation. Thus the Riemann scheme after transformation is given by

$$\left\{ \begin{matrix} \xi = b_0 & \xi = b_1 & \cdots & \xi = b_p \\ [\lambda_{0,1}]_{(m_{0,1})} & [\lambda_{1,1}]_{(m_{1,1})} & \cdots & [\lambda_{p,1}]_{(m_{p,1})} \\ \vdots & \vdots & & \vdots \\ [\lambda_{0,l_0}]_{(m_{0,l_0})} & [\lambda_{1,l_1}]_{(m_{1,l_1})} & \cdots & [\lambda_{p,l_p}]_{(m_{p,l_p})} \end{matrix} \right\}$$

Only the first row is changed.

Second we consider a gauge transformation

$$z(x) = y(x) \prod_{j=0}^{p} (x - a_j)^{\alpha_j}$$

of the unknowns. Here $\alpha_j \in \mathbb{C}$, and assume that ∞ is not in $a_0, a_1, \ldots, a_p$. If $\sum_{j=0}^{p} \alpha_j = 0$, then $x = \infty$ is a regular point also for $z(x)$, and hence the characteristic exponents at each singular point a_j are shifted by α_j. Thus we obtain the Riemann scheme

$$\left\{ \begin{matrix} x = a_0 & x = a_1 & \cdots & x = a_p \\ [\lambda_{0,1} + \alpha_0]_{(m_{0,1})} & [\lambda_{1,1} + \alpha_1]_{(m_{1,1})} & \cdots & [\lambda_{p,1} + \alpha_p]_{(m_{p,1})} \\ \vdots & \vdots & & \vdots \\ [\lambda_{0,l_0} + \alpha_0]_{(m_{0,l_0})} & [\lambda_{1,l_1} + \alpha_1]_{(m_{1,l_1})} & \cdots & [\lambda_{p,l_p} + \alpha_p]_{(m_{p,l_p})} \end{matrix} \right\} \tag{7.13}$$

for the differential equation satisfied by $z(x)$. If $\sum_{j=0}^{p} \alpha_j \neq 0$, $x = \infty$ becomes a singular point for $z(x)$, and then the column

$$\left\{ \begin{matrix} x = \infty \\ [\alpha_\infty]_{(n)} \end{matrix} \right\}$$

is added to the Riemann scheme, where we set $\alpha_\infty = -\sum_{j=0}^{p} \alpha_j$. In the case $a_0 = \infty$, we consider a gauge transformation

$$z(x) = y(x) \prod_{j=1}^{p} (x - a_j)^{\alpha_j}.$$

Then we get the same Riemann scheme (7.13) for $z(x)$, where we set $\alpha_0 = -\sum_{j=1}^{p} \alpha_j$.

When we consider a differential equation of the second order with three regular singular points, first we can send the singular points to $0, 1, \infty$ by a fractional linear transformation. Then by a gauge transformation, we can shift characteristic exponents at each singular point except ∞. In particular, we can have 0 as one of the characteristic exponents at the singular points 0, 1. The resulting Riemann scheme coincides with the Riemann scheme (7.11) of the Gauss hypergeometric differential equation.

7.3 Systems of Fuchsian Differential Equations of Normal Form

We consider a system of differential equations of the first order of rank n

$$\frac{dY}{dx} = A(x)Y, \tag{7.14}$$

where $A(x)$ is an $n \times n$ matrix with coefficients in rational functions. The singular points of the differential equation (7.14) are the poles of $A(x)$ and possibly $x = \infty$. At $x = \infty$, by using the local coordinate $t = \dfrac{1}{x}$ at $x = \infty$, we have the differential equation

$$\frac{dY}{dt} = -\frac{1}{t^2} A\left(\frac{1}{t}\right) Y.$$

Then, if $t = 0$ is a pole of $-\dfrac{1}{t^2} A\left(\dfrac{1}{t}\right)$, $x = \infty$ is a singular point of the differential equation (7.14).

If all these poles are of order 1, the singular points are regular singular, and hence the differential equation (7.14) becomes Fuchsian. Such differential equation is written as

$$\frac{dY}{dx} = \left(\sum_{j=1}^{p} \frac{A_j}{x - a_j}\right) Y, \tag{7.15}$$

where $a_1, a_2, \ldots, a_p$ are mutually distinct p points in $\mathbb{C}$, and $A_1, A_2, \ldots, A_p$ are constant $n \times n$ matrices. At $x = \infty$, we have

$$-\frac{1}{t^2} A\left(\frac{1}{t}\right) = -\frac{1}{t} \sum_{j=1}^{p} \frac{A_j}{1 - a_j t} = \frac{A_0}{t} + O(1),$$

where the residue matrix A_0 at $x = \infty$ $(t = 0)$ is given by

$$A_0 = -\sum_{j=1}^{p} A_j. \tag{7.16}$$

This implies that $x = \infty$ is at most a regular singular point. Note that $x = \infty$ becomes a regular point if $A_0 = O$. We denote ∞ by a_0.

The system (7.15) is a system of Fuchsian differential equations with regular singular points at $x = a_0, a_1, \ldots, a_p$. The system of this form is called a system of Schlesinger normal form or a Fuchsian system by literatures. In this book, we call the system (7.15) a *Fuchsian system of normal form*, as is used in Sibuya [165].

We can express a Fuchsian system of normal form by using logarithmic derivatives:

$$dY = \left(\sum_{j=1}^{p} A_j d \log(x - a_j) \right) Y.$$

By this expression, we readily see that $x = \infty$ is at most regular singular. The expression using logarithmic derivatives is important if we consider a connection to completely integrable systems which are discussed in Part II of this book.

Here we briefly explain a history concerning Fuchsian systems of normal form. Riemann posed a problem of the existence of a Fuchsian differential equation with a prescribed representation of a fundamental group as a monodromy representation. This problem is called Riemann's problem. Riemann seems to have considered the existence of a scalar differential equation, however, Schlesinger, who studied Riemann's problem intensively, considered the existence of a Fuchsian system of normal form. Then we sometimes call a Fuchsian system of normal form a Schlesinger system. In the 23 problems put forth in 1900 by Hilbert, he listed Riemann's problem as the 21st problem, so that Riemann's problem is also called Riemann-Hilbert problem. Riemann-Hilbert problem is solved affirmatively.

We note that the above explanation is based on the good exposition [145]. We also recommend [1] as a good reference.

Since Riemann's problem asks the existence of a Fuchsian system of normal form, we are also interested in whether any Fuchsian differential equation is transformed into a Fuchsian system of normal form. Kimura [105] showed that any Fuchsian differential equation can be transformed into a Fuchsian system of normal form if at most one apparent singular point is added. Bolibruch [23] gave an example of a Fuchsian differential equation which cannot be transformed into a Fuchsian system without adding an apparent singular point. We understand that, owing to Kimura's result, we may lose almost no generality if we restrict our consideration to Fuchsian systems of normal form.

Fuchsian systems of normal form behave well with Katz operations and deformation theory, and are useful in the local analysis at regular singular points. In particular, the analysis of a differential equation is reduced to the study of a tuple $(A_1, A_2, \ldots, A_p)$ of matrices.

7.3.1 Non-resonant Fuchsian Systems of Normal Form and Spectral Types

We called a square matrix non-resonant if there is no integral difference among distinct eigenvalues (Definition 4.4). We call a Fuchsian system (7.15) of normal form *non-resonant* if all residue matrices A_j ($0 \le j \le p$) are non-resonant.

Proposition 7.3 *If a Fuchsian system (7.15) of normal form is non-resonant, the local monodromy at each regular singular point* $x = a_j$ ($0 \le j \le p$) *is given by* $e^{2\pi\sqrt{-1}A_j}$.

This proposition follows directly from Theorem 4.12. We see that, in non-resonant case, the analysis becomes simple because each residue matrix corresponds directly to each local monodromy. In the following of this section, we consider non-resonant Fuchsian systems of normal form.

Let C_j be the Jordan canonical form of the residue matrix A_j of a non-resonant Fuchsian system (7.15) of normal form. Then we call the tuple

$$(C_0, C_1, \ldots, C_p)$$

the *spectral data*. The spectral data completely determines the local behaviors, and corresponds to the Riemann scheme for scalar Fuchsian differential equations.

For a matrix C of Jordan canonical form, we denote by $C^\natural$ the *spectral type* of C, which is obtained from C by forgetting the values of the eigenvalues. In order to describe the spectral types, we use the following notation.

Definition 7.1 Let A be a square matrix, and α an eigenvalue. For $j = 1, 2, \ldots$, we denote by $e_j(A; \alpha)$ the number of the Jordan cells in the Jordan canonical form of A of the eigenvalue α and of size equal to or more than j.

From this definition, we obtain the following properties for $e_j(A; \alpha)$:

$$e_1(A; \alpha) \ge e_2(A; \alpha) \ge \cdots ,$$

$$\sum_{\alpha} \sum_{j \ge 1} e_j(A; \alpha) = n,$$

where n is the size of A, and $\sum_{\alpha}$ denotes the sum over all eigenvalues of A.

Let C be a matrix of Jordan canonical form with mutually distinct eigenvalues $\lambda_1, \lambda_2, \ldots$. Then the spectral type of C is described by the tuple of the numerical data

$$((e_j(C; \lambda_1))_{j \ge 1}, (e_j(C; \lambda_2))_{j \ge 1}, \ldots).$$

For example, the spectral type of the matrix

$$
C = \begin{pmatrix}
\lambda & 1 & & & & \\
& \lambda & 1 & & & \\
& & \lambda & & & \\
& & & \lambda & & \\
& & & & \mu & 1 \\
& & & & & \mu
\end{pmatrix}
$$

is given by

$$((211), (11)).$$

Thus the spectral type is given by the numerical data

$$e = ((e_j^{(1)})_{j \geq 1}, (e_j^{(2)})_{j \geq 1}, \dots) \in (\mathbb{Z}_{\geq 0})^\infty \times (\mathbb{Z}_{\geq 0})^\infty \times \cdots$$

satisfying

$$e_j^{(i)} \geq e_{j+1}^{(i)} \quad (j \geq 1, i = 1, 2, \dots),$$

$$\sum_i \sum_{j \geq 1} e_j^{(i)} < \infty.$$

We set

$$|e| = \sum_i \sum_{j \geq 1} e_j^{(i)}.$$

When C is a diagonal matrix, the spectral type $C^\natural$ is the partition which represents the multiplicities of the eigenvalues. In this case, we write the spectral type by omitting the inner parentheses as

$$(e_1(C; \lambda_1), e_1(C; \lambda_2), \dots).$$

We call such spectral type semi-simple.

The spectral type is a notion defined for a conjugacy class of a square matrix, and then, for a conjugacy class $\mathcal{A}$ and for a matrix A contained in the conjugacy class, we define the spectral types $\mathcal{A}^\natural$ and $A^\natural$ by the spectral type $C^\natural$ of their Jordan canonical form C.

Now, for a non-resonant Fuchsian system (7.15) of normal form with spectral data $(C_0, C_1, \dots, C_p)$, we define the *spectral type* of (7.15) by the tuple

$$(C_0^\natural, C_1^\natural, \dots, C_p^\natural)$$

of the spectral types. Note that, for a non-resonant square matrix A, we have

$$\left(e^{2\pi\sqrt{-1}A}\right)^{\natural} = A^{\natural}.$$

Then, for a non-resonant Fuchsian system (7.15) of normal form, the tuple of the spectral types of the local monodromies $(L_0, L_1, \ldots, L_p)$ coincides with the spectral type of the system (7.15). We call a tuple of the spectral types of square matrices semi-simple if each matrix is semi-simple.

On the other hand, we have a condition for the eigenvalues of the residue matrices. We denote by $\lambda_{j1}, \lambda_{j2}, \ldots, \lambda_{jn}$ the eigenvalues of A_j, where we repeat the same values if there is a multiplicity. By the definition (7.16) of A_0, we have

$$\sum_{j=0}^{p} A_j = O,$$

and then we get the relation

$$\sum_{j=0}^{p}\sum_{k=1}^{n} \lambda_{jk} = 0 \tag{7.17}$$

by taking the trace. This is a relation on the characteristic exponents of the regular singular points, which corresponds to the Fuchs relation for scalar Fuchsian differential equations. Then we sometimes call this relation (7.17) the Fuchs relation.

As in the case of scalar differential equations, the condition (7.17) is a necessary condition for the existence of a Fuchsian system with the spectral data $(C_0, C_1, \ldots, C_p)$, but not a sufficient one. We study the sufficient condition in Sect. 7.6.

7.3.2 Differential Equations of Okubo Normal Form

In this section we introduce a class of Fuchsian systems of normal form called differential equations of Okubo normal form, which is found by Kenjiro Okubo and appeared also in the studies of R. Shäfke and Balser-Jurkat-Lutz. Differential equation of Okubo normal form behaves well with various integral transforms such as Riemann-Liouville transform (Euler transform) and Laplace transform, and then is a useful normal form.

Let T, A be constant $n \times n$ matrices, and assume that T is a diagonal matrix. The system of differential equations of the first order

$$(x - T)\frac{dY}{dx} = AY \qquad (7.18)$$

is called a differential equation of *Okubo normal form*. Assume that T is written in the form

$$T = \begin{pmatrix} t_1 I_{n_1} & & & \\ & t_2 I_{n_2} & & \\ & & \ddots & \\ & & & t_p I_{n_p} \end{pmatrix}, \quad t_i \neq t_j \ (i \neq j).$$

Then, by setting

$$A_j = \begin{pmatrix} O & & & \\ & \ddots & & \\ & & I_{n_j} & \\ & & & \ddots \\ & & & & O \end{pmatrix} A,$$

we can write the system (7.18) as

$$\frac{dY}{dx} = \left(\sum_{j=1}^{p} \frac{A_j}{x - t_j} \right) Y.$$

Thus a differential equation of Okubo normal form is a Fuchsian system of normal form. The residue matrix at $x = \infty$ is given by $-A$.

Here we explain Riemann-Liouville transform. Let $f(x)$ be holomorphic on a domain $D \subset \mathbb{P}^1$, and assume that, for a point $a \in \bar{D}$,

$$\lim_{\substack{x \to a \\ x \in D}} f(x) = 0$$

holds. Then, for $\lambda \in \mathbb{C}$, the integral transformation

$$(I_a^\lambda f)(x) = \frac{1}{\Gamma(\lambda)} \int_a^x f(s)(x - s)^{\lambda - 1} \, ds \qquad (7.19)$$

is called Riemann-Liouville transform, where the path from a to x is taken in D.

The followings are basic properties of Riemann-Liouville transform.

Theorem 7.4

(i) $I_q^0 = id.$
(ii) $I_a^\lambda \circ I_a^\mu = I_a^{\lambda+\mu}.$
(iii) $(I_a^{-n} f)(x) = f^{(n)}(x) \ (n \in \mathbb{Z}_{\geq 0}).$

Proof (i) is a special case of (iii) with $n = 0$. We shall prove (iii). Take $n \in \mathbb{Z}_{\geq 0}$. We cannot put $\lambda = -n$ directly into the definition (7.19), so that we consider the limit $\lambda \to -n$. At the point $s = x$ we take a regularization of the integral, which will be explained in Chap. 9. Namely, take a point c on the path of integration near x. We denote by L the path of integration from a to c. Let C be the circle centered at x of radius $|c - x|$ with initial and end point c, and we attach the positive direction on C (Fig. 7.1). Then we have

$$\int_a^x f(s)(x-s)^{\lambda-1}\, ds = \int_L f(s)(x-s)^{\lambda-1}\, ds + \frac{1}{1 - e^{2\pi\sqrt{-1}\lambda}} \int_C f(s)(x-s)^{\lambda-1}\, ds$$

(cf. Sect. 9.4). The regularization is to replace the integral of the left hand side by the right hand side.

We know that $\Gamma(\lambda)$ has a pole of order 1 at $\lambda = -n$, and has the Laurent expansion

$$\Gamma(\lambda) = \frac{(-1)^n}{n!}\frac{1}{\lambda+n} + O(1).$$

In particular we have

$$\lim_{\lambda \to -n} \frac{1}{\Gamma(\lambda)} = 0.$$

On the other hand, $1 - e^{2\pi\sqrt{-1}\lambda}$ has a zero of order 1 at $\lambda = -n$, and has a Taylor expansion

$$1 - e^{2\pi\sqrt{-1}\lambda} = -2\pi\sqrt{-1}(\lambda + n) + O((\lambda+n)^2).$$

Fig. 7.1 Paths of integration for regularization

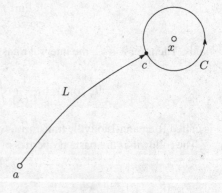

Then, if we take the limit $\lambda \to -n$ in the expression

$$(I_a^\lambda f)(x) = \frac{1}{\Gamma(\lambda)} \int_L f(s)(x-s)^{\lambda-1}\, ds + \frac{1}{\Gamma(\lambda)(1-e^{2\pi\sqrt{-1}\lambda})} \int_C f(s)(x-s)^{\lambda-1}\, ds,$$

the first term in the right hand side converges to 0, and the second term of the right hand side becomes

$$\frac{(-1)^{n+1}n!}{2\pi\sqrt{-1}} \int_C \frac{f(s)}{(x-s)^{n+1}}\, ds = f^{(n)}(x)$$

by the help of Cauchy's integral formula.

The assertion (ii) can be shown by interchanging the order of integration:

$$
\begin{aligned}
((I_a^\lambda \circ I_a^\mu)(f))(x) &= \frac{1}{\Gamma(\lambda)\Gamma(\mu)} \int_a^x \left\{ \int_a^t f(s)(t-s)^{\mu-1}\, ds \right\} (x-t)^{\lambda-1}\, dt \\
&= \frac{1}{\Gamma(\lambda)\Gamma(\mu)} \int_a^x dt \int_a^t f(s)(t-s)^{\mu-1}(x-t)^{\lambda-1}\, ds \\
&= \frac{1}{\Gamma(\lambda)\Gamma(\mu)} \int_a^x ds \int_s^x f(s)(t-s)^{\mu-1}(x-t)^{\lambda-1}\, dt \\
&= \frac{1}{\Gamma(\lambda)\Gamma(\mu)} \int_a^x f(s)\, ds \\
&\qquad \times \int_0^1 \big((x-s)u\big)^{\mu-1}\big((x-s)(1-u)\big)^{\lambda-1}(x-s)\, du \\
&= \frac{B(\lambda,\mu)}{\Gamma(\lambda)\Gamma(\mu)} \int_a^x f(s)(x-s)^{\lambda+\mu-1}\, ds \\
&= (I_a^{\lambda+\mu} f)(x),
\end{aligned}
$$

where we used the substitution of the integral variable from t to u defined by

$$u = \frac{t-s}{x-s}.$$

$\square$

We also call the transformation

$$f(x) \mapsto \int_\Delta f(s)(x-s)^{\lambda-1}\, dx$$

with parameter $\lambda \in \mathbb{C}$ Riemann-Liouville transform, where the integral path Δ is chosen appropriately. Riemann-Liouville transform is also called Euler transform.

Theorem 7.5 *A differential equation of Okubo normal form (7.18)*

$$(x - T)\frac{dY}{dx} = AY$$

is sent by Riemann-Liouville transform

$$Y(x) \mapsto Z(x) = \int_{\Delta} Y(s)(x - s)^{\lambda-1} dx$$

to a differential equation of Okubo normal form

$$(x - T)\frac{dZ}{dx} = (A + \lambda)Z$$

with A shifted to $A + \lambda$.

Proof First we show the assertion formally. We have

$$(x - T)\frac{dZ}{dx} = (x - T)(\lambda - 1) \int_{\Delta} Y(s)(x - s)^{\lambda-2} ds$$

$$= (\lambda - 1) \int_{\Delta} (x - s + s - T)Y(s)(x - s)^{\lambda-2} ds$$

$$= (\lambda - 1) \int_{\Delta} Y(s)(x - s)^{\lambda-1} ds + (\lambda - 1) \int_{\Delta} (s - T)Y(s)(x - s)^{\lambda-2} ds$$

$$= (\lambda - 1)Z(x)$$

$$+ \left[-(s - T)Y(s)(x - s)^{\lambda-1} \right]_{\partial\Delta} + \int_{\Delta} \frac{\partial}{\partial s}[(s - T)Y(s)](x - s)^{\lambda-1} ds$$

$$= (\lambda - 1)Z(x) + \int_{\Delta} Y(s)(x - s)^{\lambda-1} ds + \int_{\Delta} (s - T)\frac{dY}{ds}(s)(x - s)^{\lambda-1} ds$$

$$= \lambda Z(x) + \int_{\Delta} AY(s)(x - s)^{\lambda-1} ds$$

$$= (\lambda + A)Z(x).$$

If the path of integration Δ is taken so that

$$\left[-(s - T)Y(s)(x - s)^{\lambda-1} \right]_{\partial\Delta} = 0$$

holds, the above formal calculation is justified. For example, we can take as Δ a path connecting two regular singular points of $Y(s)$ with regularized end points.   □

Next we consider the Laplace transform. The Laplace transform is given by the integral

$$f(x) \mapsto \int_{\Delta} f(s)e^{-xs}\, ds.$$

Here the path of integration Δ is appropriately chosen. This integral transformation induces the transformation of differential operators

$$\begin{cases} \dfrac{d}{dx} & \mapsto \quad x, \\[2mm] x & \mapsto \quad -\dfrac{d}{dx}. \end{cases} \tag{7.20}$$

We can understand (7.20) as the Laplace transform for differential operators.

Theorem 7.6 *A differential equation of Okubo normal form (7.18)*

$$(x - T)\frac{dY}{dx} = AY$$

is sent by the Laplace transform

$$Y(x) \mapsto Z(x) = \int_{\Delta} Y(s)e^{-xs}\, ds$$

to a system of differential equations

$$\frac{dZ}{dx} = -\left(T + \frac{A+1}{x}\right) Z. \tag{7.21}$$

Proof Applying (7.20) formally to Okubo system (7.18), we get

$$\left(-\frac{d}{dx} - T\right)(xZ) = AZ,$$

which gives (7.21).

In order to clarify the condition for the path of integration Δ, we look at the integral:

$$
x T Z(x) = \int_\Delta T x e^{-xs} Y(s)\, ds
$$

$$
= -\left[T e^{-xs} Y(s) \right]_{\partial\Delta} + \int_\Delta T e^{-xs} \frac{dY}{dx}(s)\, ds
$$

$$
= -\int_\Delta (s - T - s) e^{-xs} \frac{dY}{ds}(s)\, ds
$$

$$
= -\int_\Delta e^{-xs} A Y(s)\, ds + \int_\Delta s e^{-xs} \frac{dY}{ds}(s)\, ds
$$

$$
= -A Z(x) + \left[s e^{-xs} Y(s) \right]_{\partial\Delta} - \int_\Delta (e^{-xs} - x s e^{-xs}) Y(s)\, ds
$$

$$
= -A Z(x) - Z(x) - x \int_\Delta \frac{\partial}{\partial x}(e^{-xs}) Y(s)\, ds
$$

$$
= -(A + 1) Z(x) - x \frac{dZ}{dx}(x),
$$

from which (7.21) follows. To justify the above calculation, we need

$$
\left[T e^{-xs} Y(s) \right]_{\partial\Delta} = \left[s e^{-xs} Y(s) \right]_{\partial\Delta} = 0,
$$

and the commutativity of the integral over Δ and the differentiation $\dfrac{d}{dx}$. These are the conditions for Δ. As in the proof of Theorem 7.5, these conditions are satisfied if we take as Δ a path connecting two regular singular points of (7.18) with regularized end points. Or, we can take $s = \infty$ as one of the end points of Δ if there exists $\delta > 0$ such that

$$
\frac{\pi}{2} + \delta \le \arg(-xs) \le \frac{3}{2}\pi - \delta
$$

holds when s goes to ∞ on Δ. $\square$

Let r be a positive integer. For constant $n \times n$ matrices $B_0, B_1, \ldots, B_r$, the system of differential equations of the first order

$$
\frac{dZ}{dx} = x^{r-1}\left(B_0 + \frac{B_1}{x} + \cdots + \frac{B_r}{x^r} \right) Z \tag{7.22}
$$

is called a differential equation of Birkhoff canonical form of Poincaré rank r. For this equation, $x = \infty$ is an irregular singular point of Poincaré rank r. A differential equation of Birkhoff canonical form is a useful normal form for the study of an irregular singular point.

We see that Eq. (7.21) in Theorem 7.6 is of Birkhoff canonical form of Poincaré rank 1. Thus, by the Laplace transform, differential equations of Okubo normal form correspond to differential equations of Birkhoff canonical form. This is one of the motivation of introducing Okubo normal form. Balser-Jurkat-Lutz [14] and Schäfke [154] showed that the Stokes multipliers of the differential equation of Birkhoff canonical form (7.21) can be directly obtained from the connection coefficients of the differential equation of Okubo normal form (7.18).

Okubo [133, 134] noticed that the normal form (7.18) is also useful for the study of Fuchsian differential equations, and built a global theory of Fuchsian differential equations. Okubo's theory is also explained in the book [111], which is interesting and useful by itself. Yokoyama continued the study along Okubo's direction, obtained many remarkable results, and established the description of connection coefficients of differential equations of Okubo normal form [188]. This result seems a terminus ad quem of the study in the direction. For Yokoyama's works, refer also to [50, 65, 186, 187] and [51].

7.4 Rigidity

7.4.1 Rigidity of Monodromy

We have seen that the local monodromies can be explicitly computed by the coefficients of the differential equation (Sect. 4.3). On the other hand, the monodromy (the global monodromy) is determined by a global analytic nature, and we have no general way to compute it. If we have an integral representation of solutions, the global monodromy may be computed (Sect. 9.7). Also, if the global monodromy is uniquely determined by the local monodromies, it is computable. We call the monodromy rigid if it is uniquely determined by the local monodromies (see Definition 7.2, below). In non-rigid case, the transcendency of the monodromy will be evaluated by the number of parameters which do not depend on the local monodromies. In this section, we shall formulate this transcendency without mentioning differential equations.

Let S be a finite subset of $\mathbb{P}^1$, and set $X = \mathbb{P}^1 \setminus S$. Take a point $b \in X$, and fix it. We consider an n dimensional anti-representation of the fundamental group $\pi_1(X, b)$

$$\rho : \pi_1(X, b) \to \mathrm{GL}(n, \mathbb{C}).$$

For a $(+1)$-loop $\gamma \in \pi_1(X, b)$ for a point $a \in S$, we see that, thanks to Proposition 5.2, the conjugacy class $[\rho(\gamma)]$ depends only on a and does not depend on the choice of a $(+1)$-loop. The conjugacy class is the local monodromy at a. We call the conjugacy class $[\rho]$ of ρ as anti-representations a *monodromy*. A monodromy is equivalent to a local system of rank n over X.

Definition 7.2 A monodromy $[\rho]$ is called *rigid* if it is uniquely determined by the local monodromies for all $a \in S$.

We consider what this definition says. Set $S = \{a_0, a_1, \ldots, a_p\}$, and take a presentation of $\pi_1(X, b)$ as in Sect. 5.3:

$$\pi_1(X, b) = \langle \gamma_0, \gamma_1, \ldots, \gamma_p \mid \gamma_0 \gamma_1 \cdots \gamma_p = 1 \rangle,$$

where γ_j is a $(+1)$-loop for a_j. If we set

$$\rho(\gamma_j) = M_j \quad (0 \le j \le p),$$

we have

$$M_p \cdots M_1 M_0 = I,$$

and ρ is determined by the tuple $(M_0, M_1, \ldots, M_p)$. The local monodromy at a_j is given by the conjugacy class $[M_j]$ of M_j. Thus, the monodromy $[\rho]$ is rigid if and only if the class $[(M_0, M_1, \ldots, M_p)]$ of the tuple $(M_0, M_1, \ldots, M_p)$ is uniquely determined by the tuple $([M_0], [M_1], \ldots, [M_p])$ of the classes. In other words, $[\rho]$ is rigid if and only if, for any other anti-representation $(N_0, N_1, \ldots, N_p)$ satisfying

$$N_j = C_j M_j C_j^{-1} \quad (0 \le j \le p)$$

for some $C_j \in \mathrm{GL}(n, \mathbb{C})$, there exists $D \in \mathrm{GL}(n, \mathbb{C})$ such that

$$N_j = D M_j D^{-1} \quad (0 \le j \le p)$$

holds.

Example 7.3 For the case $\#S = 3, n = 2$, if $[\rho]$ is irreducible, then $[\rho]$ is rigid.

Proof ρ is determined by a tuple $(M_0, M_1, M_2) \in \mathrm{GL}(2, \mathbb{C})^3$ satisfying

$$M_2 M_1 M_0 = I. \tag{7.23}$$

We shall show the assertion when M_0, M_1, M_2 are semi-simple.

By the irreducibility, none of M_0, M_1, M_2 is a scalar matrix. Then, if we set

$$M_0 \sim \begin{pmatrix} \alpha_1 & \\ & \alpha_2 \end{pmatrix}, \quad M_1 \sim \begin{pmatrix} \beta_1 & \\ & \beta_2 \end{pmatrix}, \quad M_2 \sim \begin{pmatrix} \gamma_1 & \\ & \gamma_2 \end{pmatrix},$$

we have $\alpha_1 \ne \alpha_2, \beta_1 \ne \beta_2, \gamma_1 \ne \gamma_2$. By taking the determinant of both sides of (7.23), we get $\alpha_1 \alpha_2 \beta_1 \beta_2 \gamma_1 \gamma_2 = 1$. Since M_0 is uniquely determined by M_1, M_2 using (7.23), it is enough to show the uniqueness of the equivalence class $[(M_1, M_2)]$.

By a similar transformation which sends M_1 to a diagonal matrix, we have

$$(M_1, M_2) \sim \left(\begin{pmatrix} \beta_1 & \\ & \beta_2 \end{pmatrix}, \begin{pmatrix} a & b \\ c & d \end{pmatrix} \right).$$

By the irreducibility, we have $bc \neq 0$. Then, again taking a similar transformation by the matrix

$$Q = \begin{pmatrix} 1 & \\ & c \end{pmatrix},$$

we get

$$\left(\begin{pmatrix} \beta_1 & \\ & \beta_2 \end{pmatrix}, \begin{pmatrix} a & b \\ c & d \end{pmatrix} \right) \sim \left(\begin{pmatrix} \beta_1 & \\ & \beta_2 \end{pmatrix}, \begin{pmatrix} a & bc \\ 1 & d \end{pmatrix} \right).$$

Set $bc = b_1$. From the relations

$$M_2 \sim \begin{pmatrix} \gamma_1 & \\ & \gamma_2 \end{pmatrix}, \quad M_2 M_1 \sim \begin{pmatrix} \alpha_1^{-1} & \\ & \alpha_2^{-1} \end{pmatrix},$$

we obtain

$$\begin{pmatrix} a & b_1 \\ 1 & d \end{pmatrix} \sim \begin{pmatrix} \gamma_1 & \\ & \gamma_2 \end{pmatrix}, \quad \begin{pmatrix} a\beta_1 & b_1\beta_2 \\ \beta_1 & d\beta_2 \end{pmatrix} \sim \begin{pmatrix} \alpha_1^{-1} & \\ & \alpha_2^{-1} \end{pmatrix}.$$

Then we get

$$\begin{cases} a + d = \gamma_1 + \gamma_2, \\ ad - b_1 = \gamma_1 \gamma_2, \\ \beta_1 a + \beta_2 d = \alpha_1^{-1} + \alpha_2^{-1}, \end{cases}$$

which has a unique solution (a, b_1, d) by virtue of $\beta_1 \neq \beta_2$. Thus $[(M_0, M_1, M_2)]$ is uniquely determined. □

Definition 7.3 A monodromy is called *semi-simple* if all the local monodromies are semi-simple.

Example 7.4 For the case $\#S = 3, n = 3$, if $[\rho]$ is irreducible and semi-simple with spectral type $((111), (111), (21))$, then $[\rho]$ is rigid.

Proof Let (M_0, M_1, M_2) be a tuple which represents $[\rho]$. Then we have

$$M_2 M_1 M_0 = I,$$

and may set

$$M_0 \sim \begin{pmatrix} \alpha_1 & & \\ & \alpha_2 & \\ & & \alpha_3 \end{pmatrix}, \; M_1 \sim \begin{pmatrix} \beta_1 & & \\ & \beta_2 & \\ & & \beta_3 \end{pmatrix}, \; M_2 \sim \begin{pmatrix} \gamma & & \\ & \gamma & \\ & & \gamma' \end{pmatrix},$$

where $\alpha_i \neq \alpha_j$ $(i \neq j)$, $\beta_i \neq \beta_j$ $(i \neq j)$, $\gamma \neq \gamma'$, and $\alpha_1 \alpha_2 \alpha_3 \beta_1 \beta_2 \beta_3 \gamma^2 \gamma' = 1$. By a similar transformation which sends M_1 to a diagonal matrix, we have

$$(M_1, M_2) \sim \left(\begin{pmatrix} \beta_1 & & \\ & \beta_2 & \\ & & \beta_3 \end{pmatrix}, M_2' \right).$$

Since $\mathrm{rank}(M_2' - \gamma) = 1$, we can write as

$$M_2' = \gamma + \begin{pmatrix} x \\ y \\ z \end{pmatrix} (p \; q \; r).$$

By the irreducibility, we get $xyzpqr \neq 0$. In particular, we can take $p = 1$. Moreover, taking a similar transformation by the matrix

$$Q = \begin{pmatrix} 1 & & \\ & q^{-1} & \\ & & r^{-1} \end{pmatrix},$$

we have

$$(M_1, M_2) \sim \left(\begin{pmatrix} \beta_1 & & \\ & \beta_2 & \\ & & \beta_3 \end{pmatrix}, \gamma + \begin{pmatrix} x_1 \\ y_1 \\ z_1 \end{pmatrix} (1 \; 1 \; 1) \right),$$

where $x_1 y_1 z_1 \neq 0$. Now, from the condition

$$\mathrm{tr}M_3 = 2\gamma + \gamma', \; M_2 M_1 \sim \begin{pmatrix} \alpha_1^{-1} & & \\ & \alpha_2^{-1} & \\ & & \alpha_3^{-1} \end{pmatrix},$$

we obtain the system of algebraic equations

$$\begin{cases} x_1 + y_1 + z_1 = \gamma' - \gamma, \\ \beta_1 x_1 + \beta_2 y_1 + \beta_3 z_1 = A, \\ \beta_1(\beta_2 + \beta_3)x_1 + \beta_2(\beta_1 + \beta_3)y_1 + \beta_3(\beta_1 + \beta_2)z_1 = B \end{cases}$$

for x_1, y_1, z_1, where A, B are some quantities determined by α_i, β_i, γ, γ'. Note that

$$\begin{vmatrix} 1 & 1 & 1 \\ \beta_1 & \beta_2 & \beta_3 \\ \beta_1(\beta_2+\beta_3) & \beta_2(\beta_1+\beta_3) & \beta_3(\beta_1+\beta_2) \end{vmatrix} = -\prod_{i<j}(\beta_j - \beta_i).$$

Then the system of algebraic equations is uniquely solved. Hence $[(M_0, M_1, M_2)]$ is uniquely determined, and then $[\rho]$ is rigid.   □

Let ρ be a monodromy representation, and set $\rho(\gamma_j) = M_j$ $(0 \le j \le p)$. We denote by $\mathcal{O}_j$ the conjugacy class $[M_j]$ of M_j. Namely $\mathcal{O}_j$ is the local monodromy at $x = a_j$. We set

$$\begin{aligned} \mathcal{M} &= \mathcal{M}(\mathcal{O}_0, \mathcal{O}_1, \dots, \mathcal{O}_p) \\ &= \{(N_0, N_1, \dots, N_p) \in \mathcal{O}_0 \times \mathcal{O}_1 \times \cdots \times \mathcal{O}_p ; N_p \cdots N_1 N_0 = I\}/\sim, \end{aligned}$$

where we define $(N_0, N_1, \dots, N_p) \sim (N'_0, N'_1, \dots, N'_p)$ if there exists $D \in GL(n, \mathbb{C})$ such that

$$N'_j = DN_j D^{-1} \quad (0 \le j \le p).$$

Thus, $\mathcal{M}$ is the moduli space of monodromies with prescribed local monodromies. $[\rho]$ is rigid if and only if $\mathcal{M}$ consists of a point.

Now we define the index of rigidity, which gives a simple criterion for the rigidity of monodromies. The index of rigidity corresponds to the dimension of $\mathcal{M}$ at a generic point.

Definition 7.4 The number

$$\iota = (2 - (p+1))n^2 + \sum_{j=0}^{p} \dim Z(M_j)$$

is called the *index of rigidity* for ρ, where $Z(M_j)$ denotes the centralizer of M_j.

Proposition 7.7 *The index of rigidity is an even integer.*

We can show this elementarily. A proof will be given in Sect. 7.6.

The index of rigidity gives a criterion for the rigidity. Namely the following theorem due to Katz holds.

Theorem 7.8 *If ρ is irreducible, we have $\iota \le 2$. In this case, ρ is rigid if and only if $\iota = 2$.*

We prove the theorem according to the works of Katz [100] and Völklein [179]. In the proof, we need the following Scott's lemma. Let K be a field, and V a finite dimensional vector space over K. For an element σ in $\mathrm{GL}(V)$, we set

$$F(\sigma) = \{v \in V \mid \sigma(v) = v\}.$$

For a subset Σ of $\mathrm{GL}(V)$, we set

$$F(\Sigma) = \bigcap_{\sigma \in \Sigma} F(\sigma).$$

We denote

$$d(\sigma) = \operatorname{codim} F(\sigma), \ d(\Sigma) = \operatorname{codim} F(\Sigma).$$

Moreover, for $\sigma \in \mathrm{GL}(V)$, we define $\sigma^* \in \mathrm{GL}(V^*)$ to be the unique element determined by

$$\langle \sigma(v), \sigma^*(v^*) \rangle = \langle v, v^* \rangle \quad (\forall v \in V, \ \forall v^* \in V^*),$$

where $\langle \, , \, \rangle$ is the pairing between V and the dual space V^*.

Lemma 7.9 (Scott's Lemma [158]) *Let V be a finite dimensional vector space over K, and $\sigma_1, \sigma_2, \ldots, \sigma_p$ elements in $\mathrm{GL}(V)$ satisfying*

$$\sigma_1 \sigma_2 \cdots \sigma_p = 1.$$

Then we have

$$d(\sigma_1) + \cdots + d(\sigma_p) \ge d(\Sigma) + d(\Sigma^*),$$

where $\Sigma = \{\sigma_1, \sigma_2, \ldots, \sigma_p\}$ and $\Sigma^* = \{\sigma_1^*, \sigma_2^*, \ldots, \sigma_p^*\}$.

Proof Set

$$C = \{(v_1, v_2, \ldots, v_p) \in V^p \mid v_i \in (1 - \sigma_i)V \ (1 \le i \le p)\}.$$

Define linear maps β, δ by

$$\beta : V \to C$$

$$v \mapsto ((1 - \sigma_1)v, \ldots, (1 - \sigma_p)v),$$

$$\delta : C \to V$$

$$(v_1, \ldots, v_p) \mapsto v_1 + \sigma_1 v_2 + \cdots + \sigma_1 \cdots \sigma_{p-1} v_p.$$

It is easy to see

$$\text{Im } \beta \subset \text{Ker } \delta.$$

On the other hand, we have

$$\text{Im } \delta = (1 - \sigma_1)V + \cdots + (1 - \sigma_p)V =: W$$

by using the relation

$$\sigma_1(1 - \sigma_2) = (\sigma_1 - 1 + 1)(1 - \sigma_2) = (\sigma_1 - 1)(1 - \sigma_2) + (1 - \sigma_2).$$

Now we shall show the equality

$$\dim W = d(\Sigma^*).$$

For the purpose, we recall the definition of an annihilator. For a subset A of a vector space V, the annihilator A° is defined by

$$A^\circ = \{f \in V^* \mid f|_A = 0\} \subset V^*.$$

By this definition, we have

$$(A + B)^\circ = A^\circ \cap B^\circ,$$
$$(V/A)^* \simeq A^\circ$$

for subspaces $A, B \subset V$. Using these properties, we get

$$V/W \simeq (V/W)^*$$
$$\simeq W^\circ$$
$$= \left((1 - \sigma_1)V + \cdots + (1 - \sigma_p)V\right)^\circ$$
$$= \bigcap_{i=1}^{p} \left((1 - \sigma_i)V\right)^\circ.$$

Noting that

$$\left((1 - \sigma_i)V\right)^\circ = F(\sigma_i^*),$$

we see that the last hand side of the above equality coincides with $F(\Sigma^*)$. Then we get

$$\dim W = \dim V - \dim V/W$$
$$= \operatorname{codim} W^\circ$$
$$= \operatorname{codim} F(\Sigma^*)$$
$$= d(\Sigma^*).$$

Next, it follows

$$\operatorname{Ker} \beta = F(\Sigma)$$

directly from the definition of β. Then we have

$$\dim \operatorname{Im} \beta = \operatorname{codim} \operatorname{Ker} \beta = d(\Sigma).$$

Lastly, since

$$\dim\big((1-\sigma_i)V\big) = \operatorname{codim} \operatorname{Ker} (1-\sigma_i) = d(\sigma_i),$$

we have

$$\dim C = \sum_{i=1}^{p} d(\sigma_i).$$

Combining the above results, we obtain

$$\sum_{i=1}^{p} d(\sigma_i) = \dim C$$
$$= \dim(C/\operatorname{Ker} \delta) + \dim(\operatorname{Ker} \delta/\operatorname{Im} \beta) + \dim \operatorname{Im} \beta$$
$$\geq \dim(C/\operatorname{Ker} \delta) + \dim \operatorname{Im} \beta$$
$$= \dim W + \dim \operatorname{Im} \beta$$
$$= d(\Sigma^*) + d(\Sigma).$$

$\square$

Proof of Theorem 7.8 Set

$$\rho = (M_0, M_1, \ldots, M_p).$$

Although ρ is assumed to be an anti-representation, in this proof we assume ρ is a representation, which causes no essential difference. Then we have

$$M_0 M_1 \cdots M_p = I.$$

We shall first show the former assertion, that is, $\iota \leq 2$ if ρ is irreducible. Let $(N_0, N_1, \ldots, N_p) \in \mathrm{GL}(n, \mathbb{C})^{p+1}$ be a tuple satisfying

$$N_i \sim M_i \ (0 \leq i \leq p), \ N_0 N_1 \cdots N_p = I. \tag{7.24}$$

Set $V = \mathrm{M}(n, \mathbb{C})$, and, for each i, define a linear transformation σ_i by

$$\sigma_i : V \to V$$
$$A \mapsto M_i A N_i^{-1}. \tag{7.25}$$

Then immediately we have

$$\sigma_0 \cdots \sigma_p = 1.$$

We define $F(\sigma_i), d(\sigma_i)$ as above for these σ_i. Then we have

$$d(\sigma_i) = n^2 - \dim Z(M_i).$$

In fact, if $A \in F(\sigma_i)$, we have $M_i A = A N_i$, which implies

$$M_i A C_i = A C_i M_i$$

if we use C_i such that $N_i = C_i M_i C_i^{-1}$. Then the map $A \mapsto A C_i$ becomes an isomorphism from $F(\sigma_i)$ to $\{B \in V \mid M_i B = B M_i\}$, which shows the above equality. Then we have

$$\iota = 2n^2 - \sum_{i=0}^{p} d(\sigma_i). \tag{7.26}$$

For any $B \in V$,

$$A \mapsto \mathrm{tr}(AB) \quad (A \in V)$$

becomes an element in V^*. If we identify V^* with V by this correspondence, $\sigma_i^* \in \mathrm{GL}(V^*)$ can be regarded as an element in $\mathrm{GL}(V)$

$$\sigma_i^*(B) = N_i B M_i^{-1} \quad (B \in V),$$

because, by setting $\sigma_i^*(B) = C$, we have

$$\langle \sigma_i(A), \sigma_i^*(B) \rangle = \text{tr}\,(\sigma_i(A)C)$$

$$= \text{tr}\,(M_i A N_i^{-1} C)$$

$$= \text{tr}\,(A N_i^{-1} C M_i),$$

and $\langle \sigma_i(A), \sigma_i^*(B) \rangle = \langle A, B \rangle = \text{tr}\,(AB)$.

Now we take $N_i = M_i$ ($\forall i$). Set $\Sigma = \{\sigma_0, \ldots, \sigma_p\}$, $\Sigma^* = \{\sigma_0^*, \ldots, \sigma_p^*\}$. Then by the definitions of σ_i, σ_i^*,

$$F(\Sigma) = F(\Sigma^*) = \{A \in V \mid AM_i = M_i A \ (0 \le i \le p)\}$$

holds. When $\rho = (M_0, M_1, \ldots, M_p)$ is irreducible, thanks to Schur's lemma, any matrix which commutes with every M_i is a scalar matrix. Thus $F(\Sigma) = F(\Sigma^*)$ coincides with the set of the scalar matrices, and hence is of dimension 1. Namely we have

$$d(\Sigma) = d(\Sigma^*) = n^2 - 1.$$

Then, by Scott's lemma (Lemma 7.9), we obtain

$$\iota = 2n^2 - \sum_{i=0}^{p} d(\sigma_i) \le 2n^2 - 2(n^2 - 1) = 2.$$

Second, we shall show that, when ρ is irreducible, $\iota = 2$ implies the rigidity of ρ. We consider $(N_0, N_1, \ldots, N_p)$ given at the beginning of this proof, and define σ_i by (7.25). By using the identity (7.26),

$$\sum_{i=0}^{p} d(\sigma_i) = 2n^2 - 2$$

follows from $\iota = 2$. Then applying Scott's lemma (Lemma 7.9), we get

$$2n^2 - 2 \ge d(\Sigma) + d(\Sigma^*)$$

$$= (n^2 - \dim F(\Sigma)) + (n^2 - \dim F(\Sigma^*)),$$

which implies

$$\dim F(\Sigma) + \dim F(\Sigma^*) \ge 2.$$

Thus we have $\dim F(\Sigma) > 0$ or $\dim F(\Sigma^*) > 0$. If $\dim F(\Sigma) > 0$ holds, $F(\Sigma)$ contains an element A different from O. Then, by the definition,

$$M_i A = A N_i$$

holds for every i. Since $(M_0, M_1, \ldots, M_p)$ is irreducible, A does not have 0 as an eigenvalue, and hence is invertible. Then we get

$$N_i = A^{-1} M_i A \quad (0 \leq i \leq p),$$

which shows that ρ is rigid. Similarly in the case $\dim F(\Sigma^*) > 0$, we can show that there exists B such that

$$N_i = B M_i B^{-1} \quad (0 \leq i \leq p),$$

and then ρ is rigid also in this case.

Lastly, we shall show that, when ρ is irreducible, the rigidity of ρ implies $\iota = 2$. Set $U = \mathrm{GL}(n, \mathbb{C})^{p+1}$, and define the map π by

$$\pi : U \to \mathrm{SL}(n, \mathbb{C})$$

$$(C_0, \ldots, C_p) \mapsto \prod_{i=0}^{p} (C_i M_i C_i^{-1}).$$

Set

$$G = \mathrm{SL}(n, \mathbb{C}) \times \prod_{i=0}^{p} Z(M_i).$$

Then the group G acts on U and on $\mathrm{SL}(n, \mathbb{C})$ as follows. Each $(D, Z_0, Z_1, \ldots, Z_p) \in G$ transforms elements of U and $\mathrm{SL}(n, \mathbb{C})$ by

$$U \ni (C_0, \ldots, C_p) \mapsto (D C_0 Z_0^{-1}, \ldots, D C_p Z_p^{-1}) \in U,$$

$$\mathrm{SL}(n, \mathbb{C}) \ni A \mapsto D A D^{-1} \in \mathrm{SL}(n, \mathbb{C}).$$

Then π is G-equivariant, which follows from

$$\pi(D C_0 Z_0^{-1}, \ldots, D C_p Z_p^{-1}) = D \left(\prod_{i=0}^{p} (C_i M_i C_i^{-1}) \right) D^{-1}.$$

Since $I \in \mathrm{SL}(n, \mathbb{C})$ is a fixed point of the action of G, G acts on $\pi^{-1}(I)$ by the G-equivariance.

Now we shall show that, if ρ is rigid, then G acts transitively on $\pi^{-1}(I)$. Let $(N_0, \ldots, N_p)$ be a tuple satisfying (7.24). Then there exists $(C_0, \ldots, C_p) \in U$ such that

$$N_i = C_i M_i C_i^{-1} \quad (0 \le i \le p).$$

Since $N_0 \cdots N_p = I$, we see that $(C_0, \ldots, C_p)$ belongs to $\pi^{-1}(I)$. If ρ is rigid, there exists $D \in \mathrm{SL}(n, \mathbb{C})$ such that

$$N_i = C_i M_i C_i^{-1} = D M_i D^{-1} \quad (0 \le i \le p),$$

and then $D^{-1} C_i \in Z(M_i)$ holds. If we set $D^{-1} C_i = Z_i$, $(C_0, \ldots, C_p)$ and $(I, \ldots, I)$ are connected by $(D, Z_0, \ldots, Z_p) \in G$, which implies that the action of G on $\pi^{-1}(I)$ is transitive.

Then, if ρ is rigid, we have

$$\dim G \ge \dim \pi^{-1}(I).$$

Note that

$$\dim G = \dim \mathrm{SL}(n, \mathbb{C}) + \sum_{i=0}^{p} \dim Z(M_i)$$

$$= n^2 - 1 + \sum_{i=0}^{p} \dim Z(M_i)$$

and

$$\dim \pi^{-1}(I) = (p+1)n^2 - (n^2 - 1).$$

Then, if ρ is rigid,

$$n^2 - 1 + \sum_{i=0}^{p} \dim Z(M_i) \ge (p+1)n^2 - (n^2 - 1)$$

holds. By rewriting this, we obtain $\iota \ge 2$. We already have $\iota \le 2$ by the irreducibility of ρ, and hence get $\iota = 2$. $\square$

Assume that ρ is irreducible. For the index of rigidity ι of ρ, we set

$$\alpha = 2 - \iota. \tag{7.27}$$

Then α is the dimension of $\mathcal{M}$ at a generic point. We shall explain this fact.

For simplicity, we assume that ρ is semi-simple. Let $(N_0, N_1, \ldots, N_p)$ be a representative of a generic point of $\mathcal{M}$. First, we consider how each matrix N_j is determined by the spectral type and the eigenvalues.

Lemma 7.10 *Let $(n_1, n_2, \ldots, n_q)$ be a partition of n, and $\gamma_1, \gamma_2, \ldots, \gamma_q$ mutually distinct complex numbers. A matrix with the diagonal Jordan canonical form*

$$\bigoplus_{j=1}^{q} \gamma_j I_{n_j}$$

is determined by

$$n^2 - \sum_{j=1}^{q} n_j{}^2$$

parameters. In general, such matrix A can be given by the following equalities:

$$A = \gamma_1 + \begin{pmatrix} A_1 \\ U_1 \end{pmatrix} \begin{pmatrix} I_{n'_1} & P_1 \end{pmatrix},$$

$$A_1 + P_1 U_1 = \gamma_2 - \gamma_1 + \begin{pmatrix} A_2 \\ U_2 \end{pmatrix} \begin{pmatrix} I_{n'_2} & P_2 \end{pmatrix},$$

$$A_2 + P_2 U_2 = \gamma_3 - \gamma_2 + \begin{pmatrix} A_3 \\ U_3 \end{pmatrix} \begin{pmatrix} I_{n'_3} & P_3 \end{pmatrix},$$

$$\vdots$$

$$A_{q-1} + P_{q-1} U_{q-1} = \gamma_q - \gamma_{q-1},$$

where we set

$$n'_j = n - n_1 - n_2 - \cdots - n_j$$

for each $1 \le j \le q - 1$. A_j, U_j, P_j are $n'_j \times n'_j$, $n_j \times n'_j$, and $n'_j \times n_j$ matrices, respectively, and the entries of U_j, P_j ($1 \le j \le q - 1$) are parameters.

Proof Since $\mathrm{rank}(A - \gamma_1) = n - n_1 = n'_1$, we have

$$A - \gamma_1 = \begin{pmatrix} A_1 \\ U_1 \end{pmatrix} \begin{pmatrix} I_{n'_1} & P_1 \end{pmatrix}$$

if the first n_1' columns of $A - \gamma_1$ are linearly independent. By using this expression, we get

$$(A - \gamma_1)(A - \gamma_2) = \begin{pmatrix} A_1(A_1 + P_1 U_1 - \gamma_2') & A_1(A_1 + P_1 U_1 - \gamma_2')P_1 \\ U_1(A_1 + P_1 U_1 - \gamma_2') & U_1(A_1 + P_1 U_1 - \gamma_2')P_1, \end{pmatrix}$$

where we set $\gamma_2' = \gamma_2 - \gamma_1$. Now note that

$$\operatorname{rank}((A - \gamma_1)(A - \gamma_2)) = n_2', \quad \operatorname{rank}\begin{pmatrix} A_1 \\ U_1 \end{pmatrix} = n_1'.$$

Then it follows

$$\operatorname{rank}(A_1 + P_1 U_1 - \gamma_2') = n_2'.$$

Then, if the first n_2' columns of $A_1 + P_1 U_1 - \gamma_2'$ are linearly independent, we have

$$A_1 + P_1 U_1 - \gamma_2' = \begin{pmatrix} A_2 \\ U_2 \end{pmatrix} \begin{pmatrix} I_{n_2'} & P_2 \end{pmatrix}.$$

Continuing these arguments, we obtain the equalities in Lemma. In the equalities, we can take $P_j, U_j \ (1 \le j \le q - 1)$ arbitrarily. The number of the entries is given by

$$\sum_{j=1}^{q-1} n_j' \cdot n_j + \sum_{j=1}^{q-1} n_j \cdot n_j' = 2 \sum_{i \ne j} n_i n_j$$

$$= (n_1 + \cdots + n_q)^2 - \sum_{i=1}^{q} n_i^2$$

$$= n^2 - \sum_{j=1}^{q} n_j^2.$$

$\square$

The number $\sum_{j=1}^{q} n_j^2$ which appeared in Lemma 7.10 coincides with the dimension of the centralizer $Z(A)$ of a matrix A with semi-simple spectral type $(n_1, n_2, \ldots, n_q)$. Then, for each j, N_j is determined by $n^2 - \dim Z(N_j) = n^2 - \dim Z(M_j)$ parameters. Thus, by the condition $(N_0, N_1, \ldots, N_p) \in \mathcal{O}_0 \times \mathcal{O}_1 \times \cdots \times \mathcal{O}_p$, we have

$$\sum_{j=0}^{p} (n^2 - \dim Z(M_j)) = (p+1)n^2 - \sum_{j=0}^{p} \dim Z(M_j)$$

free parameters. The condition $N_p \cdots N_1 N_0 = I$ is a system of n^2 equations, and we have already assumed that the determinants of both sides coincide. Then we have a system of $n^2 - 1$ independent equations. The equivalence relation is given by the action of $GL(n, \mathbb{C})$. Since the action by the scalar matrices is trivial, $GL(n, \mathbb{C})/\mathbb{C}^\times$ acts effectively. Thus by taking a quotient by this action, we can reduce the dimension by $n^2 - 1$. Thus the dimension of $\mathcal{M}$ is given by

$$(p+1)n^2 - \sum_{j=0}^{p} \dim Z(M_j) - 2(n^2 - 1) = 2 - \iota = \alpha.$$

This is not a rigorous proof, but an explanation. However, this viewpoint together with Lemma 5.10 will be useful in constructing a representative $(N_0, N_1, \ldots, N_p)$ for some particular cases. See [53] for further ideas of parametrizing a representative.

7.4.2 Rigidity of Fuchsian Differential Equations

The argument on the rigidity of monodromies can be applied in a similar way to the argument on the rigidity of non-resonant Fuchsian systems of normal form. We consider a Fuchsian system of normal form

$$\frac{dY}{dx} = \left(\sum_{j=1}^{p} \frac{A_j}{x - a_j} \right) Y \tag{7.28}$$

of rank n. We set $a_0 = \infty$, and define A_0 by (7.16). We assume that the system (7.28) is non-resonant. Then, for each j ($0 \le j \le p$), the conjugacy class $[A_j]$ of the residue matrix at $x = a_j$ determines the local behavior at the point. Set $[A_j] = \mathcal{O}'_j$. We define an equivalence relation on the set of Fuchsian systems of normal form by the gauge transformations

$$Y = PZ, \quad P \in GL(n, \mathbb{C})$$

by elements in $GL(n, \mathbb{C})$. Then it induces the equivalence relation on the set of tuples of residue matrices $(A_0, A_1, \ldots, A_p)$ as

$$(A_0, A_1, \ldots, A_p) \sim (P^{-1} A_0 P, P^{-1} A_1 P, \ldots, P^{-1} A_p P).$$

Then the moduli space of Fuchsian systems of normal form with prescribed local behaviors is given by

$$\mathcal{M}' = \mathcal{M}'(\mathcal{O}'_0, \mathcal{O}'_1, \dots, \mathcal{O}'_p)$$

$$= \{(B_0, B_1, \dots, B_p) \in \mathcal{O}'_0 \times \mathcal{O}'_1 \times \cdots \times \mathcal{O}'_p; \sum_{j=0}^{p} A_j = O\}/ \sim .$$

Definition 7.5 A non-resonant Fuchsian system of normal form (7.28) is called *rigid* if the moduli space $\mathcal{M}'$ consists of a point.

Definition 7.6 For a non-resonant Fuchsian system of normal form (7.28), we define the *index of rigidity* by

$$\iota = (1 - p)n^2 + \sum_{j=0}^{p} \dim Z(A_j).$$

Since we have assumed that the system (7.28) is non-resonant, the local monodromy at $x = a_j$ is given by $[e^{2\pi\sqrt{-1}A_j}]$, and A_j and $e^{2\pi\sqrt{-1}A_j}$ have the same spectral type. Then we have

$$\dim Z(e^{2\pi\sqrt{-1}A_j}) = \dim Z(A_j).$$

Thus the index of rigidity for the system (7.28) coincides with the index of rigidity for the monodromy representation of the system. In particular, the index of rigidity is an even integer.

Theorem 7.11 *Consider a non-resonant Fuchsian system of normal form (7.28). If the tuple $(A_0, A_1, \dots, A_p)$ of residue matrices is irreducible, we have $\iota \leq 2$. In this case, (7.28) is rigid if and only if $\iota = 2$.*

We can prove this in the same way as Theorem 7.8.

So far we considered the rigidity of Fuchsian systems of normal form. We can also define the rigidity for general Fuchsian differential equations. Namely, a Fuchsian differential equation is called rigid if its monodromy representation is rigid. Then it is a natural problem to find the list of all rigid Fuchsian differential equations. At this moment, we have not yet obtained the complete list. It seems that rigid Fuchsian differential equations appear sporadically. Nevertheless, several sequences of rigid Fuchsian differential equations are found. For example, Yokoyama [186] and Simpson [167] obtained such sequences. Yokoyama's list contains 8 sequences consisting of systems of Okubo normal form. The triple (M_1, M_2, M_3) of matrices we have treated in Example 5.1 in Sect. 5.4.1 is the monodromy matrices of the differential equation in Yokoyama's sequence (II*); we obtained the explicit forms of the matrices by virtue of the rigidity [48].

Another problem is to ask whether any rigid Fuchsian differential equation can be given as a Fuchsian system of normal form. This problem will be solved affirmatively by using Theorem 7.22.

If a Fuchsian system of normal form is not rigid, the moduli space $\mathcal{M}'$ has dimension $\alpha = 2 - \iota$ at a generic point, which is a similar consequence as in $\mathcal{M}$. We call a coordinate of $\mathcal{M}'$ consisting of α entries *accessory parameters*.

7.5 Middle Convolution

We shall define two operations on non-resonant Fuchsian systems of normal form. Both operations are defined as transformations of residue matrices $(A_1, A_2, \ldots, A_p)$ of the Fuchsian system

$$\frac{dY}{dx} = \left(\sum_{j=1}^{p} \frac{A_j}{x - a_j} \right) Y. \tag{7.29}$$

These operations are first formulated by Katz [100], and reformulated by Dettweiler-Reiter [34, 35]. We explain on the latter formulation.

The first operation is defined by the transformation

$$(A_1, A_2, \ldots, A_p) \mapsto (A_1 + \alpha_1, A_2 + \alpha_2, \ldots, A_p + \alpha_p),$$

where $\alpha = (\alpha_1, \alpha_2, \ldots, \alpha_p) \in \mathbb{C}^p$ is a fixed vector. We call this operation the *addition* by α, and denote it by ad_α.

The second operation is called the middle convolution. Take $\lambda \in \mathbb{C}$. For a tuple $(A_1, A_2, \ldots, A_p)$ of matrices of size n, we define a tuple $(G_1, G_2, \ldots, G_p)$ of size pn by

$$G_j = \sum_{k=1}^{p} E_{jk} \otimes (A_k + \delta_{jk}\lambda)$$

$$= \begin{pmatrix} O & \cdots & & \cdots & & \cdots & O \\ & \cdots & & \cdots & & \cdots & \\ A_1 & \cdots & A_j + \lambda & \cdots & A_p \\ & \cdots & & \cdots & & \cdots & \\ O & \cdots & & \cdots & & \cdots & O \end{pmatrix} \begin{matrix} \\ \\ (j \\ \\ \end{matrix} \tag{7.30}$$

$(1 \leq j \leq p)$, where E_{jk} denotes the matrix element of size $p \times p$, namely a $p \times p$ matrix with (j, k)-entry 1 and 0 for the other entries, and δ_{jk} denotes the Kronecker's delta. We set $V = \mathbb{C}^n$. Define subspaces $\mathcal{K}, \mathcal{L}$ of $V^p \simeq \mathbb{C}^{pn}$ by

$$
\mathcal{K} = \left\{ \begin{pmatrix} v_1 \\ \vdots \\ v_p \end{pmatrix} ; v_j \in \mathrm{Ker} A_j \ (1 \leq j \leq p) \right\},
$$

$$
\mathcal{L} = \mathrm{Ker}(G_1 + G_2 + \cdots + G_p).
$$

(7.31)

Then it is easy to see that both $\mathcal{K}$ and $\mathcal{L}$ are $(G_1, G_2, \ldots, G_p)$-invariant. Thus $(G_1, G_2, \ldots, G_p)$ induces an action of the quotient space $V^p/(\mathcal{K} + \mathcal{L})$. We denote the action by the tuple $(B_1, B_2, \ldots, B_p)$ of matrices. The transformation

$$
(A_1, A_2, \ldots, A_p) \mapsto (B_1, B_2, \ldots, B_p)
$$

(7.32)

thus obtained is called the *middle convolution* with parameter λ, and we denote it by mc_λ.

Now we explain the analytic meaning of two operations—addition and middle convolution. By the addition ad_α, the system (7.29) is transformed to the system

$$
\frac{dZ}{dx} = \left(\sum_{j=1}^{p} \frac{A_j + \alpha_j}{x - a_j} \right) Z.
$$

A solution Z of this system is given by using a solution Y of (7.29) as

$$
Z = \prod_{j=1}^{p} (x - a_j)^{\alpha_j} Y.
$$

(7.33)

Thus the addition ad_α corresponds to the gauge transformation (7.33).

Next we consider the middle convolution mc_λ. We begin with constructing solutions of the differential equation

$$
\frac{dU}{dx} = \left(\sum_{j=1}^{p} \frac{G_j}{x - a_j} \right) U
$$

(7.34)

with residue matrices $(G_1, G_2, \ldots, G_p)$ in terms of a solution Y of (7.29). For $1 \leq j \leq p$, we set

$$
W_j(x) = \frac{Y(x)}{x - a_j}.
$$

Then we have

$$\frac{dW_j(x)}{dx} = \frac{1}{x - a_j} \sum_{k=1}^{p} \frac{A_k}{x - a_k} Y(x) - \frac{Y(x)}{(x - a_j)^2}$$

$$= \frac{1}{x - a_j} \sum_{k=1}^{p} (A_k - \delta_{jk}) W_k(x),$$

and hence, by setting

$$W(x) = \begin{pmatrix} W_1(x) \\ W_2(x) \\ \vdots \\ W_p(x) \end{pmatrix},$$

we get a system of Okubo normal form

$$(x - T)\frac{dW}{dx} = (G - 1)W,$$

where

$$T = \begin{pmatrix} a_1 I_n & & & \\ & a_2 I_n & & \\ & & \ddots & \\ & & & a_p I_n \end{pmatrix}, \quad G = \begin{pmatrix} A_1 & A_2 & \cdots & A_p \\ A_1 & A_2 & \cdots & A_p \\ \vdots & \vdots & & \vdots \\ A_1 & A_2 & \cdots & A_p \end{pmatrix}.$$

Apply the Riemann-Liouville transform to $W(x)$ to get

$$U(x) = \int_\Delta W(s)(x - s)^\lambda \, ds.$$

Then, thanks to Theorem 7.5, $U(x)$ satisfies the differential equation

$$(x - T)\frac{dU}{dx} = (G + \lambda)U,$$

which is nothing but (7.34).

Let $u_1, u_2, \ldots, u_m$ be a basis of $\mathcal{K} + \mathcal{L}$, and make a basis of $\mathbb{C}^{pn}$ by adding vectors $v_{m+1}, \ldots, v_{pn}$. Set $P = (u_1, \ldots, u_m, v_{m+1}, \ldots, v_{pn})$. Since $\mathcal{K} + \mathcal{L}$ is $(G_1, G_2, \ldots, G_p)$-invariant, we have

$$G_j P = P \left(\begin{array}{c|c} * & * \\ \hline O & B_j \end{array} \right),$$

where the matrix in the right hand side is partitioned into $(m, pn-m) \times (m, pn-m)$-blocks. By the gauge transformation

$$U = P\tilde{U},$$

each residue matrix of the differential equation satisfied by $\tilde{U}$ becomes $P^{-1}G_jP$. Then, if we set

$$\tilde{U} = \left(\frac{\tilde{U}_1}{Z}\right),$$

Z satisfies the differential equation

$$\frac{dZ}{dx} = \left(\sum_{j=1}^{p} \frac{B_j}{x - a_j}\right) Z.$$

Thus, the middle convolution mc_λ corresponds to the composition of the Riemann-Liouville transform for a solution Y of (7.29) and the gauge transformation.

Under some generic condition, the middle convolution has good properties. In order to describe the properties, we use the following notation. We regard a tuple $(A_1, A_2, \ldots, A_p)$ of square matrices as a tuple of linear maps. Namely we regard

$$A = (A_1, A_2, \ldots, A_p) \in (\text{End}(V))^p$$

for some finite dimensional vector space V over $\mathbb{C}$. In other words, we regard V as an A-module. We set $V^p/(\mathcal{K} + \mathcal{L}) = mc_\lambda(V)$. Then the middle convolution with parameter λ can be regarded as an operation constructing a $(B_1, B_2, \ldots, B_p)$-module $mc_\lambda(V)$ from an A-module V. The subspace $\mathcal{K}$ of V^p is determined by the A-module V, and then we sometimes write as $\mathcal{K}_V$ in order to show the dependence on V. Similarly, we sometimes write the subspace $\mathcal{L}$ as $\mathcal{L}_V(\lambda)$, because it depends on the A-module V and the parameter λ.

We say that an A-module V satisfies the condition (M1) if

$$\bigcap_{j \neq i} \text{Ker} A_j \cap \text{Ker}(A_i + c) = \{0\} \quad (1 \leq i \leq p, \ \forall c \in \mathbb{C})$$

holds. We say that an A-module V satisfies the condition (M2) if

$$\sum_{j \neq i} \text{Im} A_j + \text{Im}(A_i + c) = V \quad (1 \leq i \leq p, \ \forall c \in \mathbb{C})$$

holds. An A-module V is said to be irreducible if any common invariant subspace for $A_1, A_2, \ldots, A_p$ is trivial.

Now we are ready to state the properties of the middle convolution.

Theorem 7.12 *Assume that an A-module V satisfies the condition (M2). Then $mc_0(V)$ is isomorphic to V.*

Theorem 7.13 *Assume that an A-module V satisfies the conditions (M1) and (M2). Then, for any λ, μ, $mc_\lambda(mc_\mu(V))$ is isomorphic to $mc_{\lambda+\mu}(V)$.*

Theorem 7.14 *Assume that an A-module V satisfies the conditions (M1) and (M2). Then, for any λ, V is irreducible if and only if $mc_\lambda(V)$ is irreducible.*

Remark 7.1 It is easy to see that, in the case $\dim V > 1$, V satisfies (M1), (M2) if it is irreducible. Then Theorem 7.14 asserts that, when $\dim V > 1$, V is irreducible if and only if $mc_\lambda(V)$ is irreducible.

Theorem 7.15 *Assume that an A-module V satisfies the condition (M2). Then the index of rigidity is invariant under the middle convolution.*

Recall that the middle convolution is analytically realized by the Riemann-Liouville transform. Then we see that the assertions in Theorems 7.12 and 7.13 correspond to the properties (i), (ii) in Theorem 7.4 of the Riemann-Liouville transform.

In order to prove these theorems, we prepare several lemmas. The following lemma is a direct consequence of the definition.

Lemma 7.16

(i)

$$\mathcal{L} = \bigcap_{i=1}^{p} \operatorname{Ker} G_i.$$

(ii) *If $\lambda \neq 0$, we have*

$$\mathcal{L} = \{{}^t(u, u, \ldots, u) \in V^p ; \left(\lambda + \sum_{i=1}^{p} A_i\right)u = 0\}.$$

If $\lambda = 0$, we have

$$\mathcal{L} = \{{}^t(u_1, u_2, \ldots, u_p) \in V^p ; \sum_{i=1}^{p} A_i u_i = 0\},$$

and in particular $\mathcal{K} \subset \mathcal{L}$.

(iii) *If $\lambda \neq 0$, we have*

$$\mathcal{K} \cap \mathcal{L} = \{0\}.$$

Then, in this case, we have $\mathcal{K} + \mathcal{L} = \mathcal{K} \oplus \mathcal{L}$.

Lemma 7.17

(i) *If W is a submodule of an A-module V, $mc_\lambda(W)$ is a submodule of $mc_\lambda(V)$.*
(ii) *If we have $V = W_1 \oplus W_2$, then $mc_\lambda(V) = mc_\lambda(W_1) \oplus mc_\lambda(W_2)$ holds.*

Proof We prove (i). Since W is A-invariant, we see that W^p is $(G_1, \ldots, G_p)$-invariant by the explicit forms of the matrices G_i. We shall show

$$\mathcal{K}_W + \mathcal{L}_W = W^p \cap (\mathcal{K}_V + \mathcal{L}_V). \tag{7.35}$$

Obviously $\mathcal{K}_W = W^p \cap \mathcal{K}_V$, $\mathcal{L}_W = W^p \cap \mathcal{L}_V$ hold. Then the inclusion $\subset$ is clear. When $\lambda = 0$, we have $\mathcal{K}_V \subset \mathcal{L}_V$, $\mathcal{K}_W \subset \mathcal{L}_W$, and then (7.35) holds. When $\lambda \neq 0$, take $w = {}^t(w_1, \ldots, w_p) \in W^p \cap (\mathcal{K}_V + \mathcal{L}_V)$. Then it can be written as

$$\begin{pmatrix} w_1 \\ \vdots \\ w_p \end{pmatrix} = \begin{pmatrix} k_1 \\ \vdots \\ k_p \end{pmatrix} + \begin{pmatrix} l \\ \vdots \\ l \end{pmatrix} \in \mathcal{K}_V + \mathcal{L}_V,$$

where $k_i \in \mathrm{Ker}A_i$, ${}^t(l, \ldots, l) \in \mathcal{L}_V$. Then we have

$$\sum_{i=1}^p A_i w_i = \sum_{i=1}^p A_i l = -\lambda l,$$

and hence $l \in W$. Then moreover we have $k_i = w_i - l \in W$, which implies $w \in \mathcal{K}_W + \mathcal{L}_W$. Thus we have shown (7.35). Now the inclusion map $W^p \to V^p$ induces

$$W^p/(\mathcal{K}_W + \mathcal{L}_W) \to V^p/(\mathcal{K}_V + \mathcal{L}_V),$$

which is injective. Then we have $mc_\lambda(W) \subset mc_\lambda(V)$.
The assertion (ii) follows directly from (i).   □

Lemma 7.18 *If an A-module V satisfies the condition (M1), then $mc_\lambda(V)$ also satisfies the condition (M1) for any λ. If an A-module V satisfies the condition (M2), then $mc_\lambda(V)$ also satisfies the condition (M2) for any λ.*

Proof We shall show the assertion for (M1). Assume that $A = (A_1, A_2, \ldots, A_p)$ satisfies (M1). Take any i, c, fix them and take

$$\bar{v} \in \bigcap_{j \neq i} \mathrm{Ker}B_j \cap \mathrm{Ker}(B_i + c).$$

It is enough to show $\bar{v} = 0$. Take a representative $v = {}^t(v_1, v_2, \ldots, v_p) \in V^p$ of $\bar{v}$. Then we have

$$G_j v \in \mathcal{K} + \mathcal{L} \quad (j \neq i), \quad (G_i + c)v \in \mathcal{K} + \mathcal{L}.$$

First we consider the case $\lambda = 0$. By Lemma 7.16 (ii), we have

$$\mathcal{L} = \{{}^t(l_1, \ldots, l_p) \, ; \, \sum_{j=1}^{p} A_j l_j = 0\},$$

and hence $\mathcal{K} \subset \mathcal{L}$. Then we have

$$G_j v \in \mathcal{L} \quad (j \neq i), \quad (G_i + c)v \in \mathcal{L}.$$

From the explicit form of the matrix G_j it follows

$$G_j v = \begin{pmatrix} 0 \\ \vdots \\ \sum_{m=1}^{p} A_m v_m \\ \vdots \\ 0 \end{pmatrix} = \begin{pmatrix} l_1 \\ \vdots \\ l_j \\ \vdots \\ l_p \end{pmatrix} \in \mathcal{L},$$

and then we have $l_m = 0$ $(m \neq j)$, which implies $A_j l_j = 0$ by the fact ${}^t(l_1, l_2, \ldots, l_p) \in \mathcal{L}$. Thus we obtain

$$\sum_{m=1}^{p} A_m v_m \in \mathrm{Ker} A_j \quad (j \neq i).$$

On the other hand, we also have

$$(G_i + c)v = \begin{pmatrix} cv_1 \\ \vdots \\ cv_i + \sum_{m=1}^{p} A_m v_m \\ \vdots \\ cv_p \end{pmatrix} = \begin{pmatrix} l_1' \\ \vdots \\ l_i' \\ \vdots \\ l_p' \end{pmatrix} \in \mathcal{L}.$$

By using this relation, we get

$$c \sum_{m=1}^{p} A_m v_m = \sum_{m=1}^{p} A_m (c v_m)$$

$$= A_i \left(l_i' - \sum_{m=1}^{p} A_m v_m \right) + \sum_{j \neq i} A_j l_j'$$

$$= -A_i \sum_{m=1}^{p} A_m v_m,$$

which shows

$$(A_i + c) \left(\sum_{m=1}^{p} A_m v_m \right) = 0.$$

Thus we have

$$\sum_{m=1}^{p} A_m v_m \in \bigcap_{j \neq i} \mathrm{Ker} A_j \cap \mathrm{Ker}(A_i + c),$$

and the right hand side is $\{0\}$ by the condition (M1). Then we have

$$\sum_{m=1}^{p} A_m v_m = 0,$$

which implies $v \in \mathcal{L}$. This shows $\bar{v} = 0$.

Next we assume $\lambda \neq 0$. Then, owing to Lemma 7.16 (ii), we have

$$\mathcal{L} = \{{}^t(l, \ldots, l) \, ; \, \left(\lambda + \sum_{m=1}^{p} A_m \right) l = 0 \}.$$

For any $j \neq i$, we have

$$G_j v = \begin{pmatrix} 0 \\ \vdots \\ \lambda v_j + \sum_{m=1}^{p} A_m v_m \\ \vdots \\ 0 \end{pmatrix} = \begin{pmatrix} k_1' \\ \vdots \\ k_j' \\ \vdots \\ k_p' \end{pmatrix} + \begin{pmatrix} l' \\ \vdots \\ l' \\ \vdots \\ l' \end{pmatrix},$$

where ${}^t(k'_1, \ldots, k'_p) \in \mathcal{K}$, ${}^t(l', \ldots, l') \in \mathcal{L}$. Then we have $l' = -k'_m \in \mathrm{Ker} A_m$ $(m \neq j)$, from which

$$0 = \left(\lambda + \sum_{m=1}^{p} A_m\right) l' = (\lambda + A_j) l'$$

follows. Thus we obtain

$$l' \in \bigcap_{m \neq j} \mathrm{Ker} A_m \cap \mathrm{Ker}(A_j + \lambda) = \{0\},$$

which implies $l' = 0$. Then we have, for every $j \neq i$,

$$\lambda v_j + \sum_{m=1}^{p} A_m v_m = k'_j \in \mathrm{Ker} A_j. \tag{7.36}$$

Moreover, we also have

$$(G_i + c)v = \begin{pmatrix} cv_1 \\ \vdots \\ (c + \lambda)v_i + \sum_{m=1}^{p} A_m v_m \\ \vdots \\ cv_p \end{pmatrix} = \begin{pmatrix} k_1 \\ \vdots \\ k_i \\ \vdots \\ k_p \end{pmatrix} + \begin{pmatrix} l \\ \vdots \\ l \\ \vdots \\ l \end{pmatrix},$$

where ${}^t(k_1, \ldots, k_p) \in \mathcal{K}$, ${}^t(l, \ldots, l) \in \mathcal{L}$. Then we have

$$cv_j = k_j + l \quad (j \neq i), \tag{7.37}$$

$$(c + \lambda)v_i + \sum_{m=1}^{p} A_m v_m = k_i + l. \tag{7.38}$$

If we set

$$\sum_{m=1}^{p} A_m v_m = w,$$

we obtain from (7.36) and (7.37)

$$\lambda A_j v_j + A_j w = 0, \quad c A_j v_j = A_j l,$$

respectively, and hence we get

$$A_j(cw + \lambda l) = 0$$

for $j \neq i$. Now we assume $c \neq 0$. Then, by using (7.38), (7.37), we have

$$
\begin{aligned}
A_i(cw + \lambda l) &= c(A_i l - (c + \lambda)A_i v_i) + \lambda A_i l \\
&= (c + \lambda)A_i l - c(c + \lambda)\Big(w - \sum_{j \neq i} A_j v_j\Big) \\
&= (c + \lambda)A_i l - c(c + \lambda)\Big(w - \frac{1}{c}\sum_{j \neq i} A_j l\Big) \\
&= (c + \lambda)A_i l - c(c + \lambda)w + (c + \lambda)(-\lambda l - A_i l) \\
&= -(c + \lambda)(cw + \lambda l),
\end{aligned}
$$

and hence

$$(A_i + c + \lambda)(cw + \lambda l) = 0.$$

Thus we get

$$cw + \lambda l \in \bigcap_{j \neq i} \operatorname{Ker} A_j \cap \operatorname{Ker}(A_i + c + \lambda) = \{0\},$$

which shows

$$w = -\frac{\lambda}{c} l.$$

If moreover $c + \lambda \neq 0$, by using (7.37), (7.38), we have

$$v_j = \frac{1}{c} k_j + \frac{1}{c} l \quad (j \neq i),$$

$$v_i = \frac{1}{c + \lambda}(k_i + l + \frac{\lambda}{c} l) = \frac{1}{c + \lambda} k_i + \frac{1}{c} l.$$

Then we get $v \in \mathcal{K} + \mathcal{L}$, which implies $\bar{v} = 0$. If $c \neq 0$ and $c + \lambda = 0$, by using (7.36), (7.37), we have

$$\lambda v_j + w = k'_j, \quad -\lambda v_j = k_j + l$$

for $j \neq i$, which shows

$$w - l \in \mathrm{Ker} A_j.$$

We also have $w - l \in \mathrm{Ker} A_i$ by (7.38), and then

$$w - l \in \bigcap_{j=1}^{p} \mathrm{Ker} A_j = \{0\},$$

namely $w = l$. Set

$$v' = v + \frac{1}{\lambda} \begin{pmatrix} l \\ \vdots \\ l \end{pmatrix} = v + \frac{1}{\lambda} \begin{pmatrix} w \\ \vdots \\ w \end{pmatrix}.$$

If we write $v' = {}^{t}(v'_1, \ldots, v'_p)$, we have

$$v'_j = v_j + \frac{1}{\lambda} l \in \mathrm{Ker} A_j$$

for $j \neq i$. On the other hand, from

$$l = w$$

$$= \sum_{m=1}^{p} A_m \left(v'_m - \frac{1}{\lambda} l \right)$$

$$= A_i v'_i - \frac{1}{\lambda} \left(\sum_{m=1}^{p} A_m \right) l$$

$$= A_i v'_i - \frac{1}{\lambda} (-\lambda l),$$

we obtain $A_i v'_i = 0$. Thus we have

$$v'_i \in \mathrm{Ker} A_i,$$

which implies $v' \in \mathcal{K}$. Then also in this case, we have $\bar{v} = 0$. It remains to check the case $c = 0$. In this case, we have

$$\lambda v_j + w \in \mathrm{Ker} A_j$$

for every j. Then

$$A_j w = -\lambda A_j v_j$$

holds. Summing up this equality for all j, we get

$$\left(\sum_{j=1}^{p} A_j\right) w = -\lambda \sum_{j=1}^{p} A_j v_j = -\lambda w.$$

Thus we get

$$\left(\lambda + \sum_{j=1}^{p} A_j\right) w = 0,$$

which shows ${}^t(w, \ldots, w) \in \mathcal{L}$. Then we have $v \in \mathcal{K} + \mathcal{L}$, and hence $\bar{v} = 0$.

The assertion for (M2) can be shown easily by the definition.   □

Proof of Theorem 7.12 By $\lambda = 0$, we have

$$\mathcal{L} = \{{}^t(v_1, \ldots, v_p) ;\ \sum_{i=1}^{p} A_i v_i = 0\}$$

and $\mathcal{K} \subset \mathcal{L}$. Then $mc_0(V) = V^p/\mathcal{L}$ holds. Define the map

$$\varphi : V^p \to V$$

by

$$\begin{pmatrix} v_1 \\ \vdots \\ v_p \end{pmatrix} \mapsto \sum_{i=1}^{p} A_i v_i.$$

Then, by virtue of (M2), φ becomes surjective. Since we have

$$\mathrm{Ker}\varphi = \mathcal{L},$$

φ induces an isomorphism

$$mc_0(V) = V^p/\mathcal{L} \simeq V.$$

Moreover, for each i, we readily see that

$$\varphi \circ G_i = A_i \circ \varphi$$

holds. Thus the isomorphism is an isomorphism between $(B_1, \ldots, B_p)$-module $mc_0(V)$ and $(A_1, \ldots, A_p)$-module V.   □

Proof of Theorem 7.13 The assertion holds if $\lambda = 0$ or $\mu = 0$ by virtue of Theorem 7.12. Then we may assume $\lambda\mu \neq 0$. Set $mc_\mu(V) = M$. Then we have

$$M = V^p/(\mathcal{K}_V + \mathcal{L}_V(\mu)),$$

$$mc_\lambda(mc_\mu(V)) = M^p/(\mathcal{K}_M + \mathcal{L}_M).$$

We shall describe $mc_\lambda(mc_\mu(V))$ as a quotient space of $(V^p)^p$, and then construct the isomorphism in the theorem.

Define G_j by (7.30) for $1 \leq j \leq p$. We use the following notation:

$$\mathcal{K}_1 = \mathcal{K}_V, \quad \mathcal{L}_1 = \mathcal{L}_V(\mu), \quad \mathcal{L}_1' = \mathcal{L}_V(\lambda + \mu),$$

$$\mathcal{K}_2 = \mathcal{K}_{V^p}, \quad \mathcal{L}_2 = \mathcal{L}_{V^p}(\lambda).$$

Here $\mathcal{K}_1, \mathcal{L}_1, \mathcal{L}_1'$ are subspaces of V^p, and $\mathcal{K}_2, \mathcal{L}_2$ are subspaces of $(V^p)^p$. First we shall show

$$\mathcal{K}_M = (\mathcal{K}_2 + \mathcal{K}_1{}^p + \mathcal{L}_1{}^p)/(\mathcal{K}_1{}^p + \mathcal{L}_1{}^p), \tag{7.39}$$

$$\mathcal{L}_M = (\mathcal{L}_2 + \mathcal{K}_1{}^p + \mathcal{L}_1{}^p)/(\mathcal{K}_1{}^p + \mathcal{L}_1{}^p). \tag{7.40}$$

By the definition, $\bar{v} = {}^t(\bar{v}_1, \ldots, \bar{v}_p) \in \mathcal{K}_M$ means

$$B_i\bar{v}_i = 0 \quad (\forall i).$$

If $\bar{v} \in M^p$ is represented by $v = {}^t(v_1, \ldots, v_p) \in \mathcal{K}_2$, we have

$$G_iv_i = 0 \quad (\forall i),$$

and hence $\bar{v} \in \mathcal{K}_M$. Thus the inclusion $\supset$ of (7.39) holds. Take $\bar{v} = {}^t(\bar{v}_1, \ldots, \bar{v}_p) \in \mathcal{K}_M$, and let $v = {}^t(v_1, \ldots, v_p) \in (V^p)^p$ be a representative of $\bar{v}$. Then we have

$$G_iv_i \in \mathcal{K}_1 + \mathcal{L}_1 \quad (\forall i).$$

Then, in the same argument as the proof of Lemma 7.18, we obtain

$$G_iv_i \in \mathcal{K}_1 \quad (\forall i)$$

by using (M1). Hence, for each i, we have

$$G_iv_i = {}^t(0, \ldots, k_i, \ldots, 0), \quad k_i \in \mathrm{Ker}A_i.$$

Now set

$$v_i' = v_i - \frac{1}{\mu}{}^t(0, \ldots, k_i, \ldots, 0).$$

Then we have $G_i v_i' = 0$, and hence ${}^t(v_1', \ldots, v_p') \in \mathcal{K}_2$. Thus we have shown $v \in \mathcal{K}_2 + \mathcal{K}_1{}^p$. This implies the inclusion $\subset$ of (7.39), and then (7.39) holds.

We consider the relation (7.40). Since we assume $\lambda \neq 0$, any element in $\mathcal{L}_M$ is written as $\bar{v} = (\bar{v}_0, \ldots, \bar{v}_0)$ by $\bar{v}_0 \in M$ satisfying

$$\left(\lambda + \sum_{i=1}^p B_i\right)\bar{v}_0 = 0.$$

If $v \in \mathcal{L}_2$, we have $v = {}^t(v_0, \ldots, v_0)$ and

$$\left(\lambda + \sum_{i=1}^p G_i\right)v_0 = 0,$$

and hence the element in M^p represented by v belongs to $\mathcal{L}_M$. Hence the inclusion $\supset$ of (7.40) holds. Take any $\bar{v} = {}^t(\bar{v}_0, \ldots, \bar{v}_0) \in \mathcal{L}_M$. We can take as a representative of $\bar{v}$ an element $v \in (V^p)^p$ of the form $v = {}^t(v_0, \ldots, v_0)$. Then we have

$$\left(\lambda + \sum_{i=1}^p G_i\right)v_0 \in \mathcal{K}_1 + \mathcal{L}_1,$$

which can be written as

$$\left(\lambda + \sum_{i=1}^p G_i\right)v_0 = \left[\lambda + \mu + \begin{pmatrix} A_1 & \cdots & A_p \\ & \cdots & \\ A_1 & \cdots & A_p \end{pmatrix}\right]v_0 = k + l$$

with $k \in \mathcal{K}_1, l \in \mathcal{L}_1$. By the definitions of $\mathcal{K}_1, \mathcal{L}_1$, we get

$$\left(\lambda + \sum_{i=1}^p G_i\right)k = (\lambda + \mu)k, \quad \left(\lambda + \sum_{i=1}^p G_i\right)l = \lambda l.$$

If $\lambda + \mu \neq 0$, by setting

$$v_0' = v_0 - \frac{1}{\lambda + \mu}k - \frac{1}{\lambda}l,$$

we have

$$\left(\lambda + \sum_{i=1}^{p} G_i\right) v_0' = 0,$$

and hence ${}^t(v_0', \ldots, v_0') \in \mathcal{L}_2$. Namely we can take as another representative of $\bar{v} \in \mathcal{L}_M$ an element in $\mathcal{L}_2$. If $\lambda + \mu = 0$, we have

$$\begin{pmatrix} A_1 \cdots A_p \\ \cdots \\ A_1 \cdots A_p \end{pmatrix} v_0 = k + l.$$

Then, if we set $k = {}^t(k_1, \ldots, k_p)$, we have $k_1 = \cdots = k_p$, and hence

$$k_1 = \cdots = k_p \in \bigcap_{i=1}^{p} \mathrm{Ker} A_i = \{0\}.$$

Thus we get $k = 0$. Then if we set

$$v_0' = v_0 - \frac{1}{\lambda} l,$$

we have

$$\left(\lambda + \sum_{i=1}^{p} G_i\right) v_0' = 0,$$

which shows ${}^t(v_0', \ldots, v_0') \in \mathcal{L}_2$. Then also in this case, we can take as another representative of $\bar{v} \in \mathcal{L}_M$ an element in $\mathcal{L}_2$. Thus the inclusion $\subset$ of (7.40) holds, and hence (7.40).

Therefore we get

$$mc_\lambda(mc_\mu(V)) = (V^p)^p/(\mathcal{K}_2 + \mathcal{L}_2 + \mathcal{K}_1{}^p + \mathcal{L}_1{}^p).$$

Next we define the map φ by

$$\varphi : (V^p)^p \to V^p$$

$$\begin{pmatrix} v_1 \\ \vdots \\ v_p \end{pmatrix} \mapsto \sum_{i=1}^{p} G_i v_i.$$

We shall show that φ induces the map

$$\bar{\varphi} : mc_\lambda(mc_\mu(V)) \to mc_{\lambda+\mu}(V).$$

For the purpose, it is enough to show

$$\varphi(\mathcal{K}_2 + \mathcal{L}_2 + \mathcal{K}_1{}^p + \mathcal{L}_1{}^p) \subset \mathcal{K}_1 + \mathcal{L}_1'.$$

Take any $v = {}^t(v_1, \ldots, v_p) \in \mathcal{K}_1{}^p$. Then $v_i \in \mathcal{K}_1$ and $\mathcal{K}_1$ is $(G_1, \ldots, G_p)$-invariant, from which $G_i v_i \in \mathcal{K}_1$ follows. Thus we have

$$\varphi(v) = \sum_{i=1}^{p} G_i v_i \in \mathcal{K}_1.$$

Take any $v = {}^t(v_1, \ldots, v_p) \in \mathcal{L}_1{}^p$. Since

$$v_i \in \mathcal{L}_1 = \bigcap_{j=1}^{p} \mathrm{Ker}\, G_j,$$

we have $\varphi(v) = 0$. Thus we have shown $\varphi(\mathcal{K}_1{}^p + \mathcal{L}_1{}^p) \subset \mathcal{K}_1$. It is immediate from the definition that $\varphi(v) = 0$ holds for $v \in \mathcal{K}_2$. Take any $v \in \mathcal{L}_2$. Then we have $v = {}^t(v_0, \ldots, v_0)$, and

$$\left(\lambda + \sum_{i=1}^{p} G_i\right) v_0 = \left[\lambda + \mu + \begin{pmatrix} A_1 \cdots A_p \\ \cdots \\ A_1 \cdots A_p \end{pmatrix}\right] v_0 = 0. \tag{7.41}$$

Hence $v_0 \in \mathcal{L}_1'$, and we obtain $\varphi(v) \in \mathcal{L}_1'$ from $\varphi(v) = -\lambda v_0$. Thus we see that $\bar{\varphi}$ is well-defined.

For $v = {}^t(v_1, \ldots, v_p) \in (V^p)^p$, we have

$$\varphi(v) = G_1 v_1 + \cdots + G_p v_p$$

$$= \begin{pmatrix} A_1 + \mu & A_2 \cdots A_p \\ O & \cdots\cdots O \\ & \cdots\cdots \\ O & \cdots\cdots O \end{pmatrix} v_1 + \cdots + \begin{pmatrix} O \cdots & \cdots & O \\ \cdots & \cdots \\ O \cdots & \cdots & O \\ A_1 \cdots & A_{p-1} & A_p + \mu \end{pmatrix} v_p.$$

Then, by the condition (M2), any element in V^p can be realized by $\varphi(v)$ if we take $v_1, \ldots, v_p$ appropriately. Thus φ is surjective, and hence $\bar{\varphi}$ is also surjective.

Lastly, we shall show that $\bar{\varphi}$ is injective. Since $\bar{\varphi}$ is surjective, it is enough to show that $mc_\lambda(mc_\mu(V))$ and $mc_{\lambda+\mu}(V)$ have the same dimension. By the definition,

$\mathcal{L}_1{}^p \subset \mathcal{K}_2$ immediately follows. Since we assume $\lambda \neq 0$, we have $\mathcal{K}_2 + \mathcal{L}_2 = \mathcal{K}_2 \oplus \mathcal{L}_2$. Then we get

$$\dim(\mathcal{K}_2 + \mathcal{L}_2 + \mathcal{K}_1{}^p + \mathcal{L}_1{}^p)$$
$$= \dim \mathcal{K}_2 + \dim \mathcal{L}_2 + \dim \mathcal{K}_1{}^p - \dim(\mathcal{K}_2 \cap \mathcal{K}_1{}^p) - \dim(\mathcal{L}_2 \cap \mathcal{K}_1{}^p).$$

For $v = {}^t(v_1, \ldots, v_p) \in \mathcal{K}_2$, we have $v_i \in \mathrm{Ker} G_i$ $(1 \leq i \leq p)$, and by setting $v_i = {}^t(v_{i1}, \ldots, v_{up})$, we get

$$0 = G_i v_i = \begin{pmatrix} 0 \\ \vdots \\ \sum_{j=1}^{p} A_j v_{ij} + \mu v_{ii} \\ \vdots \\ 0 \end{pmatrix}.$$

Thus v_i belongs to the kernel of the linear map

$$V^p \to V$$

$$\begin{pmatrix} u_1 \\ \vdots \\ u_p \end{pmatrix} \mapsto \sum_{j \neq i} A_j u_j + (A_i + \mu) u_i.$$

By (M2), the rank of this linear map is equal to $\dim V$, and then the dimension of the kernel is $\dim V^p - \dim V$. Then we have

$$\dim \mathcal{K}_2 = p(\dim V^p - \dim V) = p^2 \dim V - p \dim V.$$

By the assumption $\lambda \neq 0$, any elements in $\mathcal{L}_2$ is written as $v = {}^t(v_0, \ldots, v_0)$, where v_0 satisfies (7.41). Then we have $v_0 \in \mathcal{L}_1'$, and hence

$$\dim \mathcal{L}_2 = \dim \mathcal{L}_1'.$$

For ${}^t(v_1, \ldots, v_p) \in \mathcal{K}_2 \cap \mathcal{K}_1{}^p$, we have

$$v_i \in \mathcal{K}_1 \cap \mathrm{Ker} G_i \quad (\forall i).$$

If we write $v_i = {}^t(v_{i1}, \ldots, v_{ip})$, we get

$$0 = G_i v_i = {}^t(0, \ldots, \lambda v_{ii}, \ldots, 0),$$

from which $v_{ii} = 0$ follows. Thus we have

$$\dim(\mathcal{K}_2 \cap \mathcal{K}_1{}^p) = \sum_{i=1}^{p} \left(\sum_{j \neq i} \dim \operatorname{Ker} A_j \right)$$

$$= \dim \mathcal{K}_1{}^p - \dim \mathcal{K}_1.$$

Any element in $\mathcal{L}_2 \cap \mathcal{K}_1{}^p$ can be written as ${}^t(v_0, \ldots, v_0)$ with $v_0 \in \mathcal{K}_1$ satisfying

$$0 = \left(\lambda + \sum_{i=1}^{p} G_i \right) v_0 = (\lambda + \mu) v_0.$$

If $\lambda + \mu \neq 0$, we have $v_0 = 0$, from which $\mathcal{L}_2 \cap \mathcal{K}_1{}^p = \{0\}$ follows. If $\lambda + \mu = 0$, $v_0 \in \mathcal{K}_1$ implies ${}^t(v_0, \ldots, v_0) \in \mathcal{L}_2$, and hence we have

$$\dim(\mathcal{L}_2 \cap \mathcal{K}_1{}^p) = \dim \mathcal{K}_1.$$

Now assume $\lambda + \mu \neq 0$. In this case we have $\mathcal{K}_1 + \mathcal{L}_1' = \mathcal{K}_1 \oplus \mathcal{L}_1'$, and then, by combining with the above results, we get

$\dim(V^p)^p / (\mathcal{K}_2 + \mathcal{L}_2 + \mathcal{K}_1{}^p + \mathcal{L}_1{}^p)$

$= p^2 \dim V - ((p^2 \dim V - p \dim V) + \dim \mathcal{L}_1' + \dim \mathcal{K}_1{}^p - (\dim \mathcal{K}_1{}^p - \dim \mathcal{K}_1))$

$= p \dim V - \dim \mathcal{K}_1 - \dim \mathcal{L}_1'$

$= \dim V^p / (\mathcal{K}_1 + \mathcal{L}_1').$

Assume $\lambda + \mu = 0$. Then, by noting $\mathcal{K}_1 \subset \mathcal{L}_1'$, we get

$$\dim(V^p)^p / (\mathcal{K}_2 + \mathcal{L}_2 + \mathcal{K}_1{}^p + \mathcal{L}_1{}^p)$$

$$= p^2 \dim V - ((p^2 \dim V - p \dim V) + \dim \mathcal{L}_1' + \dim \mathcal{K}_1{}^p$$

$$- (\dim \mathcal{K}_1{}^p - \dim \mathcal{K}_1) - \dim \mathcal{K}_1)$$

$$= p \dim V - \dim \mathcal{L}_1'$$

$$= \dim V^p / (\mathcal{K}_1 + \mathcal{L}_1').$$

Thus in both cases we have

$$\dim mc_\lambda(mc_\mu(V)) = \dim mc_{\lambda+\mu}(V),$$

which shows that $\bar{\varphi}$ is injective.

Thus we have shown that $\bar{\varphi}$ is an isomorphism. For each i, we set

$$
H_i = \begin{pmatrix} O & \cdots & \cdots & \cdots & O \\ & \cdots & \cdots & \cdots & \\ G_1 & \cdots & G_i + \lambda & \cdots & G_p \\ & \cdots & \cdots & \cdots & \\ O & \cdots & \cdots & \cdots & O \end{pmatrix} \quad (i,
$$

$$
G_i' = \begin{pmatrix} O & \cdots & & & \cdots & O \\ & \cdots & & \cdots & & \\ A_1 & \cdots & A_i + \lambda + \mu & \cdots & A_p \\ & \cdots & & \cdots & & \\ O & \cdots & & & \cdots & O \end{pmatrix} \quad (i.
$$

Then we can directly check that

$$
\varphi \circ H_i = G_i' \circ \varphi \quad (\forall i)
$$

holds. This shows that $\bar{\varphi}$ is an isomorphism of modules.   □

Proof of Theorem 7.14 We assume that V satisfies (M1), (M2), and is irreducible. Then, thanks to Lemma 7.18, $mc_\lambda(V)$ also satisfies (M1), (M2). Let M be a minimal submodule of $mc_\lambda(V)$ different from $\{0\}$. Then we see that M is irreducible. Since $mc_\lambda(V)$ satisfies (M1), M also satisfies (M1). Moreover, by noting that M is irreducible, we see that M satisfies (M2). We consider the middle convolution with parameter $-\lambda$, and set

$$
W = mc_{-\lambda}(M).
$$

Then, by Theorems 7.12 and 7.13, W is a subspace of V, and by Lemma 7.17, W is a submodule of V. Since V is irreducible, $W = V$ or $W = \{0\}$ holds. Then again by Theorems 7.12 and 7.13, $M = mc_\lambda(W) = mc_\lambda(V)$ or $M = \{0\}$ holds, which implies that $mc_\lambda(V)$ is irreducible.

The converse assertion is reduced to the above argument by Theorems 7.12 and 7.13.   □

To show Theorem 7.15, we look at the change of the Jordan canonical form of each residue matrix A_j by the middle convolution. The change itself is an important result, and then we describe it as a theorem.

Theorem 7.19 *Let V be an $(A_1, A_2, \ldots, A_p)$-module satisfying (M2). Assume that $\lambda \neq 0$, and set*

$$
mc_\lambda(A_1, A_2, \ldots, A_p) = (B_1, B_2, \ldots, B_p).
$$

(i) *Fix each $1 \leq i \leq p$. If the Jordan canonical form of A_i contains a Jordan cell $J(\alpha; m)$, the Jordan canonical form of B_i contains the Jordan cell $J(\alpha + \lambda; m')$, where*

$$m' = \begin{cases} m & (\alpha \neq 0, -\lambda), \\ m - 1 & (\alpha = 0), \\ m + 1 & (\alpha = -\lambda). \end{cases}$$

The Jordan canonical form of B_i is built of these Jordan cells together with some number of the Jordan cells $J(0; 1)$.

(ii) *We set*

$$A_0 = -\sum_{i=1}^{p} A_i, \quad B_0 = -\sum_{i=1}^{p} B_i.$$

Then, if the Jordan canonical form of A_0 contains a Jordan cell $J(\alpha; m)$, the Jordan canonical form of B_0 contains the Jordan cell $J(\alpha - \lambda; m')$, where

$$m' = \begin{cases} m & (\alpha \neq 0, \lambda), \\ m + 1 & (\alpha = 0), \\ m - 1 & (\alpha = \lambda). \end{cases}$$

The Jordan canonical form of B_0 is built of these Jordan cells together with some number of the Jordan cells $J(-\lambda; 1)$.

Lemma 7.20 *Let V be an $(A_1, \ldots, A_p)$-module satisfying (M2).*

(i) *For each $1 \leq i \leq p$, the linear map*

$$\phi_i : \mathrm{Im}(G_i|_{mc_\lambda(V)}) \to \mathrm{Im} A_i$$

$${}^t(0, \ldots, w_i, \ldots, 0) \mapsto A_i w_i$$

is an isomorphism. Moreover we have

$$\phi_i \circ G_i = (A_i + \lambda) \circ \phi_i.$$

(ii) *The linear map*

$$\phi_0 : \mathrm{Im}\big((G_0 + \lambda)|_{mc_\lambda(V)}\big) \to \mathrm{Im}(A_0 - \lambda)$$

$${}^t(w_1, \ldots, w_1) \mapsto (A_0 - \lambda)w_1$$

is an isomorphism. Moreover we have

$$\phi_0 \circ G_0 = (A_0 - \lambda) \circ \phi_0.$$

Proof We prove (i). For any $v = {}^t(v_1, \ldots, v_p) \in V^p$, $k = {}^t(k_1, \ldots, k_p) \in \mathcal{K}$, $l \in \mathcal{L}$, we have

$$G_i(v + k + l) \doteq \begin{pmatrix} 0 \\ \vdots \\ \sum_{j \neq i} A_j v_j + (A_i + \lambda)v_i + \lambda k_i \\ \vdots \\ 0 \end{pmatrix} =: \begin{pmatrix} 0 \\ \vdots \\ w_i + \lambda k_i \\ \vdots \\ 0 \end{pmatrix}.$$

Since $k_i \in \mathrm{Ker} A_i$, the value of ϕ_i does not depend on k, l. Thus ϕ_i is well-defined. If $A_i w_i = 0$, we have $w_i \in \mathrm{Ker} A_i$, and hence ${}^t(0, \ldots, w_i, \ldots, 0) \in \mathcal{K}$. This proves that ϕ_i is injective. By the assumption (M2),

$$w_i = \sum_{j \neq i} A_j v_j + (A_i + \lambda)v_i$$

can be any element in V, which shows that ϕ_i is surjective. From the relation

$$\phi_i \circ G_i(G_i(v)) = \phi_i \left(G_i \begin{pmatrix} 0 \\ \vdots \\ w_i \\ \vdots \\ 0 \end{pmatrix} \right) = \phi_i \left(\begin{pmatrix} 0 \\ \vdots \\ (A_i + \lambda)w_i \\ \vdots \\ 0 \end{pmatrix} \right) = (A_i + \lambda)A_i w_i,$$

we obtain $\phi_i \circ G_i = (A_i + \lambda) \circ \phi_i$.

Next we prove (ii). Since $\phi_0((G_0 + \lambda)(\mathcal{K} + \mathcal{L})) = 0$, ϕ_0 is well-defined. ϕ_0 is injective, because $(A_0 - \lambda)w_1 = 0$ implies ${}^t(w_1, \ldots, w_1) \in \mathcal{L}$. By the assumption (M2),

$$w_1 = -\sum_{j=1}^p A_j v_j$$

can take any element in V, which shows that ϕ_0 is surjective. By using the above w_1, we get

$$\phi_0 \circ G_0((G_0 + \lambda)v) = \phi_0 \left(\begin{pmatrix} (A_0 - \lambda)w_1 \\ \vdots \\ (A_0 - \lambda)w_1 \end{pmatrix} \right) = (A_0 - \lambda)^2 w_1 = (A_0 - \lambda) \circ \phi_0((G_0 + \lambda)v),$$

which shows $\phi_0 \circ G_0 = (A_0 - \lambda) \circ \phi_0$.   □

Here we recall basic facts on the Jordan canonical form of a matrix. The Jordan canonical form of a matrix A contains the Jordan cell $J(\alpha; m)$ if and only if there exists x such that

$$(A - \alpha)^m x = 0, \ (A - \alpha)^{m-1} x \neq 0, \ x \notin \text{Im}(A - \alpha).$$

In this case,

$$x, (A - \alpha)x, \ldots, (A - \alpha)^{m-1} x$$

are linearly independent, and make a part of the basis which sends A to the Jordan canonical form.

Proof of Theorem 7.19 We prove (i). Fix one $1 \leq i \leq p$. Since $B_i = G_i|_{mc_\lambda(V)}$, the eigenvalues of B_i are contained in the eigenvalues of G_i. By the definition (7.30) of G_i, we see that the eigenvalues of G_i consists of the eigenvalues of $A_i + \lambda$ and 0 of multiplicity $(p-1)n$. Then we are going to trace the change of each Jordan cell in the Jordan canonical form of A_i in order to find the Jordan canonical form of B_i.

First we consider the case $\alpha \neq 0, -\lambda$. Since the Jordan canonical form of A_i contains the Jordan cell $J(\alpha, m)$, there exists x such that

$$(A_i - \alpha)^m x = 0, \ (A_i - \alpha)^{m-1} x \neq 0, \ x \notin \text{Im}(A_i - \alpha).$$

Set

$$x_1 = x, \ x_2 = (A_i - \alpha)x, \ \ldots, \ x_m = (A_i - \alpha)^{m-1} x.$$

Then $x_1, x_2, \ldots, x_m$ are linearly independent, and make a partial basis corresponding to $J(\alpha, m)$. We shall show $x_j \in \text{Im} A_i$ $(1 \leq j \leq m)$. Expanding $(A_i - \alpha)^m x = 0$, we get

$$(-\alpha)^m x = A(-A^{m-1} - \cdots - m(-\alpha)^{m-1})x,$$

which implies $x \in \text{Im} A_i$ by virtue of $\alpha \neq 0$. If we write $x = A_i u$, we have $(A_i - \alpha)^j x = A_i(A_i - \alpha)^j u$, and hence $x_j \in \text{Im} A_i$. Since we assumed (M2), we can apply Lemma 7.20, and hence there exists an isomorphism

$$\phi : \mathrm{Im}(G_i|_{mc_\lambda(V)}) \to \mathrm{Im}A_i.$$

Then the elements

$$y_1 = \phi^{-1}(x_1), \ldots, y_m = \phi^{-1}(x_m)$$

are linearly independent in $mc_\lambda(V)$. Note that ϕ satisfies

$$\phi \circ G_i = (A_i + \lambda) \circ \phi.$$

Then we have

$$(\phi \circ G_i)(y_j) = (A_i + \lambda)(A_i - \alpha)^{j-1}x$$

$$= (A_i - \alpha)^j x + (\alpha + \lambda)(A_i - \alpha)^{j-1}x,$$

and hence

$$(G_i - (\alpha + \lambda))y_j = y_{j+1}$$

holds for $1 \le j \le m - 1$. For $j = m$, from $(A_i - \alpha)^m x = 0$ we obtain

$$(G_i - (\alpha + \lambda))y_m = 0.$$

On the other hand, if there exists y_0 such that $y_1 = (G_i - (\alpha + \lambda))y_0$, we have

$$(\phi \circ G_i)(y_0) = A_i\phi(y_0) + \lambda\phi(y_0),$$

$$(\phi \circ G_i)(y_0) = \phi(y_1 + (\alpha + \lambda)y_0) = x - (\alpha + \lambda)\phi(y_0),$$

and hence

$$x = (A_i - \alpha)\phi(y_0) \in \mathrm{Im}(A_i - \alpha).$$

This contradicts the assumption on x. Then we have $y_1 \notin \mathrm{Im}(G_i - (\alpha + \lambda))$. Hence, by the partial basis $y_1, \ldots, y_m$, the Jordan canonical form of B_i contains the Jordan cell $J(\alpha + \lambda; m)$.

Next we consider the case $\alpha = 0$. A partial basis corresponding to the Jordan cell $J(0; m)$ of A_i is given by

$$x_1 = x, \ x_2 = A_i x, \ \ldots, \ x_m = A_i{}^{m-1}x,$$

where x is so chosen that

$$A_i{}^m x = 0, \; A_i{}^{m-1} x \neq 0, \; x \notin \mathrm{Im} A_i$$

hold. Since $x_2, \ldots, x_m \in \mathrm{Im} A_i$, the elements

$$y_2 = \phi^{-1}(x_2), \; y_3 = \phi^{-1}(x_3), \; \ldots, \; y_m = \phi^{-1}(x_m)$$

belong to $\mathrm{Im} G_i$ and are linearly independent in $mc_\lambda(V)$. Moreover, they satisfy

$$(G_i - \lambda) y_i = y_{i+1} \quad (2 \leq i \leq m - 1),$$
$$(G_i - \lambda) y_m = 0.$$

If there exists y_1 such that $y_2 = (G_i - \lambda) y_1$, we have $y_1 \neq 0$ by $y_2 \neq 0$, and obtain

$$y_1 = \frac{1}{\lambda} (G_i y_1 - y_2) \in \mathrm{Im} G_i$$

by $\lambda \neq 0$. Then, by setting

$$u = \phi(y_1),$$

we have $u \in \mathrm{Im} A_i$, $u \neq 0$. Note that

$$A_i x = \phi(y_2) = \phi((G_i - \lambda) y_1) = A_i \phi(y_1) = A_i u.$$

Then, by setting $x - u = z$, we get $z \in \mathrm{Ker} A_i$. We decompose V into a direct sum by the basis which gives the Jordan canonical form of A_i, and set

$$V = \langle x_1, x_2, \ldots, x_m \rangle \oplus W.$$

According to this decomposition, we write

$$u = u' + u'', \; z = z' + z'',$$

where $u', z' \in \langle x_1, x_2, \ldots, x_m \rangle$. Then we have $x = u' + z'$, and obtain $z' \in \langle x_m \rangle$ from $z' \in \mathrm{Ker} A_i$ and $u' \in \langle x_2, \ldots, x_m \rangle$ from $u' \in \mathrm{Im} A_i$. Thus we have

$$x_1 = x = u' + z' \in \langle x_2, \ldots, x_m \rangle,$$

which is a contradiction. Then there does not exist y_1, and hence $y_2 \notin \mathrm{Im}(G_i - \lambda)$. Therefore, by the partial basis $y_2, \ldots, y_m$, the Jordan canonical form of B_i contains the Jordan cell $J(\lambda; m - 1)$.

Lastly we consider the case $\alpha = -\lambda$. A partial basis corresponding to the Jordan cell $J(-\lambda; m)$ of A_i is given by

$$x_1 = x, \ x_2 = (A_i + \lambda)x, \ \ldots, \ x_m = (A_i + \lambda)^{m-1}x,$$

where x is so chosen that

$$(A_i + \lambda)^m x = 0, \ (A_i + \lambda)^{m-1}x \neq 0, \ x \notin \mathrm{Im}(A_i + \lambda)$$

hold. Similarly as in the cases $a \neq 0, -\lambda$, we obtain $x_j \in \mathrm{Im}A_i$ $(1 \leq j \leq m)$ from $(A_i + \lambda)^m x = 0$. By using the isomorphism ϕ, we define

$$y_1 = \phi^{-1}(x_1), \ \ldots, \ y_m = \phi^{-1}(x_m).$$

Then $y_1, \ldots, y_m$ are elements in $\mathrm{Im}G_i$, and linearly independent in $mc_\lambda(V)$. Since $y_1 \in \mathrm{Im}G_i$, there exists $y_0 \in V^p$ such that

$$y_1 = G_i y_0.$$

Then we may write

$$y_1 = {}^t(0, \ldots, 0, y_{1i}, 0, \ldots, 0).$$

If $y_0 \in \mathcal{K} + \mathcal{L}$, we get $y_{1i} \in \mathrm{Ker}A_i$, which causes the contradiction

$$x_1 = \phi(y_1) = A_i y_{1i} = 0.$$

Then y_0 is different from 0 in $mc_\lambda(V)$. Moreover, if there exists y_{-1} such that $y_0 = G_i y_{-1}$, we may write

$$y_0 = {}^t(0, \ldots, 0, y_{0i}, 0, \ldots, 0),$$

and have $y_{1i} = (A_i + \lambda)y_{0i}$. Then we get

$$x_1 = \phi(y_1) = A_i y_{1i} = (A_i + \lambda)A_i y_{0i},$$

which contradicts $x \notin \mathrm{Im}(A_i + \lambda)$. Therefore, by the partial basis $y_0, y_1, \ldots, y_m$, the Jordan canonical form of B_i contains the Jordan cell $J(0; m+1)$.

Next we prove (ii). Since $B_0 = G_0|_{mc_\lambda(V)}$, the eigenvalues of B_0 are contained in the eigenvalues of G_0. By

$$G_0 = -\lambda - \begin{pmatrix} A_1 & A_2 & \cdots & A_p \\ & \cdots\cdots & \\ A_1 & A_2 & \cdots & A_p \end{pmatrix},$$

the eigenvalues of G_0 are the eigenvalues of $A_0 - \lambda$ and $-\lambda$ of multiplicity $(p-1)n$. Then we can show the assertion in a similar way as (i), and we omit the proof. □

Remark 7.2 If $\lambda \neq 0$, we have $\mathcal{K} + \mathcal{L} = \mathcal{K} \oplus \mathcal{L}$, and hence the size of B_i is given by

$$np - (\dim \mathcal{K} + \dim \mathcal{L}). \tag{7.42}$$

Thus the Jordan canonical form of B_i $(0 \leq i \leq p)$ are completely determined by Theorem 7.18.

Proposition 7.21 *Let J be a Jordan matrix with only one eigenvalue. For $j = 1, 2, 3 \ldots$, we denote by e_j the number of the Jordan cells in J of size equal to or greater than j. Then we have*

$$\dim Z(J) = \sum_{j \geq 1} e_j{}^2.$$

Proof It is reduce to the case $J(\alpha; m) \oplus J(\alpha; n)$, and in this case, we see that the dimension of the centralizer is given by

$$m + n + 2\min\{m, n\}.$$

□

Proof of Theorem 7.15 The assertion clearly holds if $\lambda = 0$ by Theorem 7.12. Then we assume $\lambda \neq 0$, and set

$$mc_\lambda(A_1, \ldots, A_p) = (B_1, \ldots, B_p).$$

Set

$$\dim \mathcal{K} = k, \ \dim \mathcal{L} = l.$$

For each i $(1 \leq i \leq p)$, we set

$$\dim \mathrm{Ker} A_i = k_i,$$

and then we have

$$k = \sum_{i=1}^{p} k_i.$$

If we set $\dim V = n$, the dimension of $mc_\lambda(V)$ becomes $pn - k - l$. From Theorem 7.19, we readily obtain the following relations. For $1 \leq i \leq p$, we have

$$e_j(B_i; \alpha + \lambda) = e_j(A_i; \alpha) \quad (\alpha \neq 0, -\lambda),$$
$$e_j(B_i; \lambda) = e_{j+1}(A_i; 0),$$
$$e_j(B_i; 0) = e_{j-1}(A_i; -\lambda) \quad (j > 1),$$

(7.43)

and for B_0,

$$e_j(B_0; \alpha - \lambda) = e_j(A_0; a) \quad (\alpha \neq 0, \lambda),$$
$$e_j(B_0; 0) = e_{j+1}(A_0; \lambda),$$
$$e_j(B_0; -\lambda) = e_{j-1}(A_0; 0) \quad (j > 1).$$

(7.44)

Note that the dimension of the generalized eigenspace of a square matrix A of the eigenvalue α is given by

$$\sum_{j \geq 1} e_j(A; \alpha),$$

and that

$$k_i = \dim \mathrm{Ker} A_i = e_1(A_i; 0).$$

Then we can express $\dim mc_\lambda(V)$ in terms of $e_j(B_i; \beta)$, and get

$$e_1(B_i; 0) = (p - 1)n - k - l + k_i.$$

Similarly, by noting

$$l = \dim \mathcal{L} = e_1(A_0; \lambda),$$

we get

$$e_1(B_0; -\lambda) = (p - 1)n - k.$$

Now we apply Proposition 7.21. For $1 \leq i \leq p$, we have

$$\dim Z(B_i) = ((p - 1)n - (k + l))^2 + 2k_i((p - 1)n - (k + l)) + \dim Z(A_i),$$

and for B_0,

$$\dim Z(B_0) = ((p - 1)n - (k + l))^2 + 2l((p - 1)n - (k + l)) + \dim Z(A_0).$$

Putting these into the definition of the index of rigidity, we get the invariance. $\square$

We write the results of Theorem 7.19 explicitly in the semi-simple case. Assume that $(A_0, A_1, \ldots, A_p)$ is semi-simple. The result of the middle convolution

$$mc_\lambda(A_0, A_1, \ldots, A_p) = (B_0, B_1, \ldots, B_p)$$

is also semi-simple if and only if $-\lambda$ is not an eigenvalue of any A_i $(1 \le i \le p)$ and 0 is not an eigenvalue of A_0. We assume these conditions.

For each i, we denote the spectral type of A_i by

$$A_i^\natural = (e_{i1}, e_{i2}, \ldots, e_{iq_i}).$$

We assume that, for $1 \le i \le p$, e_{i1} denotes the multiplicity of the eigenvalue 0 of A_i, and for $i = 0$, e_{01} denotes the multiplicity of the eigenvalue λ of A_0. By this convention, we may have $e_{i1} = 0$. Set

$$g = \sum_{i=0}^{p} e_{i1}.$$

Then we have

$$\dim mc_\lambda(V) = pn - g,$$

and moreover, by Theorem 7.19 and the proof of Theorem 7.15, we obtain

$$B_i^\natural = (e_{i1} + (p-1)n - g, e_{i2}, \ldots, e_{iq_i}). \tag{7.45}$$

Set

$$g - (p-1)n = d.$$

Then we obtain the change of the spectral type as follows. We also write the eigenvalues. For $1 \le i \le p$, we have

$$\begin{pmatrix} 0 & \alpha_{i2} & \cdots & \alpha_{iq_i} \\ e_{i1} & e_{i2} & \cdots & e_{iq_i} \end{pmatrix} \xrightarrow{mc_\lambda} \begin{pmatrix} 0 & \alpha_{i2} + \lambda & \cdots & \alpha_{iq_i} + \lambda \\ e_{i1} - d & e_{i2} & \cdots & e_{iq_i} \end{pmatrix},$$

and for $i = 0$,

$$\begin{pmatrix} \lambda & \alpha_{02} & \cdots & \alpha_{0q_0} \\ e_{01} & e_{02} & \cdots & e_{0q_0} \end{pmatrix} \xrightarrow{mc_\lambda} \begin{pmatrix} -\lambda & \alpha_{02} - \lambda & \cdots & \alpha_{0q_0} - \lambda \\ e_{01} - d & e_{02} & \cdots & e_{0q_0} \end{pmatrix}.$$

Example 7.5 We shall see the changes of the spectral types

$$(A_0^\natural, A_1^\natural, A_2^\natural) \to (B_0^\natural, B_1^\natural, B_2^\natural)$$

given by (7.45) in several cases with $p = 2$.

(i) The case $n = 3$, $(A_0^\natural, A_1^\natural, A_2^\natural) = ((111), (111), (111))$. Since $g = 1+1+1 = 3$, we have $d = 0$, and hence there is no change of the spectral type.

(ii) The case $n = 3$, $(A_0^\natural, A_1^\natural, A_2^\natural) = ((21), (111), (111))$. We have $d = 1$ by $g = 2+1+1 = 4$. Then we get the change of the spectral type

$$((21), (111), (111)) \to ((11), (11), (11)).$$

Here we omit 0 in the partitions, because 0 means the vanishing of the corresponding eigenspace.

(iii) The case $n = 3$, $(A_0^\natural, A_1^\natural, A_2^\natural) = ((21), (21), (111))$. In this case, the index of rigidity is given by

$$\iota = (1 - 2) \cdot 3^2 + ((2^2 + 1^2) + (2^2 + 1^2) + (1^2 + 1^2 + 1^2)) = 4,$$

which implies, by virtue of Theorem 7.11, that (A_0, A_1, A_2) is reducible if it exists.

(iv) The case $n = 5$, $(A_0^\natural, A_1^\natural, A_2^\natural) = ((311), (311), (11111))$. In this case, the index of rigidity is $\iota = 2$, and then a necessary condition for the existence of an irreducible tuple is satisfied. We have $d = 2$ by $g = 3+3+1 = 7$. Applying (7.45) formally, we get

$$((311), (311), (11111)) \to ((111), (111), (-11111)),$$

in which the dimension -1 of an eigenspace has no meaning. Then we can conclude that there is no tuple (A_0, A_1, A_2) of this spectral type.

As explained in the above examples, the existence of a tuple $(A_0, A_1, \ldots, A_p)$ with a prescribed spectral type and its irreducibility can be determined by the index of rigidity and the change of the spectral types by the middle convolution. We shall show an algorithm for such criterion in the next section. For the rigid case, the criterion becomes very simple, and is one of the main results of the Katz theory. In order to describe this result, we note the properties of the addition.

We have shown several properties of the middle convolution. Also for the addition, similar properties can be easily shown. Namely, we have

$$ad_0 = \mathrm{id}, \quad ad_\alpha \circ ad_\beta = ad_{\alpha+\beta}.$$

Moreover, for any $\alpha \in \mathbb{C}^p$, ad_α leaves the dimension and irreducibility of V invariant. Also ad_α leaves the spectral type of $(A_0, A_1, \ldots, A_p)$, and hence the index of rigidity, invariant.

Now we are ready to state one of the main results of Katz.

Theorem 7.22 *Any irreducible rigid tuple $(A_0, A_1, \ldots, A_p)$ can be obtained from a tuple of rank 1 by a finite iteration of the additions and the middle convolutions.*

Proof Let n be greater than 1, and $(A_0, A_1, \ldots, A_p)$ an irreducible tuple of rank (size) n. For each i, we denote by m_i the maximum of the dimensions of the eigenspaces of A_i. Then we have

$$\dim Z(A_i) \leq m_i{}^2 \times \frac{n}{m_i} = nm_i.$$

Since the index of rigidity is 2, we have

$$2 = (1 - p)n^2 + \sum_{i=0}^{p} \dim Z(A_i) \leq (1 - p)n^2 + \sum_{i=0}^{p} nm_i,$$

and hence

$$\sum_{i=0}^{p} m_i \geq (p - 1)n + \frac{2}{n} > (p - 1)n.$$

Let α_i be an eigenvalue of A_i such that the dimension of the eigenspace is m_i for each i, and operate the addition by $(-\alpha_1, \ldots, -\alpha_p)$. We set

$$ad_{(-\alpha_1, \ldots, -\alpha_p)}(A_1, \ldots, A_p) = (B_1, \ldots, B_p), \quad B_0 = -\sum_{i=1}^{p} B_i.$$

Then we have

$$\dim \mathrm{Ker}\, B_i = m_i \quad (1 \leq i \leq p),$$

and the dimension of the eigenspace of B_0 of the eigenvalue $\alpha_0 + \sum_{i=1}^{p} \alpha_i = \sum_{i=0}^{p} \alpha_i$ is m_0. Now we operate the middle convolution with parameter $\lambda = \sum_{i=0}^{p} \alpha_i$. Then the dimension of $mc_\lambda(V)$ becomes

$$pn - \sum_{i=1}^{p} m_i - m_0 < pn - (p - 1)n = n,$$

which is strictly less than n. □

Remark 7.3 In Theorem 7.22, we assumed that $(A_0, A_1, \ldots, A_p)$ is irreducible, and then it satisfies (M2). Hence we can apply Theorem 7.12 to see that the middle convolution mc_0 with parameter 0 is an isomorphism, which induces $\dim mc_0(V) = \dim V$. Then we see that the parameter $\lambda = \sum_{i=0}^{p} \alpha_i$ of the middle convolution in the above proof is different from 0.

Example 7.6 We give the list of the spectral types of all irreducible semi-simple rigid tuples of rank 2, 3, 4. For rank 2, we have only

$$((11), (11), (11)).$$

For rank 3, we have two tuples

$$((21), (111), (111)), \ ((21), (21), (21), (21)).$$

For rank 4, we have six tuples

$$((31), (1111), (1111)), \ ((22), (211), (1111)), \ ((211), (211), (211)),$$

$$((31), (31), (22), (211)), \ ((31), (22), (22), (22)),$$

$$((31), (31), (31), (31), (31)).$$

These spectral types are obtained by using the algorithm which we will introduce in the next section.

By Katz's theorem (Theorem 7.22), we see that any rigid Fuchsian system of normal form (7.15) can be reduced to a differential equation of rank 1

$$\frac{dy}{dx} = \left(\sum_{j=1}^{p} \frac{\alpha_j}{x - a_j} \right) y. \tag{7.46}$$

The last differential equation is solved by

$$y(x) = \prod_{j=1}^{p} (x - a_j)^{\alpha_j}. \tag{7.47}$$

Thus various information of rigid differential equations will be explicitly obtained if we know how the information are transmitted by the addition and the middle convolution. Actually, we can describe monodromy, connection coefficients, power series expansions of solutions, irreducibility conditions and so on for rigid differential equations. Here we shall explain the description of integral representations of solutions [59]. For the other information, please refer to [137].

Consider an irreducible rigid Fuchsian system of normal form

$$\frac{dY}{dx} = \left(\sum_{j=1}^{p} \frac{A_j}{x - a_j} \right) Y. \tag{7.48}$$

Thanks to Theorem 7.22, this system is obtained from a differential equation (7.46) of rank 1 by a finite iteration of additions and middle convolutions. The addition and the middle convolution are analytically realized as a gauge transformation and Riemann-Liouville transform, respectively. Then, by using the explicit form (7.47) of a solution of the differential equation (7.46) as a seed, we get an integral representation of solutions of the system (7.48).

Theorem 7.23 *Any irreducible rigid Fuchsian system of normal form has an integral representation of solutions*

$$Y(x) = Q \prod_{j=1}^{p} (x - a_j)^{\alpha_{q+1,j}} \int_{\Delta} \prod_{i=1}^{q} \prod_{j \neq 1}^{p} (t_i - a_j)^{\alpha_{ij}} \prod_{i=1}^{q-1} (t_i - t_{i+1})^{\lambda_i} (t_q - x)^{\lambda_q} \, \eta, \tag{7.49}$$

where

$$\alpha_{ij} \in \mathbb{C} \ (1 \leq i \leq q+1, \ 1 \leq j \leq p), \quad \lambda_i \in \mathbb{C} \ (1 \leq i \leq q),$$

Q is a linear transformation with constant coefficients, η a vector of twisted q-forms in the variable $(t_1, \ldots, t_q)$ and Δ a twisted cycle in $(t_1, \ldots, t_q)$-space.

Remark 7.4 We will introduce in Sect. 9.9 the notions of twisted q-forms and twisted cycles.

Proof Consider a chain of additions and middle convolutions which transforms the differential equation (7.46) of rank 1 to the system (7.48). Note that the middle convolution is a combination of a tensor product (7.30) which makes the size of residue matrices p times and a restriction to a quotient space. Then the chain of additions ("ad"s) and middle convolutions is decomposed into a finer chain

$$\cdots \rightarrow \text{ad} \rightarrow \text{tensor} \rightarrow \text{restriction} \rightarrow \text{ad} \rightarrow \text{tensor} \rightarrow \text{restriction} \rightarrow \cdots .$$

We shall show that this finer chain is replaced by another chain

$$\cdots \rightarrow \text{ad} \rightarrow \text{tensor} \rightarrow \text{ad} \rightarrow \text{tensor} \rightarrow \cdots \rightarrow \text{restriction} \tag{7.50}$$

which takes restriction once in the last step. In order to show this, it is enough to show that a chain

$$\text{tensor} \rightarrow \text{restriction} \rightarrow \text{ad} \rightarrow \text{tensor} \rightarrow \text{restriction} \tag{7.51}$$

can be replaced by a chain

$$\text{tensor} \rightarrow \text{ad} \rightarrow \text{tensor} \rightarrow \text{restriction} \tag{7.52}$$

which makes the same result. Take a tuple $(A_1, A_2, \ldots, A_p)$ of matrices, and construct $(G_1, G_2, \ldots, G_p)$ by the tensor product (7.30). The result $(B_1, B_2, \ldots, B_p)$ of the middle convolution mc_λ is determined by

$$P^{-1} G_j P = \left(\begin{array}{c|c} * & * \\ \hline O & B_j \end{array} \right).$$

Operate the addition by $(\alpha_1, \alpha_2, \ldots, \alpha_p) \in \mathbb{C}^p$ to this result, and then take a tensor product. Namely we have the tuple $(H_1, H_2, \ldots, H_p)$ defined by

$$H_j = \left(\begin{array}{ccccc} & & O & & \\ B_1 + \alpha_1 & \cdots & B_j + \alpha_j + \mu & \cdots & B_p + \alpha_p \\ & & O & & \end{array} \right).$$

Lastly define $(C_1, C_2, \ldots, C_p)$ by

$$Q^{-1} H_j Q = \left(\begin{array}{c|c} * & * \\ \hline O & C_j \end{array} \right).$$

Thus we obtain the result of the chain (7.51). On the other hand, we operate the addition by $(\alpha_1, \alpha_2, \ldots, \alpha_p) \in \mathbb{C}^p$ to the tuple $(G_1, G_2, \ldots, G_p)$, and then take a tensor product. Thus we get $(K_1, K_2, \ldots, K_p)$ defined by

$$K_i = \left(\begin{array}{ccccc} & & O & & \\ G_1 + \alpha_1 & \cdots & G_j + \alpha_j + \mu & \cdots & G_p + \alpha_p \\ & & O & & \end{array} \right).$$

Then we have

$$
\begin{pmatrix} P & & \\ & \ddots & \\ & & P \end{pmatrix}^{-1} K_j \begin{pmatrix} P & & \\ & \ddots & \\ & & P \end{pmatrix}
$$

$$
= \left(\begin{array}{c|c|c|c|c|c|c} \multicolumn{7}{c}{O} \\ \hline * & * & & * & * & & * & * \\ \hline O & B_1 + \alpha_1 & \cdots & O & B_j + \alpha_j + \mu & \cdots & O & B_p + \alpha_p \\ \hline \multicolumn{7}{c}{O} \end{array} \right).
$$

Therefore, by taking an appropriate permutation matrix S, we can obtain

$$
S^{-1} \begin{pmatrix} P & & \\ & \ddots & \\ & & P \end{pmatrix}^{-1} K_j \begin{pmatrix} P & & \\ & \ddots & \\ & & P \end{pmatrix} S = \left(\begin{array}{c|c} * & * \\ \hline O & H_j \end{array} \right).
$$

Then, by setting

$$
R = \begin{pmatrix} P & & \\ & \ddots & \\ & & P \end{pmatrix} S \begin{pmatrix} I & \\ & Q \end{pmatrix},
$$

we have

$$
R^{-1} K_j R = \left(\begin{array}{c|c} * & * \\ \hline O & C_j \end{array} \right).
$$

Hence, by this R, we have the same result by the chain (7.52) and the chain (7.51).

Now we study the change of solutions by the chain of operations (7.50). As explained in the beginning of this section, solutions of the result of the middle convolution mc_λ to a Fuchsian system of normal form with a solution $U(x)$ are obtained in the following way. First we take the Riemann-Liouville transform of the vector

$$
\hat{U}(x) = \begin{pmatrix} \dfrac{U(x)}{x - a_1} \\ \vdots \\ \dfrac{U(x)}{x - a_p} \end{pmatrix}
$$

to get

$$V(x) = \int_{\Delta} \hat{U}(t)(t-x)^{\lambda} \, dt,$$

which is a solution of the result of the tensor product. Then, by a linear transformation P which gives the restriction to the quotient space, we get

$$W(x) = PV(x),$$

which is a solution of the result of mc_{λ}. We start from the differential equation (7.46) of rank 1 whose solution is given by (7.47). Then we get a solution

$$\int_{\delta_1} \prod_{j=1}^{p} (t_1 - a_j)^{\alpha_{1j}} (t_1 - x)^{\lambda_1} \eta_1,$$

$$\eta_1 = {}^t (dt_1/(t_1 - a_1), dt_1/(t_1 - a_2), \dots, dt_1/(t_1 - a_p))$$

of the differential equation obtained by the convolution (tensor product) with parameter λ_1, where we have set $\alpha_{1j} = \alpha_j$. Operating the addition, we get

$$\prod_{j=1}^{p} (x - a_j)^{\alpha_{2j}} \int_{\delta_1} \prod_{j=1}^{p} (t_1 - a_j)^{\alpha_{1j}} (t_1 - x)^{\lambda_1} \eta_1,$$

and operating the next tensor product with parameter λ_2, we get

$$\int_{\delta_2} \prod_{i=1}^{2} \prod_{j=1}^{p} (t_i - a_j)^{\alpha_{ij}} (t_1 - t_2)^{\lambda_1} (t_2 - x)^{\lambda_2} \eta_2,$$

$$\eta_2 = \left(\frac{dt_1 \wedge dt_2}{(t_1 - a_k)(t_2 - a_l)} \right)_{(k,l) \in \{1,\dots,p\}^2}.$$

Continuing these operations and taking a linear transformation in the last step, we arrive at the representation (7.49). □

Here we give a proof of Proposition 7.7 which asserts that the index of rigidity is an even integer.

Proof of Proposition 7.7 We use that notation in the proof of Theorem 7.15. In particular, we use Definition 7.6 for the index of rigidity. By Proposition 7.21, we have

$$\dim Z(A_i) = \sum_{\alpha} \sum_{j \geq 1} e_j(A_i; \alpha)^2,$$

where α runs all distinct eigenvalues of A_i. We see

$$\sum_{j \geq 1} e_j(A_i; \alpha)^2 \equiv \sum_{j \geq 1} e_j(A_i; \alpha) \quad (\text{mod } 2),$$

and the right hand side is equal to the dimension of the generalized eigenspace of A_i for the eigenvalue α. Then we have

$$\dim Z(A_i) \equiv \sum_{\alpha} \sum_{j \geq 1} e_j(A_i; \alpha) = n \quad (\text{mod } 2).$$

Thus, if n is even, ι is also even. If n is odd, ι is still even by

$$\iota \equiv (1 - p)n^2 + (p + 1)n \equiv (1 - p) + (p + 1) = 2 \quad (\text{mod } 2).$$

$\square$

7.5.1 Middle Convolution for Monodromy

We have defined and analyzed the addition and the middle convolution for Fuchsian systems of normal form. We can also define these operations to monodromies, and then obtain similar properties. For details, we recommend to refer to [34]. Here we overview the definition and main results.

Let $(M_1, M_2, \ldots, M_p) \in \mathrm{GL}(n, \mathbb{C})^p$ be a tuple of non-singular matrices. For $(\lambda_1, \lambda_2, \ldots, \lambda_p) \in (\mathbb{C}^\times)^p$, we define the map

$$(M_1, M_2, \ldots, M_p) \mapsto (\lambda_1 M_1, \lambda_2 M_2, \ldots, \lambda_p M_p),$$

and call it the *multiplication* by $(\lambda_1, \lambda_2, \ldots, \lambda_p)$. This is an analogue of the addition, and is related to the addition in the following way. Consider a Fuchsian system of normal form

$$\frac{dY}{dx} = \left(\sum_{j=1}^{p} \frac{A_j}{x - a_j} \right) Y, \tag{7.53}$$

and let $(M_1, M_2, \ldots, M_p)$ be a tuple of matrices determining the monodromy representation. Namely, in the monodromy representation with respect to a fundamental system of solutions $\mathcal{Y}(x)$, M_j is the circuit matrix for a $(+1)$-loop for $x = a_i$. For the tuple $(A_1, A_2, \ldots, A_p)$ of the residue matrices of (7.53), we operate the addition to obtain $(A_1 + \alpha_1, A_2 + \alpha_2, \ldots, A_p + \alpha_p)$. Then we have the fundamental matrix solution

$$\mathcal{Y}(x) \prod_{j=1}^{p} (x - a_j)^{\alpha_j}$$

for the Fuchsian system of normal form with these new residue matrices. The monodromy representation with respect to this fundamental system of solutions is given by

$$(e^{2\pi \sqrt{-1} \alpha_1} M_1, e^{2\pi \sqrt{-1} \alpha_2} M_2, \ldots, e^{2\pi \sqrt{-1} \alpha_p} M_p),$$

which is the result of the multiplication of $(M_1, M_2, \ldots, M_p)$.

Next we define the middle convolution. Take a parameter $\lambda \in \mathbb{C}^\times$. For a tuple $(M_1, M_2, \ldots, M_p) \in GL(n, \mathbb{C})^p$, we define the tuple $(G_1, G_2, \ldots, G_p)$ of size p times by

$$G_j = \begin{pmatrix} I_n & O & \cdots & \cdots & & O & O \\ & \ddots & & & & & \\ & & \ddots & & & & \\ M_1 - I_n & \cdots & M_{j-1} - I_n & \lambda M_j & \lambda(M_{j+1} - I_n) & \cdots & \lambda(M_p - I_n) \\ & & & & \ddots & & \\ & & & & & \ddots & \\ O & O & \cdots & \cdots & & O & I_n \end{pmatrix} \quad (j.$$

Define the subspaces $\mathcal{K}, \mathcal{L}$ of $\mathbb{C}^{pn}$ by

$$\mathcal{K} = \left\{ \begin{pmatrix} v_1 \\ \vdots \\ v_p \end{pmatrix} ; v_j \in \mathrm{Ker}(M_j - I_n) \ (1 \le j \le p) \right\},$$

$$\mathcal{L} = \bigcap_{j=1}^{p} \mathrm{Ker}(G_j - I).$$

Then we see that these subspaces are $(G_1, G_2, \ldots, G_p)$-invariant. Hence the tuple $(G_1, G_2, \ldots, G_p)$ acts on the quotient space $\mathbb{C}^{pn}/(\mathcal{K} + \mathcal{L})$, and we denote by $(N_1, N_2, \ldots, N_p)$ the action. The operation

$$(M_1, M_2, \ldots, M_p) \mapsto (N_1, N_2, \ldots, N_p)$$

is called the *middle convolution* with parameter λ, and is denoted by MC_λ.

Generically, the middle convolution of Fuchsian system and the middle convolution of the monodromy are compatible. Let $(M_1, M_2, \ldots, M_p)$ be a tuple of matrices which determines the monodromy representation of the system (7.53). We operate the middle convolution with parameter ℓ to the Fuchsian system (7.53). Then the result $(N_1, N_2, \ldots, N_p)$ of the middle convolution with parameter $e^{2\pi\sqrt{-1}\ell}$ of $(M_1, M_2, \ldots, M_p)$ can be regarded as a tuple which determines the monodromy representation of the Fuchsian system of normal form obtained by the middle convolution mc_ℓ. This compatibility is proved by Dettweiler and Reiter [35].

Conditions analogous to (M1), (M2) for an $(M_1, M_2, \ldots, M_p)$-module V are also defined. Under these conditions, similar assertions as Theorems 7.12–7.15 hold. As the consequences, we have

$$MC_1(V) \simeq V,$$

$$MC_\lambda(MC_\mu(V)) \simeq MC_{\lambda\mu}(V),$$

and that the middle convolution MC_λ leaves the irreducibility and the index of rigidity invariant. Please refer to [34] for the precise assertions.

Moreover, we have the result corresponding to Theorem 7.22 which is the main result of this section.

Theorem 7.24 *Any irreducible rigid tuple* $(M_1, M_2, \ldots, M_p)$ *of elements in* $GL(n, \mathbb{C})$ *can be obtained from a tuple of elements in* $GL(1, \mathbb{C})$ *by a finite iteration of the multiplications and the middle convolutions.*

7.6 Problem of Existence of Fuchsian Differential Equations

It is natural to ask the existence of a monodromy representation with prescribed local monodromies. We formulate the problem as to ask the existence of a tuple $(M_0, M_1, \ldots, M_p)$ of matrices satisfying

$$M_p \cdots M_1 M_0 = I$$

such that, for a given tuple $(L_0, L_1, \ldots, L_p)$ of conjugacy classes in $GL(n; \mathbb{C})$, $M_j \in L_j$ holds for $0 \le j \le p$. Also, the problem to ask the existence of an irreducible tuple $(M_0, M_1, \ldots, M_p)$ is called the Deligne-Simpson problem.

As an additive analogue, we can consider the additive Dligne-Simpson problem, which asks, for a given tuple $(C_0, C_1, \ldots, C_p)$ of conjugacy classes in $M(n; \mathbb{C})$, the existence of an irreducible tuple of matrices $(A_0, A_1, \ldots, A_p)$ satisfying

$$A_0 + A_1 + \cdots + A_p = O \tag{7.54}$$

such that $A_j \in C_j$ holds for $0 \le j \le p$. This problem can be regarded as a problem of asking the existence of a Fuchsian system of normal form with the residue matrices $(A_0, A_1, \ldots, A_p)$. Thanks to the Riemann-Hilbert correspondence, these two problems are almost equivalent. Precisely speaking, there remains a problem of transforming a Fuchsian differential equation to a Fuchsian system of normal form, which we have discussed in the beginning of Sect. 7.3.

A conjugacy class is determined by the spectral type and the eigenvalues. Kostov [116] and Crawley-Boevey [30] independently solved the Deligne-Simpson problem, and, in the solution of Crawley-Boevey, the spectral type plays an essential role. Then in this section we reformulate the Deligne-Simpson problem as a problem on the spectral types, and introduce the result of Crawley-Boevey. Namely, we ask the condition on the tuple of spectral types such that there exists an irreducible tuple $(A_0, A_1, \ldots, A_p)$ of matrices satisfying (7.54) with the spectral types. This problem can be solved by the help of the addition and the middle convolution. We shall give an explicit algorithm.

First, we introduce the result of Crawley-Boevey. Let $(C_0, C_1, \ldots, C_p)$ be a tuple of conjugacy classes in $M(n; \mathbb{C})$. With this tuple, we associate a Kac-Moody root system as follows. For each i $(0 \le i \le p)$, take $\xi_{i,1}, \xi_{i,2}, \ldots, \xi_{i,d_i} \in \mathbb{C}$ such that

$$\prod_{j=1}^{d_i} (A_i - \xi_{i,j}) = O$$

holds, where A_i is any matrix in C_i. The set $\{\xi_{i,j}\}$ should contain all eigenvalues of C_i, and may contain other values. We define $r_{i,j}$ by

$$r_{i,j} = \operatorname{rank} \prod_{k=1}^{j} (A_i - \xi_{i,k}).$$

Clearly

$$n = r_{i,0} \ge r_{i,1} \ge \cdots \ge r_{i,d_i} = 0 \tag{7.55}$$

holds. If ξ is an eigenvalue of C_i, and if the indices j such that $\xi = \xi_{i,j}$ are labelled as $j_1 < j_2 < \cdots$, we have

$$r_{i,j_k-1} - r_{i,j_k} = e_k(A_i; \xi). \tag{7.56}$$

We take the set

$$I = \{0\} \cup \{[i, j] ; 0 \le i \le p, \ 1 \le j \le d_i\}$$

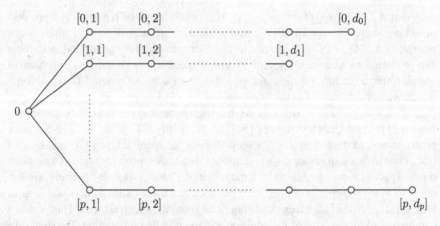

Fig. 7.2 Dynkin diagram for $(C_0, C_1, \ldots, C_p)$

as the set of vertices, and consider the star shaped Dynkin diagram given in Fig. 7.2. Using this Dynkin diagram, we define the generalized Cartan matrix $C = (C_{uv})_{u,v \in I}$ by

$$
C_{uv} = \begin{cases} 2 & (u = v), \\ -1 & (u \neq v, \ u \text{ and } v \text{ are connected by an edge}), \\ 0 & (\text{otherwise}). \end{cases}
$$

We denote by ϵ_v the unit vector corresponding to a vertex $v \in I$, and define the reflection $s_v : \mathbb{Z}^I \to \mathbb{Z}^I$ with respect to ϵ_v by

$$
s_v(\alpha) = \alpha - ({}^t\epsilon_v C\alpha)\epsilon_v \quad (\alpha \in \mathbb{Z}^I). \tag{7.57}
$$

These reflections generate the Weyl group $W = \langle s_v \mid v \in I \rangle$. Define the fundamental region B by

$$
B = \{\alpha \in (\mathbb{Z}_{\geq 0})^I \setminus \{0\} \, ; \, \text{support of } \alpha \text{ is connected}, \, {}^t\epsilon_v C\alpha \leq 0 \, (\forall v \in I)\}.
$$

We denote by Δ_{re} the set of images of ϵ_v $(v \in I)$ by the action of W, and call the elements real roots. Also we denote by Δ_{im}^+ the set of images of elements of B by the action of W, set

$$
\Delta_{\mathrm{im}} = \Delta_{\mathrm{im}}^+ \cup -\Delta_{\mathrm{im}}^+,
$$

and call the elements imaginary roots. The set

$$
\Delta = \Delta_{\mathrm{re}} \cup \Delta_{\mathrm{im}}
$$

is called Kac-Moody root system, and we call an elements in Δ a root. Set

$$\Delta^+ = \Delta \cap (\mathbb{Z}_{\geq 0})^I,$$

and call the elements positive roots. It is known that

$$\Delta = \Delta^+ \cup -\Delta^+$$

holds. Define a map $q : \mathbb{Z}^I \to \mathbb{Z}$ by

$$q(\alpha) = 1 - \frac{1}{2}{}^t\alpha C\alpha \quad (\alpha \in \mathbb{Z}^I). \tag{7.58}$$

Since $q(-\alpha) = q(\alpha), q(s_v(\alpha)) = q(\alpha)$ hold, q takes a non-negative value for any element in Δ, and takes 0 only for real roots.

For $\lambda \in \mathbb{C}^I$, we set

$$\Delta_\lambda^+ = \{\alpha \in \Delta^+ ; \lambda \cdot \alpha := \sum_{v \in I} \lambda_v \alpha_v = 0\}.$$

Moreover, we set

$$\Sigma_\lambda = \{\alpha \in \Delta_\lambda^+ ; \text{if we write } \alpha = \beta^{(1)} + \beta^{(2)} + \cdots \text{ as a sum}$$

$$\text{of more than one elements in } \Delta_\lambda^+, \text{ we have}$$

$$q(\alpha) > q(\beta^{(1)}) + q(\beta^{(2)}) + \cdots \}.$$

By using these notation, we can state the result of Crawley-Boevey.

Theorem 7.25 *Let* $(C_0, C_1, \ldots, C_p)$ *be a tuple of conjugacy classes in* $\mathrm{M}(n; \mathbb{C})$, *and define* $\xi_{i,j}$ *and* $r_{i,j}$ *as above. Define* $\lambda \in \mathbb{C}^I$ *and* $\alpha \in \mathbb{Z}^I$ *by*

$$\lambda_0 = -\sum_{i=0}^p \xi_{i,1}, \quad \lambda_{[i,j]} = \xi_{i,j} - \xi_{i,j+1},$$

$$\alpha_0 = n, \quad \alpha_{[i,j]} = r_{i,j}.$$

Then there exists an irreducible tuple $(A_0, A_1, \ldots, A_p)$ *satisfying (7.54) such that* $A_i \in C_i$ $(0 \leq i \leq p)$ *if and only if* $\alpha \in \Sigma_\lambda$. *Moreover, if* α *is a real root, the solution* $(A_0, A_1, \ldots, A_p)$ *is unique up to conjugation, and if* α *is an imaginary root, there exist infinitely many mutually non-conjugate solutions.*

This theorem is proved by reducing to the existence of representations of quivers. We do not go into the proof. Please refer to [29, 30].

We shall reexamine this beautiful theorem from our viewpoint. In Theorem 7.25 we see that, for a tuple $(C_0, C_1, \ldots, C_p)$ of conjugacy classes, λ represents the eigenvalues and α represents the spectral types. The relation $\lambda \cdot \alpha = 0$ among λ

and α represents the Fuchs relation (7.17), which is necessary for the existence of a solution satisfying (7.54). Suppose that a solution exists. Then the result of an addition to the solution is also irreducible and is a solution for the tuple of conjugacy classes obtained by shifting each conjugacy class by a scalar. Thus we understand that the essential part of the problem depends not on the eigenvalues but on the spectral types. Hence we look at α.

We examine how the spectral type of $(C_0, C_1, \ldots, C_p)$ is related to α. First we note that

$$r_{i,1} = \mathrm{rank}(A_i - \xi_{i,1}) = n - \dim \mathrm{Ker}(A_i - \xi_{i,1}) = n - e_1(A_i; \xi_{i,1}).$$

The integer $r_{i,2}$ is obtained from $r_{i,1}$ by subtracting the difference of the rank of $(A_i - \xi_{i,1})$ and the rank of $(A_i - \xi_{i,1})(A_i - \xi_{i,2})$. If $\xi_{i,2} = \xi_{i,1}$, thanks to (7.56) the difference is equal to $e_2(A_i; \xi_{i,1})$, and if $\xi_{i,2} \neq \xi_{i,1}$, the difference is equal to $\dim \mathrm{Ker}(A_i - \xi_{i,2}) = e_1(A_i; \xi_{i,2})$. In a similar way, we see that the integer $r_{i,j}$ is obtained from $r_{i,j-1}$ by subtracting $e_k(A_i; \xi_{i,j})$, where k is such a number that $k-1$ is the number of $\xi_{i,j}$ appearing in $\xi_{i,1}, \ldots, \xi_{i,j-1}$.

Then we examine how the action of the Weyl group changes α. First we consider the reflection s_v with respect to the vertex $v = [i, j]$. Set $s_v(\alpha) = \beta$. By the definition (7.57), we see that $\alpha, \beta \in \mathbb{Z}^I$ coincide except the $[i, j]$-entry, which is given by

$$\beta_{[i,j]} = \alpha_{[i,j]} - (-\alpha_{[i,j-1]} + 2\alpha_{[i,j]} - \alpha_{[i,j+1]})$$

$$= \alpha_{[i,j-1]} - (\alpha_{[i,j]} - \alpha_{[i,j+1]})$$

$$= r_{i,j-1} - (r_{i,j} - r_{i,j+1}).$$

By noting $r_{i,j} = r_{i,j-1} - (r_{i,j-1} - r_{i,j})$, we understand that the reflection $s_v = s_{[i,j]}$ permutes the order of $\xi_{i,j}$ and $\xi_{i,j+1}$. Next we consider the reflection s_0 with respect to the vertex 0. We also set $s_0(\alpha) = \beta$. By the definition, we see that only the 0-entry is changed, and is given by

$$\beta_0 = \alpha_0 - \left(2\alpha_0 - \sum_{i=0}^{p} \alpha_{[i,1]}\right)$$

$$= \sum_{i=0}^{p} \alpha_{[i,1]} - \alpha_0$$

$$= \sum_{i=0}^{p} r_{i,1} - n$$

$$= \sum_{i=0}^{p} (n - e_1(A_i; \xi_{i,1})) - n$$

$$= pn - \sum_{i=0}^{p} e_1(A_i; \xi_{i,1}).$$

Set $\beta_0 = n'$. Since $e_1(A_i; \xi_{i,1}) = \dim \mathrm{Ker}(A_i - \xi_{i,1})$, we notice that, by (7.42), n' looks similar to the rank after the middle convolution. To establish the similarity explicitly, we operate the addition. Choose any i_0 in $\{0, 1, \ldots, p\}$, and take a transformation

$$A_i \mapsto A_i - \xi_{i,1} = A_i'$$

for $i \neq i_0$. Then A_{i_0} is transformed by

$$A_{i_0} \mapsto A_{i_0} + \sum_{i \neq i_0} \xi_{i,1} = A_{i_0}'.$$

If we set

$$\mu = \sum_{i=0}^{p} \xi_{i,1},$$

we have

$$\mathrm{Ker}(A_{i_0} - \xi_{i_0,1}) = \mathrm{Ker}(A_{i_0}' - \mu).$$

Then we operate the middle convolution with parameter μ. Since

$$e_1(A_i; \xi_{i,1}) = \dim \mathrm{Ker} A_i' \ (i \neq i_0), \quad e_1(A_{i_0}; \xi_{i_0,1}) = \dim \mathrm{Ker}(A_{i_0}' - \mu),$$

we see that the result

$$(B_0, B_1, \ldots, B_p) = mc_\mu(A_0', A_1', \ldots, A_p')$$

becomes of rank n'. We shall show that the entry $\beta_{[i,j]} = \alpha_{[i,j]} = r_{i,j}$ corresponding to a vertex $v = [i, j]$ comes from $(B_0, B_1, \ldots, B_p)$. For the spectral types of $(A_0', A_1', \ldots, A_p')$ and $(B_0, B_1, \ldots, B_p)$, we have relations corresponding to (7.43) and (7.44). For $i \neq i_0$, we have

$$e_1(B_i; 0) = n' - \sum_{j \geq 2} e_j(B_i; 0) - \sum_{\eta \neq 0} \sum_{j \geq 1} e_j(B_i; \eta)$$

$$= n' - \sum_{j \geq 1} e_j(A_i'; -\mu) - \sum_{j \geq 2} e_j(A_i'; 0) - \sum_{\xi \neq 0, -\mu} \sum_{j \geq 1} e_j(A_i'; \xi)$$

$$= n' - \sum_{j \geq 2} e_j(A_i'; 0) - \sum_{\xi \neq 0} \sum_{j \geq 1} e_j(A_i'; \xi)$$

$$= n' - (n - e_1(A_i'; 0)),$$

which shows

$$\beta_{[i,1]} = \alpha_{[i,1]} = n - e_1(A_i'; 0) = n' - e_1(B_i; 0).$$

Thus $\beta_{[i,1]}$ can be regarded as a value coming from B_i. For $j \geq 2$, we see that $\beta_{[i,j]}$ is obtained from $\beta_{[i,j-1]}$ by subtracting some $e_k(A_i'; \xi)$, and that any $e_k(A_i'; \xi)$ other than $e_1(A_i'; 0)$ coincides with some $e_{k'}(B_i; \eta)$. Then also in this case $\beta_{[i,j]}$ can be regarded as a value coming from B_i. Also for $i = i_0$, we have

$$\beta_{[i_0,1]} = \alpha_{[i_0,1]} = n - e_1(A_{i_0}'; \mu) = n' - e_1(B_{i_0}; -\mu),$$

and hence $\beta_{[i_0,j]}$ can be regarded as a value coming from B_{i_0}.

Therefore we obtain the following result.

Theorem 7.26 *The action of the Weyl group has the following meanings. The reflection $s_{[i,j]}$ with respect to a vertex $[i, j]$ coincides with the permutation of $\xi_{i,j}$ and $\xi_{i,j+1}$. The reflection s_0 with respect to the vertex 0 coincides with the composition of the operations to a tuple $(A_0, A_1, \ldots, A_p)$ of matrices, that are the addition*

$$A_i \mapsto A_i - \xi_{i,1} \quad (i \neq i_0)$$

and the middle convolution with parameter

$$\mu = \sum_{i=0}^{p} \xi_{i,1},$$

where i_0 can be chosen arbitrarily from $\{0, 1, \ldots, p\}$.

Now we consider the meaning of $q(\alpha)$ which is defined by (7.58) and appears in Σ_λ.

Proposition 7.27 *We have*

$$^t\alpha C\alpha = \iota,$$

where ι is the index of rigidity.

Proof Since $\alpha_{[i,j]}$ is obtained from n by subtracting j $e_k(A_i; \xi_{i,l})$'s, we may set

$$\alpha_{[i,j]} = n - e_{i,1} - e_{i,2} - \cdots - e_{i,j},$$

where $e_{i,1}, e_{i,2} \ldots$ is a sequence of elements in $\{e_k(A_i; \xi_{i,l}); \ k \geq 1, l \geq 1\}$ in some order. Then the assertion is shown as follows:

$$
{}^t\alpha C\alpha = \sum_{u,v \in I} \alpha_u \alpha_v {}^t\epsilon_u C\epsilon_v
$$

$$
= 2{\alpha_0}^2 - 2\sum_i \alpha_0 \alpha_{[i,1]} + 2\sum_{i,j} {\alpha_{[i,j]}}^2 - 2\sum_{i,j} \alpha_{[i,j]}\alpha_{[i,j+1]}
$$

$$
= 2\alpha_0\left(\alpha_0 - \sum_i \alpha_{[i,1]}\right) + 2\sum_{i,j} \alpha_{[i,j]}(\alpha_{[i,j]} - \alpha_{[i,j+1]})
$$

$$
= 2n\left(n - \sum_i (n - e_{i,1})\right) + 2\sum_{i,j}(n - e_{i,1} - \cdots - e_{i,j})e_{i,j+1}
$$

$$
= 2\left(-pn^2 + n\sum_i e_{i,1}\right) + 2n\sum_{i,j} e_{i,j+1} - 2\sum_{i,j}\sum_{k=1}^{j} e_{i,k}e_{i,j+1}
$$

$$
= -2pn^2 + 2n\sum_{i,j} e_{i,j} - 2\sum_i \sum_{k<l} e_{i,k}e_{i,l}
$$

$$
= -2pn^2 + 2(p+1)n^2 + \sum_{i=0}^{p}\left(\sum_j {e_{i,j}}^2 - \left(\sum_j e_{i,j}\right)^2\right)
$$

$$
= 2n^2 + \sum_{i=0}^{p}\sum_{j\geq 1} {e_{i,j}}^2 - (p+1)n^2
$$

$$
= (1-p)n^2 + \sum_{i=0}^{p}\sum_{j\geq 1} {e_{i,j}}^2
$$

$$
= \iota.
$$

□

By this proposition, we get

$$
q(\alpha) = 1 - \frac{1}{2}\iota = \frac{1}{2}(2 - \iota),
$$

where $2 - \iota$ in the right hand side is the number of accessory parameters, or the dimension of the moduli space. It is known that the number of accessory parameters for a scalar Fuchsian differential equation is a half of the number for the corresponding Fuchsian system of normal form. Thus $q(\alpha)$ represents the number of accessory parameters for a scalar Fuchsian differential equation.

We extract from Theorem 7.25 the assertion for spectral types. In other words, we are going to describe the condition on such a spectral type that there exists an irreducible tuple $(A_0, A_1, \ldots, A_p)$ of the spectral type with appropriate sets of eigenvalues.

For the purpose, we slightly change the definition of the above Kac-Moody root system. Since we will start from $\alpha \in \mathbb{Z}^I$, we cannot give d_i's in advance. Then we set

$$I_d = \{0\} \cup \{[i, j]; \, 0 \le i \le p, \, 1 \le j \le d\}$$

and

$$I = \bigcup_{d \ge 1} I_d.$$

For this new I, we can define the generalized Cartan matrix C, the fundamental region B, reflections s_v and the root system in exactly the same manner. We can also define I so that it does not depend on p, however, here we understand that p is an arbitrarily fixed non-negative integer.

If $\alpha \in \mathbb{Z}^I$ comes from a tuple of conjugacy classes, we have

$$\alpha_0 \ge \alpha_{[i,1]} \ge \alpha_{[i,2]} \ge \cdots \quad (0 \le i \le p)$$

by (7.56). Then we denote by $\mathbb{Z}_o^I$ the subset of the elements in $\mathbb{Z}^I$ satisfying this condition.

Definition 7.7 We call the spectral type $(A_0^{\natural}, A_1^{\natural}, \ldots, A_p^{\natural})$ of any irreducible tuple $(A_0, A_1, \ldots, A_p)$ of matrices satisfying (7.54) *irreducibly realizable*. We also call $\alpha \in \mathbb{Z}_o^I$ obtained from this tuple *irreducibly realizable*.

Theorem 7.28 $\alpha \in \mathbb{Z}_o^I$ *is irreducibly realizable if and only if, $q(\alpha) \ne 1$ and α is a positive root, or $q(\alpha) = 1$, α is a positive root and the greatest common divisor of all entries of α is 1. In this case, α is a real root if and only if the corresponding irreducible tuple of matrices is rigid.*

Proof By Theorem 7.25, α is irreducibly realizable if and only if there exists $\lambda \in \mathbb{C}^I$ such that $\alpha \in \Sigma_\lambda$. Assume that, for $\alpha \in \mathbb{Z}_o^I \cap \Delta^+$, there exists a decomposition

$$\alpha = \beta^{(1)} + \beta^{(2)} + \cdots + \beta^{(k)}$$

into a sum of positive roots. If the conditions

$$\lambda \cdot \beta^{(i)} = 0 \quad (1 \le i \le k) \tag{7.59}$$

on λ are not induced from the condition

$$\lambda \cdot \alpha = 0, \tag{7.60}$$

there exists λ which satisfies (7.60) and does not satisfy (7.59). For such λ, we have $\alpha \in \Sigma_\lambda$, and hence α is irreducibly realizable. Otherwise, for each i there exists $c_i \in \mathbb{C}$ such that $\beta^{(i)} = c_i \alpha$. Then we have

$$\alpha = \beta^{(1)} + \beta^{(2)} + \cdots + \beta^{(k)} = m\beta$$

for some integer m greater than 1. We consider this case. Take λ satisfying (7.60). By the definition (7.58), we have

$$q(\alpha) - mq(\beta) = (m - 1)(mq(\beta) - (m + 1)). \tag{7.61}$$

If $q(\beta) = 0$, we have $q(\alpha) < 0$, which is impossible. If $q(\beta) = 1$, the right hand side of (7.61) becomes negative, which implies $\alpha \notin \Sigma_\lambda$. Note that $q(\beta) = 1$ if and only if $q(\alpha) = 1$. If $q(\beta) > 1$, the right hand side of (7.61) becomes positive, which shows $\alpha \in \Sigma_\lambda$. Summing up the above, we see that, when $q(\alpha) \neq 1$, α is irreducibly realizable if and only if it is a positive root, and when $q(\alpha) = 1$, α is irreducibly realizable if and only if it is a positive root and $\alpha = m\beta$ does not hold for any $m \geq 2$ and for any β.   □

We have characterized the irreducibly realizable spectral types in terms of root systems. By using this characterization, we have an algorithm for the criterion of the irreducible realizability.

Consider a spectral type

$$(e^{(0)}, e^{(1)}, \ldots, e^{(p)}), \quad e^{(i)} = ((e_j^{(i;1)})_{j\geq 1}, (e_j^{(i;2)})_{j\geq 1}, \ldots) \ (0 \leq i \leq p), \tag{7.62}$$

where each $e^{(i)}$ gives a spectral type of one matrix, and satisfies

$$\left|e^{(i)}\right| = \sum_k \sum_{j\geq 1} e_j^{(i;k)} = n.$$

Step I Compute the index of rigidity

$$\iota = (1 - p)n^2 + \sum_{i=0}^p \sum_k \sum_{j\geq 1} (e_j^{(i;k)})^2.$$

If $\iota \geq 4$, the spectral type cannot be irreducible even if it is realizable by Theorem 7.11. If $\iota = 0$ and if the greatest common divisor of all $e_j^{(i;k)}$ is greater than 1, the spectral type is not irreducibly realizable by Theorem 7.28. Otherwise, go to the next step.

Step II For each i, choose k such that $e_1^{(i;k)}$ is the maximum among $e_1^{(i;j)}$. By permuting the indices if necessary, we may assume that $k = 1$. Compute

$$d = \sum_{i=0}^{p} e_1^{(i;1)} - (p-1)n.$$

If

$$d \le 0,$$

the spectral type is irreducibly realizable (by the reason which will be explained later). Suppose that $d > 0$. If there exists i such that

$$e_1^{(i;1)} < d,$$

the spectral type is not irreducibly realizable. Otherwise, we have

$$e_1^{(i;1)} \ge d > 0 \quad (0 \le i \le p).$$

Then go to the next step.

Step III Define a new spectral type $(e^{(0)'}, e^{(1)'}, \ldots, e^{(p)'})$ by

$$e^{(i)'} = ((e_1^{(i;1)} - d), (e_2^{(i;1)}, e_3^{(i;1)}, \ldots), (e_j^{(i;2)})_{j \ge 1}, (e_j^{(i;3)})_{j \ge 1}, \ldots) \quad (0 \le i \le p).$$

Then we have

$$n' = \left| e^{(i)'} \right| = n - d \quad (0 \le i \le p).$$

If $n' = 1$, this new spectral type, and hence the original spectral type (7.62), is irreducibly realizable and rigid. If $n' > 1$, go to Step II with this new spectral type.

In Step II, the conclusion is irreducibly realizable, not irreducibly realizable, or $d > 0$. In the last case $d > 0$, we go to Step III, and have lower rank spectral type because $n' = n - d$. Thus this algorithm stops in a finite time.

Now we explain the reason why the spectral type is irreducibly realizable if $d \le 0$ in Step II. We look at the relation between this algorithm and the Kac-Moody root systems. Suppose that $n > 1$, and consider $\alpha \in \mathbb{Z}^I$ which corresponds to the spectral type (7.62). Then we may set

$$\alpha_0 = n, \quad \alpha_{[i,j]} = n - e_{i,1} - e_{i,2} - \cdots - e_{i,j},$$

where $e_{i,1}, e_{i,2}, \ldots$ is a permutation of $e_j^{(i;k)}$ $(k = 1, 2, \ldots; j \geq 1)$ under the condition that $e_{j_2}^{(i;k)}$ appears later than $e_{j_1}^{(i;k)}$ if $j_1 < j_2$. By operating reflections $s_{[i,j]}$ if necessary, we may assume

$$e_{i,1} \geq e_{i,2} \geq \cdots \quad (0 \leq i \leq p). \tag{7.63}$$

In particular, we have

$$e_{i,1} = e_1^{(i;1)}.$$

Then we have

$$^t\epsilon_0 C\alpha = 2\alpha_0 - \sum_{i=0}^{p} \alpha_{[i,1]}$$

$$= 2n - \sum_{i=0}^{p}(n - e_{i,1})$$

$$= \sum_{i=0}^{p} e_1^{(i;1)} - (p-1)n$$

$$= d,$$

and

$$^t\epsilon_{[i,1]}C\alpha = -\alpha_0 + 2\alpha_{[i,1]} - \alpha_{[i,2]} = e_{i,2} - e_{i,1},$$

$$^t\epsilon_{[i,j]}C\alpha = -\alpha_{[i,j-1]} + 2\alpha_{[i,j]} - \alpha_{[i,j+1]} = e_{i,j+1} - e_{i,j} \quad (j > 1).$$

Thus, if $d \leq 0$, taking (7.63) into account, we get

$$^t\epsilon_v C\alpha \leq 0 \quad (v \in I),$$

which implies that α belongs to the fundamental region B. Namely α is a root. Therefore, thanks to Theorem 7.28, α is irreducibly realizable, and hence so is the spectral type (7.62).

Here we note that the operation to obtain the spectral type $(e^{(0)'}, e^{(1)'}, \ldots, e^{(p)'})$ from the spectral type (7.62) in Step III is the reflection s_0 with respect to the vertex 0, which can be regarded as a composition of an addition and an middle convolution as is explained in Theorem 7.26. The Katz theorem (Theorem 7.22) corresponds to the case that this algorithm stops by $n = 1$ in Step III.

Example 7.7

(i) The spectral type $((111), (111), (111))$ in Example 7.5 (i) is irreducibly realizable because $d = 0$.

(ii) The spectral type $((21), (111), (111))$ in Example 7.5 (ii) has $d = 1$, and hence can be sent to a new spectral type $((11), (11), (11))$. The new spectral type also has $d = 1$, and can be sent to $((1), (1), (1))$, which is irreducibly realizable and rigid because the rank is 1.

(iii) Consider the spectral type $(((21)1), ((21)1), ((111)1))$, whose index of rigidity is 0. We have $d = 5 - 4 = 1$, and hence it is sent to

$$((111), (111), (0(11)1)) = ((111), (111), ((11)1)).$$

For this new spectral type, we have $d = 0$, and hence it is irreducibly realizable. We realize this operation of the spectral types as a transformation for tuples of matrices of Jordan canonical forms:

$$\left(\begin{pmatrix} \lambda_1 & 1 & & \\ & \lambda_1 & & \\ & & \lambda_1 & \\ & & & \lambda_2 \end{pmatrix}, \begin{pmatrix} \mu_1 & 1 & & \\ & \mu_1 & & \\ & & \mu_1 & \\ & & & \mu_2 \end{pmatrix}, \begin{pmatrix} \nu_1 & 1 & & \\ & \nu_1 & 1 & \\ & & \nu_1 & \\ & & & \nu_2 \end{pmatrix} \right)$$

$$\rightarrow \left(\begin{pmatrix} \lambda_1' & & \\ & \lambda_2' & \\ & & \lambda_3' \end{pmatrix}, \begin{pmatrix} \mu_1' & & \\ & \mu_2' & \\ & & \mu_3' \end{pmatrix}, \begin{pmatrix} \nu_1' & 1 & \\ & \nu_1' & \\ & & \nu_2' \end{pmatrix} \right).$$

Example 7.8 In Example 7.6, we gave a list of irreducible rigid semi-simple spectral types of rank 2, 3, 4. We can check that these spectral types are irreducibly realizable by applying the above algorithm. We have one more spectral type of rank 4 with the index of rigidity $\iota = 2$:

$$((31), (31), (31), (1111)).$$

For this spectral type, we have $d = 3 + 3 + 3 + 1 - 2 \times 4 = 2$, and $e_1 = 1 < 2 = d$ for (1111). Then it is not irreducibly realizable.

We have studied the problem of the existence of a Fuchsian system of normal form with a prescribed spectral type. We can also consider the problem of the existence of a scalar Fuchsian differential equation with a prescribe spectral type. The last problem is formulated and solved by Oshima. He has a similar result as Theorem 7.28, however, the proof is completely independent from one by Crawley-Boevey.

Oshima called the spectral types with $d \leq 0$ *basic*, and showed that the number of the basic spectral types with any fixed index of rigidity is finite. We can regard a basic spectral type as a representative of smallest rank in a equivalence class under additions and middle convolutions. The basic spectral types of index of rigidity 0 are listed by Kostov [115]; there are four basic spectral types

$$((11), (11), (11), (11)), \ ((111), (111), (111)),$$
$$((22), (1111), (1111)), \ ((33), (222), (111111)).$$

(7.64)

We give only semi-simple ones in the list, however, by putting inner parentheses we get any non-semi-simple basic spectral type. For example, from the second spectral type in the list, we obtain non-semi-simple ones

$$(((11)1), ((11)1), ((11)1)), \ (((11)1), (111), (111)), \ (((111)), ((11)1), (111)).$$

The basic spectral types of index of rigidity -2 are listed by Oshima; there are 13 basic spectral types

$$((11), (11), (11), (11), (11)), \ ((21), (21), (111), (111)),$$
$$((22), (22), (22), (211)), \ ((31), (22), (22), (1111)),$$
$$((211), (1111), (1111)), \ ((32), (11111), (11111)),$$
$$((221), (221), (11111)), \ ((33), (2211), (111111)),$$
$$((222), (222), (2211)), \ ((44), (2222), (22211)),$$
$$((44), (332), (11111111)), \ ((55), (3331), (22222)),$$
$$((66), (444), (2222211)).$$

(7.65)

Also for this list, we can get any basic non-semi-simple spectral type. The basic spectral types of index of rigidity -4 are also obtained by Oshima. Please refer to [138, 139] for details.

Chapter 8
Deformation Theory

8.1 Isomonodromic Deformation

We consider a system of Fuchsian ordinary differential equations

$$\frac{dY}{dx} = A(x)Y, \qquad (8.1)$$

where $A(x)$ is an $n \times n$-matrix with rational functions in x as entries. Without loss of generality, we may assume the point ∞ is one of the regular singular points of (8.1). Let $t_0 = \infty, t_1, t_2, \ldots, t_p$ be the set of the regular singular points, and set

$$t = (t_1, t_2, \ldots, t_p).$$

The parameters $t_1, t_2, \ldots, t_p$ appear in $A(x)$ as positions of the singular points. We assume that the coefficients of the rational functions of the entries of $A(x)$ also depend analytically on t. Then we use $A(x, t)$ instead of $A(x)$, and write the system (8.1) as

$$\frac{dY}{dx} = A(x, t)Y. \qquad (8.2)$$

Now we set

$$\mathbb{P}^1 \setminus \{t_0, t_1, \ldots, t_p\} = D_t,$$

and take a point $x_0 \in D_t$. For each j $(1 \leq j \leq p)$, we take a $(+1)$-loop Γ_j in D_t for t_j with base point x_0. As we noted in Chap. 5, the homotopy classes $[\Gamma_j]$ generate the fundamental group $\pi_1(D_t, x_0)$:

$$\pi_1(D_t, x_0) = \langle [\Gamma_1], [\Gamma_2], \dots, [\Gamma_p] \rangle.$$

In this chapter, however, we do not take the homotopy class, but consider the loop Γ_j itself.

Let t be fixed. Then for each fundamental system of solutions $\mathcal{Y}(x, t)$ of (8.2), the circuit matrices M_j $(1 \leq j \leq p)$ with respect to $\mathcal{Y}(x, t)$ are determined by

$$\Gamma_{j_*} \mathcal{Y}(x, t) = \mathcal{Y}(x, t) M_j.$$

Since M_j's are determined for each fixed t, we understand that they depend on t. Now we vary t in a small neighborhood, while Γ_j's are fixed. We take the neighborhood of t so small that Γ_j's are a generators of the fundamental group $\pi_1(D_t, x_0)$ for any t in the neighborhood (Fig. 8.1). Then we come to the following definition.

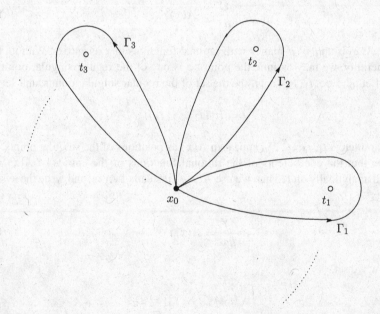

Fig. 8.1 $(+1)$-loops Γ_j

Definition 8.1 A fundamental system of solutions $\mathcal{Y}(x, t)$ is said to be *isomonodromic* if the circuit matrices M_j ($1 \leq j \leq p$) with respect to the fundamental system $\mathcal{Y}(x, t)$ do not vary by a variation of t in a small neighborhood of the original position.

If t appears only as positions of the singular points in $A(x, t)$, the system (8.2) is rigorously determined by fixing each t, and hence the monodromy representation is definitely determined. Then there may be no possibility of existing isomonodromic fundamental systems of solutions. If the coefficients of the entries of $A(x, t)$ also depend on t in a suitable way, we may have a chance to get an isomonodromic fundamental systems of solutions. This is our problem.

Theorem 8.1 *The Fuchsian system of differential equations (8.2) has an isomonodromic fundamental system of solutions analytic with respect to t in a small neighborhood (as specified before) if and only if there exist $n \times n$-matrix functions $B_j(x, t)$ ($1 \leq j \leq p$) rational in x such that the system of partial differential equations*

$$
\begin{cases}
\dfrac{\partial Y}{\partial x} = A(x, t)Y, \\[2mm]
\dfrac{\partial Y}{\partial t_j} = B_j(x, t)Y & (1 \leq j \leq p)
\end{cases}
\tag{8.3}
$$

becomes completely integrable.

The system (8.3) is called completely integrable if all the partial differential equations are compatible and if the dimension of the space of solutions is finite. We shall discuss on completely integrable systems in detail in Part II of this book.

Proof Assume that there exists an isomonodromic fundamental system of solutions $\mathcal{Y}(x, t)$ of (8.2). Then the circuit matrix M_k determined by

$$
\Gamma_{k*}\mathcal{Y}(x, t) = \mathcal{Y}(x, t)M_k
$$

does not depend on $t = (t_1, t_2, \ldots, t_p)$. Taking the partial derivative of both sides with respect to t_j, we get

$$
\Gamma_{k*}\frac{\partial \mathcal{Y}}{\partial t_j}(x, t) = \frac{\partial}{\partial t_j}\left(\Gamma_{k*}\mathcal{Y}(x, t)\right)
$$

$$
= \frac{\partial}{\partial t_j}\left(\mathcal{Y}(x, t)M_k\right)
$$

$$
= \frac{\partial \mathcal{Y}}{\partial t_j}(x, t)M_k.
$$

If we set

$$B_j(x, t) = \frac{\partial \mathcal{Y}}{\partial t_j}(x, t)\mathcal{Y}(x, t)^{-1} \tag{8.4}$$

for each j, we have

$$\Gamma_{k*}B_j(x, t) = \frac{\partial \mathcal{Y}}{\partial t_j}(x, t)M_k\Big(\mathcal{Y}(x, t)M_k\Big)^{-1}$$

$$= B_j(x, t).$$

This implies that $B_j(x, t)$ is single-valued on D_t as a function in x. Since $B_j(x, t)$ is obtained from a solutions of the Fuchsian ordinary differential equations, the singular points as a function in x are $x = t_0, t_1, \ldots, t_p$ all of which are at most regular singular points. Then, thanks to Theorem 4.1, the singular points are at most poles, and hence $B_j(x, t)$ is a rational function in x. The t_j-equation in the system (8.3) of partial differential equations is satisfied by $\mathcal{Y}(x, t)$, which is just the definition of $B_j(x, t)$. Then the system (8.3) has a fundamental system of solutions $\mathcal{Y}(x, t)$. It follows that the system (8.3) is completely integrable.

Conversely, we assume that the system (8.3) of partial differential equations is completely integrable. Then, as we shall show later in Part II (Theorem 11.1), there exists a fundamental system of solutions $\mathcal{Y}(x, t)$ of (8.3). Define the matrix M_k by

$$\Gamma_{k*}\mathcal{Y}(x, t) = \mathcal{Y}(x, t)M_k.$$

Differentiating the left hand side with respect to t_j, we get

$$\frac{\partial}{\partial t_j}\Big(\Gamma_{k*}\mathcal{Y}(x, t)\Big) = \Gamma_{k*}\frac{\partial \mathcal{Y}}{\partial t_j}(x, t)$$

$$= \Gamma_{k*}\Big(B_j(x, t)\mathcal{Y}(x, t)\Big)$$

$$= B_j(x, t)\mathcal{Y}(x, t)M_k,$$

and differentiation of the right hand side with respect to t_j yields

$$\frac{\partial}{\partial t_j}\Big(\mathcal{Y}(x, t)M_k\Big) = B_j(x, t)\mathcal{Y}(x, t)M_k + \mathcal{Y}(x, t)\frac{\partial M_k}{\partial t_j}.$$

Thus, noting that $\mathcal{Y}(x, t)$ is invertible, we have

$$\frac{\partial M_k}{\partial t_j} = O.$$

This implies that the fundamental system of solutions $\mathcal{Y}(x,t)$ of (8.3) is isomonodromic. □

Definition 8.2 The *isomonodromic deformation* of the system (8.2) is to determine $A(x,t)$ as a function in t so that the system (8.2) has a isomonodromic fundamental system of solutions.

By Theorem 8.1, we see that the isomonodromic deformation is a problem to determine $A(x,t)$ so that the ordinary differential equation (8.2) is compatible with the partial differential equations with respect to the positions of the singular points t. Thus the isomonodromic deformation is sometimes said to be a holonomic deformation, because completely integrable systems are called holonomic systems.

8.2 Deformation of Fuchsian Systems of Normal Form

We consider the isomonodromic deformation of Fuchsian systems of normal form

$$\frac{dY}{dx} = \left(\sum_{j=1}^{p} \frac{A_j}{x - t_j} \right) Y, \tag{8.5}$$

where A_j $(1 \leq j \leq p)$ is constant with respect to x and depends on $t = (t_1, t_2, \ldots, t_p)$. Set $t_0 = \infty$, and set

$$A_0 = - \sum_{j=1}^{p} A_j.$$

The problem is to determine A_j as a function in t so that the system (8.5) has an isomonodromic fundamental system of solutions.

In the following of this chapter, we assume that $A_0, A_1, \ldots, A_p$ are non-resonant. We first note the following fact.

Theorem 8.2 *Assume that $A_0, A_1, \ldots, A_p$ are non-resonant. Then, if the system (8.5) of differential equations has an isomonodromic fundamental system of solutions, the conjugacy classes of $A_0, A_1, \ldots, A_p$ do not depend on t.*

Proof Let M_j be the circuit matrix with respect to the isomonodromic fundamental system of solutions. Since it does not depend on t, so is its conjugacy class. If A_j is non-resonant, we have

$$M_j \sim e^{2\pi \sqrt{-1} A_j},$$

and hence the conjugacy class of $e^{2\pi \sqrt{-1} A_j}$ does not depend on t. Since A_j depends continuously on t, the conclusion follows. □

The following theorem and corollary are the main result in this section.

Theorem 8.3

(i) *Consider the Fuchsian system of normal form (8.5). We assume that A_j $(0 \le j \le p)$ are non-resonant. Then the system (8.5) has an isomonodromic fundamental system of solutions if and only if there exists an invertible matrix Q which is constant with respect to x such that, by setting*

$$QA_jQ^{-1} = A'_j \quad (1 \le j \le p), \tag{8.6}$$

the system of partial differential equations

$$
\begin{cases}
\dfrac{\partial Z}{\partial x} = \left(\displaystyle\sum_{j=1}^{p} \dfrac{A'_j}{x - t_j} \right) Z, \\[4ex]
\dfrac{\partial Z}{\partial t_j} = -\dfrac{A'_j}{x - t_j} Z \quad (1 \le j \le p)
\end{cases}
\tag{8.7}
$$

becomes completely integrable.

(ii) *The system (8.7) of partial differential equations is completely integrable if and only if*

$$
\begin{cases}
\dfrac{\partial A'_i}{\partial t_i} = -\displaystyle\sum_{\substack{j=1 \\ j \ne i}}^{p} \dfrac{[A'_i, A'_j]}{t_i - t_j} \quad (1 \le i \le p), \\[4ex]
\dfrac{\partial A'_i}{\partial t_j} = \dfrac{[A'_i, A'_j]}{t_i - t_j} \quad (1 \le i, j \le p, \ j \ne i)
\end{cases}
\tag{8.8}
$$

hold.

Remark 8.1

(i) As we have noted in the previous section, the complete integrability will be treated in Part II. Please have a look if you want to know in more detail.

(ii) The above system (8.7) of partial differential equations and its integrability condition (8.8) can be written in terms of exterior derivative as follows:

$$dZ = \left(\sum_{j=1}^{p} A'_j d \log(x - t_j) \right) Z, \tag{8.9}$$

$$dA'_i = -\sum_{j \ne i} [A'_i, A'_j] d \log(t_i - t_j) \quad (1 \le i \le p). \tag{8.10}$$

These are called Pfaffian systems or total differential equations. The system (8.9) is a linear Pfaffian system, and the system (8.10) a non-linear Pfaffian system.

Proof We readily obtain the integrability condition (8.8) of the system (8.7) by the argument in Chap. 11 in Part II. Then we are going to show the assertion (i).

As we have shown in Theorem 8.1, an isomonodromic fundamental system of solutions exists if and only if there exist the matrices $B_j(x, t)$ such that the system (8.3) becomes completely integrable. In our case of Fuchsian systems of normal form, the matrices $B_j(x, t)$ can be obtained explicitly in the following manner.

Let $\mathcal{Y}(x, t)$ be an isomonodromic fundamental system of solutions of (8.5), and define $B_j(x, t)$ for each j ($1 \leq j \leq p$) by (8.4). By the definition, $B_j(x, t)$ is a rational function in x, and the positions of the poles are contained in $\{t_0, t_1, \ldots, t_p\}$. We shall compute the Laurent series expansion at $x = t_k$.

First we study the Laurent series expansion of $B_j(x, t)$ at $x = t_j$. By Corollary 4.6, we see that there exists a fundamental system of solutions

$$\mathcal{Y}_j(x, t) = F_j(x, t)(x - t_j)^{\Lambda_j}$$

at $x = t_j$, where Λ_j is the Jordan canonical form of A_j and $F_j(x, t)$ is a convergent power series of the form

$$F_j(x, t) = P_j + \sum_{m=1}^{\infty} F_{jm}(t)(x - t_j)^m$$

with an invertible matrix P_j satisfying

$$P_j^{-1} A_j P_j = \Lambda_j.$$

Thanks to Theorem 8.2, the matrix Λ_j does not depend on t. The isomonodromic solution $\mathcal{Y}(x, t)$ is connected to this solution $\mathcal{Y}_j(x, t)$ as

$$\mathcal{Y}(x, t) = \mathcal{Y}_j(x, t) C_j,$$

where C_j is the connection matrix. Put this into the definition (8.4) of $B_j(x, t)$, and we have

$$B_j(x, t) = \frac{\partial}{\partial t_j} \left(F_j(x - t_j)^{\Lambda_j} C_j \right) \left(F_j(x - t_j)^{\Lambda_j} C_j \right)^{-1}$$

$$= \left(\frac{\partial F_j}{\partial t_j}(x - t_j)^{\Lambda_j} C_j - F_j \Lambda_j (x - t_j)^{\Lambda_j - 1} C_j + F_j(x - t_j)^{\Lambda_j} \frac{\partial C_j}{\partial t_j} \right)$$

$$\times C_j^{-1}(x - t_j)^{-\Lambda_j} F_j^{-1}$$

$$= \frac{\partial F_j}{\partial t_j} F_j^{-1} - F_j \Lambda_j (x - t_j)^{-1} F_j^{-1} + F_j(x - t_j)^{\Lambda_j} \frac{\partial C_j}{\partial t_j} C_j^{-1}(x - t_j)^{-\Lambda_j} F_j^{-1}.$$

Note that $B_j(x, t)$ is single valued around $x = t_j$, and so are the first and the second terms in the last hand side of the above equation. Thus the third term should be single valued around $x = t_j$. We consider the third term in detail. We decompose Λ_j into the direct sum

$$\Lambda_j = \begin{pmatrix} \Lambda_{j1} & & \\ & \ddots & \\ & & \Lambda_{jq} \end{pmatrix},$$

where Λ_{jk} has only one eigenvalue λ_{jk} such that $\lambda_{jk} \neq \lambda_{jl}$ for $k \neq l$. We partition $\partial C_j / \partial t_j \cdot C_j^{-1}$ into the same block sizes as Λ_j:

$$\frac{\partial C_j}{\partial t_j} C_j^{-1} = \begin{pmatrix} C_{11} & \cdots & C_{1q} \\ \vdots & & \vdots \\ C_{q1} & \cdots & C_{qq} \end{pmatrix},$$

here C_{kl} has the same number of rows as Λ_{jk} and the same number of columns as Λ_{jl}. Then the (k, l)-block of $(x - t_j)^{\Lambda_j} \partial C_j / \partial t_j \cdot C_j^{-1}(x - t_j)^{-\Lambda_j}$ is $(x - t_j)^{\Lambda_{jk}} C_{kl}(x - t_j)^{-\Lambda_{jl}}$, which has a factor $(x - t_j)^{\lambda_{jk} - \lambda_{jl}}$. If $k \neq l$, by the non-resonance we have $\lambda_{jk} - \lambda_{jl} \notin \mathbb{Z}$, and hence the factor is not single-valued around $x = t_j$. Thus we should have $C_{kl} = O$ for $k \neq l$. Now we look at a diagonal block C_{kk}. If Λ_{jk} has a Jordan cell of size greater than 1, $(x - t_j)^{\Lambda_{jk}}$ has an entry with the term $\log(x - t_j)$. Then by the single-valuedness, C_{kk} should be commutative with Λ_j, so that the factor $(x - t_j)^{\Lambda_{jk}}$ is cancelled by its inverse. Otherwise, there is no logarithmic term. Thus, in any case, the third term of the right hand side of the above equality becomes holomorphic at $x = t_j$. Hence the principal part of the Laurent series expansion of $B_j(x, t)$ at $x = t_j$ appears only in the second term. The second term can be written as

$$- F_j \Lambda_j (x - t_j)^{-1} F_j^{-1}$$
$$= -(P_j + F_{j1}(x - t_j) + \cdots) \Lambda_j (x - t_j)^{-1} (P_j + F_{j1}(x - t_j) + \cdots)^{-1}$$
$$= -P_j \Lambda_j P_j^{-1} (x - t_j)^{-1} + \cdots$$
$$= -\frac{A_j}{x - t_j} + \cdots,$$

and then the principal part is just $-A_j / (x - t_j)$.

The Laurent series expansion of $B_j(x, t)$ at $x = t_k$ ($k \neq j, 0$) can be computed in a similar way by using the fundamental system of solutions

$$\mathcal{Y}_k(x, t) = F_k(x, t)(x - t_k)^{\Lambda_k}$$

at $x = t_k$, and, since we have

$$\frac{\partial}{\partial t_j}(x - t_k)^{\Lambda_k} = O,$$

there appear no principal part. Thus $B_j(x, t)$ is holomorphic at $x = t_k$. In a similar way, we can show that $B_j(x, t)$ is also holomorphic at $x = \infty$. In conclusion, we have

$$B_j(x, t) = -\frac{A_j}{x - t_j} + B_j^0(t),\tag{8.11}$$

where $B_j^0(t)$ is constant in x^1.

Next we are going to delete the term $B_j^0(t)$ by some gauge transformation. Since $\mathcal{Y}(x, t)$ is isomonodromic, for any invertible matrix $Q(t)$, the circuit matrices for

$$\mathcal{Z}(x, t) = Q(t)\mathcal{Y}(x, t)$$

do not depend on t. Namely $\mathcal{Z}(x, t)$ is isomonodromic. The system of differential equations satisfied by $\mathcal{Z}(x, t)$ is just

$$\frac{\partial \mathcal{Z}}{\partial x} = \left(\sum_{j=1}^{p}\frac{A'_j}{x - t_j}\right)\mathcal{Z},$$

where we set $QA_j Q^{-1} = A'_j$. For each j, we set

$$B'_j(x, t) = \frac{\partial \mathcal{Z}}{\partial t_j}\mathcal{Z}^{-1}.$$

Then by using (8.11), we get

$$B'_j(x, t) = \frac{\partial Q}{\partial t_j}Q^{-1} + Q\left(-\frac{A_j}{x - t_j} + B_j^0\right)Q^{-1}.$$

Now we want to take Q so that

$$\frac{\partial Q}{\partial t_j}Q^{-1} + QB_j^0 Q^{-1} = O$$

---

[1]Bolobruch [24, 25] called the Pfaffian systems with coefficients (8.11) *unnormalized Schlesinger deformations*.

holds. Namely we want to get a solution $Q(t)$ of the system

$$\frac{\partial Q}{\partial t_j} = -Q B_j^0(t) \quad (1 \le j \le p) \tag{8.12}$$

of partial differential equations. The system (8.12) is completely integrable by the following reason. We rewrite (8.11) by using the definition of B_j as

$$\frac{\partial \mathcal{Y}}{\partial t_j} \mathcal{Y}^{-1} = -\frac{A_j}{x - t_j} + B_j^0,$$

and differentiate the both sides with respect to t_k to get

$$\frac{\partial^2 \mathcal{Y}}{\partial t_k \partial t_j} \mathcal{Y}^{-1} - \frac{\partial \mathcal{Y}}{\partial t_j} \mathcal{Y}^{-1} \frac{\partial \mathcal{Y}}{\partial t_k} \mathcal{Y}^{-1} = -\frac{\frac{\partial A_j}{\partial t_k}}{x - t_j} + \frac{\partial B_j^0}{\partial t_k}.$$

The second term of the left hand side is $-B_j(x,t)B_k(x,t)$. We also have the equation where j and k are interchanged. Since the first term in the left hand side is symmetric in j and k, we get

$$B_j(x,t)B_k(x,t) - \frac{\frac{\partial A_j}{\partial t_k}}{x - t_j} + \frac{\partial B_j^0}{\partial t_k} = B_k(x,t)B_j(x,t) - \frac{\frac{\partial A_k}{\partial t_j}}{x - t_k} + \frac{\partial B_k^0}{\partial t_j}.$$

By using (8.11), we can rewrite this equation as

$$\left(-\frac{A_j}{x - t_j} + B_j^0\right)\left(-\frac{A_k}{x - t_k} + B_k^0\right) - \frac{\frac{\partial A_j}{\partial t_k}}{x - t_j} + \frac{\partial B_j^0}{\partial t_k}$$

$$= \left(-\frac{A_k}{x - t_k} + B_k^0\right)\left(-\frac{A_j}{x - t_j} + B_j^0\right) - \frac{\frac{\partial A_k}{\partial t_j}}{x - t_k} + \frac{\partial B_k^0}{\partial t_j}.$$

Now taking the limit $x \to \infty$ in both sides, we obtain

$$B_j^0 B_k^0 + \frac{\partial B_j^0}{\partial t_k} = B_k^0 B_j^0 + \frac{\partial B_k^0}{\partial t_j},$$

which is the completely integrability condition for the system (8.12). Thus the system (8.12) is completely integrable, and then we can take its fundamental matrix of solutions Q. Define $\mathcal{Z}$ by using this Q, then we obtain

$$B_j'(x,t) = -\frac{A_j'}{x - t_j}$$

for $1 \leq j \leq p$. This shows that, if there exists an isomonodromic fundamental system of solutions for (8.5), the system (8.7) of partial differential equations is completely integrable.

Conversely, if, for some invertible matrix Q, $A'_1, A'_2, \ldots, A'_p$ satisfy the system (8.8), where we set $A'_j = Q A_j Q^{-1}$ $(1 \leq j \leq p)$, then the system (8.7) of partial differential equations becomes completely integrable, and hence a fundamental system of solutions $\mathcal{Z}(x, t)$ exists. Set

$$\mathcal{Y}(x, t) = Q^{-1} \mathcal{Z}(x, t).$$

Then $\mathcal{Y}(x, t)$ becomes a fundamental system of solutions for (8.5), and satisfies the completely integrable system obtained from (8.7) by the transformation. This shows that $\mathcal{Y}(x, t)$ is isomonodromic. □

Corollary 8.4 *The local solution*

$$\mathcal{Y}(x, t) = F_0(x, t) x^{-\Lambda_0}$$

of the system of differential equations (8.5) at infinity is isomonodromic, if and only if the system of partial differential equations

$$\begin{cases} \dfrac{\partial Y}{\partial x} = \left(\displaystyle\sum_{j=1}^{p} \dfrac{A_j}{x - t_j} \right) Y, \\[4mm] \dfrac{\partial Y}{\partial t_j} = -\dfrac{A_j}{x - t_j} Y \quad (1 \leq j \leq p) \end{cases}$$

is completely integrable. The completely integrability condition is given by

$$\begin{cases} \dfrac{\partial A_i}{\partial t_i} = -\displaystyle\sum_{\substack{j=1 \\ j \neq i}}^{p} \dfrac{[A_i, A_j]}{t_i - t_j} \quad (1 \leq i \leq p), \\[4mm] \dfrac{\partial A_i}{\partial t_j} = \dfrac{[A_i, A_j]}{t_i - t_j} \quad (1 \leq i, j \leq p, \ j \neq i). \end{cases} \tag{8.13}$$

Proof Construct $B_j(x, t)$ by using this $\mathcal{Y}(x, t)$. Then we see $B_j^0(x, t) = O$ for any j. □

Definition 8.3 The system of differential equations (8.13), which is obtained by Schlesinger [157], is called the *Schlesinger system*.

The Schlesinger system is a system of algebraic differential equations of the tuple $(A_1, A_2, \ldots, A_p)$ of the residue matrices, however, for the tuple, a very transcendental condition that the local fundamental system of solutions at $x = \infty$ is isomonodromic is assumed. On the other hand, for the system (8.8) in Theorem 8.3, no such transcendental condition is assumed, while the tuple $(A_1, A_2, \ldots, A_p)$ itself is not the unknown. Then we want to derive a system of differential equations such that $(A_1, A_2, \ldots, A_p)$ is the unknown and that no transcendental extra condition is assumed.

First we note that, for any solution $(A'_1, A'_2, \ldots, A'_p)$ of the system (8.8), the conjugacy class of each A'_j is invariant thanks to Theorem 8.2. Then we may regard the system (8.8) as a system on the moduli space $\mathcal{M}'$ which is introduced in Sect. 7.4.2. Moreover, since the accessory parameters give a coordinate of $\mathcal{M}'$, they can be regarded as the unknowns of the system (8.8). We call the system (8.8) the *deformation equation* when we regard it as a system of differential equations for the accessory parameters. To write down the deformation equation by taking accessory parameters explicitly is an important problem. We shall discuss this problem later in Sect. 8.4.

The idea to rewrite the system (8.8) is to use the new unknown functions

$$\mathrm{tr}(A_i A_j) \quad (i \neq j).$$

If $(A_1, \ldots, A_p)$ and $(A'_1, \ldots, A'_p)$ are related as (8.6) by some $Q \in \mathrm{GL}(n; \mathbb{C})$, then we have

$$\mathrm{tr}(A_i A_j) = \mathrm{tr}(A'_i A'_j),$$

so that we get the same equation for these new unknowns if we start from (8.13) instead of (8.8). The idea of using these unknowns seems to come from the following fact.

Theorem 8.5 (Procesi [143]) *Let K be a field of characteristic 0. We define the action of $\mathrm{GL}(n, K)$ on the space $\mathrm{End}(K^n)^p$ by*

$$Q \cdot (A_1, A_2, \ldots, A_p) = (Q A_1 Q^{-1}, Q A_2 Q^{-1}, \ldots, Q A_p Q^{-1})$$

$$((A_1, A_2, \ldots, A_p) \in \mathrm{End}(K^n)^p, \ Q \in \mathrm{GL}(n, K)).$$

Then the ring of the invariant polynomials with respect to this action is generated by

$$\mathrm{tr}(A_{i_1} A_{i_2} \cdots A_{i_k}) \quad (1 \leq k \leq 2^n - 1)$$

over K.

Now we are going to derive a system of differential equations for these new unknowns from the Schlesinger system (8.13). By a simple calculation, we obtain

$$
\begin{cases}
\dfrac{\partial}{\partial t_i} \operatorname{tr}(A_i A_j) = -\sum_{k \neq i,j} \dfrac{\operatorname{tr}([A_i, A_k]A_j)}{t_i - t_k}, \\[2ex]
\dfrac{\partial}{\partial t_i} \operatorname{tr}(A_j A_k) = \dfrac{\operatorname{tr}([A_i, A_j]A_k)}{t_i - t_j} + \dfrac{\operatorname{tr}(A_j[A_i, A_k])}{t_i - t_k},
\end{cases}
\tag{8.14}
$$

where i, j, k are mutually distinct. If the new functions $\operatorname{tr}([A_i, A_j]A_k)$ etc. are expressed as functions of $\operatorname{tr}(A_i A_j)$ $(i \neq j)$, the system (8.14) is a closed system for the unknowns $\operatorname{tr}(A_i A_j)$ $(i \neq j)$. Otherwise, we differentiate these new functions and use (8.13) to get

$$
\begin{cases}
\dfrac{\partial}{\partial t_i} \operatorname{tr}([A_i, A_j]A_k) = -\sum_{l \neq i} \dfrac{\operatorname{tr}([[A_i, A_l], A_j]A_k)}{t_i - t_l} \\[2ex]
\qquad + \dfrac{\operatorname{tr}([A_i, [A_i, A_j]]A_k)}{t_i - t_j} + \dfrac{\operatorname{tr}([A_i, A_j][A_i, A_k])}{t_i - t_k}, \\[2ex]
\dfrac{\partial}{\partial t_i} \operatorname{tr}([A_j, A_k]A_l) = \dfrac{\operatorname{tr}([[A_i, A_j], A_k]A_l)}{t_i - t_j} + \dfrac{\operatorname{tr}([A_j, [A_i, A_k]]A_l)}{t_i - t_k} \\[2ex]
\qquad + \dfrac{\operatorname{tr}([A_j, A_k][A_i, A_l])}{t_i - t_l}.
\end{cases}
\tag{8.15}
$$

If the new functions $\operatorname{tr}([[A_i, A_j], A_k]A_l)$ etc. are not expressed as functions of already appeared unknowns, we continue a similar process. Thus we obtain a system (H) of infinitely many equations for infinitely many unknowns from the Schlesinger system (8.13). However, we know that the deformation equation is an equation for the accessory parameters whose number is bounded, and hence these infinitely many unknowns should be functionally dependent. Thus the system (H) is equivalent to a system of finitely many differential equations for finitely many unknowns. We call the system (H) the *Hitchin system*. Hitchin [72] studied the system (H) when the system (8.14) becomes a closed system.

Thus we obtained the following assertion.

Theorem 8.6 *The non-resonant normalized Fuchsian system (8.5) has an isomonodromic fundamental system of solutions if and only if the conjugacy classes of the tuple of the residue matrices $(A_0, A_1, \ldots, A_p)$ do not depend on t and the tuple becomes a solution of the Hitchin system (H).*

8.3   Deformation and Middle Convolution

Deformation of a non-resonant normalized Fuchsian system is described as a system of differential equations on the moduli space $\mathcal{M}'$ to which the equivalence class of the tuple $(A_0, A_1, \ldots, A_p)$ of the residue matrices belong. On the other hand, the addition and the middle convolution introduced in Sect. 7.4 can be naturally regarded as operations between the moduli spaces. We noted that these two operations do not change the index of rigidity, and hence the dimension of the moduli spaces. This implies that they do not change the number of the unknowns of the deformation equation. Then it is natural to ask whether these two operations leave the deformation equation itself invariant. The answer is affirmative, which we shall show by using the Hitchin system [58].

Theorem 8.7 *Additions and middle convolutions with parameters independent of t leave the Hitchin system invariant.*

Proof By the process of deriving the Hitchin system from the Schlesinger system, we see that the unknowns of the Hitchin system consist the set $\{\operatorname{tr} U \; ; \; U \in \mathcal{U}\}$, where the infinite set $\mathcal{U}$ is obtained from the set of the products $A_i A_j$ of distinct residue matrices by recursive replacements of each matrix A_i by some $[A_k, A_l]$. Then, by this construction, any element of $\mathcal{U}$ has one of the forms

$$A_i A_j, \quad [U_1, U_2] A_i, \quad [U_1, U_2][U_3, U_4] \quad (U_1, U_2, U_3, U_4 \in \mathcal{U}). \tag{8.16}$$

First we show that the Hitchin system is invariant by addition. Let $\alpha = (\alpha_1, \alpha_2, \ldots, \alpha_p)$ be a tuple of constants independent of t, and consider the addition

$$ad_\alpha(A_1, A_2, \ldots, A_p) \mapsto (A_1 + \alpha_1, A_2 + \alpha_2, \ldots, A_p + \alpha_p). \tag{8.17}$$

We look at the change of the traces of the elements in (8.16). First we have

$$\operatorname{tr}(A_i A_j) \mapsto \operatorname{tr}((A_i + \alpha_i)(A_j + \alpha_j))$$
$$= \operatorname{tr}(A_i A_j) + \alpha_i \operatorname{tr} A_j + \alpha_j \operatorname{tr} A_i + \alpha_i \alpha_j.$$

Since $(A_1, A_2, \ldots, A_p)$ is a solution of the Hitchin system, the conjugacy class of each A_i is independent of t, and so is the trace. Then we have

$$\frac{\partial}{\partial t_k} \operatorname{tr}((A_i + \alpha_i)(A_j + \alpha_j)) = \frac{\partial}{\partial t_k} \operatorname{tr}(A_i A_j).$$

This shows the invariance, because $\operatorname{tr}(A_i A_j)$ appears in the Hitchin system only in the left hand side. For the other elements in (8.16), in U_1, U_2 the matrix A_j appears always in the form $[A_j, A_k]$, and then U_1, U_2 are not changed by the addition (8.17).

Then we have

$$\mathrm{tr}([U_1, U_2]A_i) \mapsto \mathrm{tr}([U_1, U_2](A_i + \alpha_i)$$
$$= \mathrm{tr}([U_1, U_2]A_i) + \alpha_i \mathrm{tr}[U_1, U_2]$$
$$= \mathrm{tr}([U_1, U_2]A_i),$$

and hence we obtain the invariance. Similar holds for $\mathrm{tr}([U_1, U_2][U_3, U_4])$, and then we have proved the invariance of the Hitchin system by the addition (8.17).

Next we show that the Hitchin system is invariant by the middle convolution mc_λ. We use the notations (7.30), (7.31), (7.32) for the middle convolution. First we consider the change of unknowns of the Hitchin system by the operation

$$(A_1, A_2, \ldots, A_p) \mapsto (G_1, G_2, \ldots, G_p). \tag{8.18}$$

Note that the unknowns of the Hitchin system are the traces of sums or differences of the products

$$A_i A_j A_k \cdots A_l.$$

If we replace the product by $G_i G_j G_k \cdots G_l$, we see that there appears non-zero block row only in the i-th position, in which the diagonal (i, i)-block is

$$\tilde{A}_j \tilde{A}_k \cdots \tilde{A}_l \tilde{A}_i,$$

where $\tilde{A}_i$ denotes A_i or $A_i + \lambda$. Then we have

$$\mathrm{tr}(G_i G_j G_k \cdots G_l) = \mathrm{tr}(\tilde{A}_j \tilde{A}_k \cdots \tilde{A}_l \tilde{A}_i) = \mathrm{tr}(\tilde{A}_i \tilde{A}_j \tilde{A}_k \cdots \tilde{A}_l).$$

In a similar way in the proof for addition, we can replace $\tilde{A}_i$ in the right hand side by A_i. This proves that the Hitchin system is invariant under the operation (8.18). Next we consider the change by the operation

$$(G_1, G_2, \ldots, G_p) \mapsto (B_1, B_2, \ldots, B_p). \tag{8.19}$$

Take a basis $u_1, u_2, \ldots, u_m$ of $\mathcal{K} + \mathcal{L}$, and then take $v_{m+1} \ldots, v_{pn}$ so that the union makes a basis of V^p. Set

$$Q = (u_1, u_2, \ldots, u_m, v_{m+1}, \ldots, v_{pn}).$$

Then we have

$$Q^{-1} G_i Q = \begin{pmatrix} C_i & * \\ O & B_i \end{pmatrix}$$

for $1 \leq i \leq p$, so that

$$\mathrm{tr}(G_i G_j G_k \cdots G_l) = \mathrm{tr}(B_i B_j B_k \cdots B_l) + \mathrm{tr}(C_i C_j C_k \cdots C_l)$$

holds. We can take the basis of $\mathcal{K} + \mathcal{L}$ in the following way. If $u_j \in \mathcal{K}$, there exists l such that

$$u_j = \begin{pmatrix} 0 \\ \vdots \\ 0 \\ w \\ 0 \\ \vdots \\ 0 \end{pmatrix} \quad (l, \quad w \in \mathrm{Ker} A_l.$$

Then we have

$$G_i u_j = \delta_{il} \lambda u_j.$$

If $u_j \in \mathcal{L}$, we always have

$$G_i u_j = 0.$$

These imply that every C_i is a diagonal matrix, whose diagonal entries are λ in the j-th position for j with $u_j \in \mathrm{Ker} A_i$ and 0 for the other positions. In particular, we have $C_i C_j = O$ if $i \neq j$. Then, if there are distinct indices among $i, j, k, \ldots, l$, which is always the case for the unknowns of the Hitchin system, we have

$$C_i C_j C_k \cdots C_l = O.$$

Thus we get

$$\mathrm{tr}(G_i G_j G_k \cdots G_l) = \mathrm{tr}(B_i B_j B_k \cdots B_l),$$

which shows that the Hitchin system is invariant by the operation (8.19). This completes the proof. □

8.4 Painlevé Equation and Garnier System

Thanks to the result of the previous section, we see that, in order to obtain all deformation equations of Fuchsian systems, it is enough to obtain the deformation equations of Fuchsian systems of basic spectral types. For the rank $n = 2$ case, we consider the spectral type

$$((11), (11), \ldots, (11)) \qquad\qquad (8.20)$$

for $p + 1$ points. For this spectral type, we have

$$d = (p + 1) \times 1 - (p - 1) \times 2 = 3 - p,$$

so that it is basic if $p \geq 3$. In this section we derive the deformation equation for this basic spectral type. When $p = 3$, the deformation equation is called the *Painlevé equation*, and when $p \geq 4$, it is called the *Garnier system*.

In order to derive the deformation equation, we need to choose the accessory parameters as a coordinate of the moduli space. Since the index of rigidity is $\iota = 6 - 2p$, the number of the accessory parameters is $2 - \iota = 2p - 4$. There may be several ways to choose the accessory parameters. In this section we use the following lemma from linear algebra, which will be also useful for other cases.

Lemma 8.8 *Let A, B be a generic pair of $n \times n$-matrices. The definition of generic will be explained in the proof. Then there exists $P \in \mathrm{GL}(n, \mathbb{C})$ such that $P^{-1}AP$ is upper triangular and $P^{-1}BP$ lower triangular.*

Proof Let $u_1, u_2, \ldots, u_n$ be a set of linearly independent generalized eigenvectors of A satisfying

$$Au_i \in \langle u_1, \ldots, u_i \rangle \quad (1 \leq i \leq n),$$

and $v_1, v_2, \ldots, v_n$ a set of linearly independent generalized eigenvectors of B satisfying

$$Bv_i \in \langle v_i, \ldots, v_n \rangle \quad (1 \leq i \leq n).$$

In these basis, A (resp. B) becomes upper (resp. lower) triangular Jordan canonical form. For each i, we denote by β_i the eigenvalue of the generalized eigenspace of B containing v_i. We assume that $u_1, \ldots, u_{n-1}, v_n$ are linearly independent.

We shall prove that, for $i = 2, 3, \ldots, n - 1$, there exists u_i' such that

$$u_i' \in \langle u_1, \ldots, u_i \rangle,$$
$$v_i \in \langle u_i', \ldots, u_{n-1}', v_n \rangle,$$
$$Bu_i' \in \beta_i u_i' + \langle u_{i+1}', \ldots, u_{n-1}', v_n \rangle.$$

Assume that we have obtained $u'_{i+1}, \ldots, u'_{n-1}$, and that $u_1, \ldots, u_i, u'_{i+1}, \ldots,$ u'_{n-1}, v_n are linearly independent. Then there exist $c_1, c_2, \ldots, c_n \in \mathbb{C}$ such that

$$v_i = c_1 u_1 + \cdots + c_i u_i + c_{i+1} u'_{i+1} + \cdots + c_{n-1} u'_{n-1} + c_n v_n.$$

By setting

$$u'_i = c_1 u_1 + \cdots + c_i u_i,$$

we get

$$
\begin{aligned}
Bu'_i &= B(v_i - c_{i+1} u'_{i+1} - \cdots - c_{n-1} u'_{n-1} - c_n v_n) \\
&\in \beta_i v_i + \langle v_{i+1}, \ldots, v_n \rangle + \langle u'_{i+1}, \ldots, u'_{n-1}, v_n \rangle \\
&= \beta_i (u'_i + \langle u'_{i+1}, \ldots, u'_{n-1}, v_n \rangle) + \langle u'_{i+1}, \ldots, u'_{n-1}, v_n \rangle \\
&= \beta_i u'_i + \langle u'_{i+1}, \ldots, u'_{n-1}, v_n \rangle,
\end{aligned}
$$

which shows that this u'_i satisfies the condition. We assume that $u_1, u'_2, \ldots, u'_{n-1}, v_n$ thus obtained are linearly independent. Then we see that

$$Au'_i \in \langle u_1, u'_2, \ldots, u'_i \rangle, \quad Bu'_i \in \langle u'_i, \ldots, u'_{n-1}, v_n \rangle$$

hold. Hence by setting

$$P = (u_1, u'_2, \ldots, u'_{n-1}, v_n),$$

the similar transformation by P sends A to an upper triangle matrix and B a lower triangle matrix.

In the above, we have assumed several linear independence of tuples of vectors. We call A, B a generic pair if these assumptions hold.   □

Consider the tuple

$$(A_0, A_1, \ldots, A_p), \quad \sum_{i=0}^{p} A_i = O$$

of matrices of spectral type (8.20). By using Lemma 8.8, we may assume that A_0 is upper triangular and A_1 lower triangular. By operating addition, we can send one of the eigenvalues of A_i $(1 \le i \le p)$ to 0. We denote the remaining eigenvalue by θ_i. Thus we have

$$A_0 = \begin{pmatrix} \kappa_1 & a \\ 0 & \kappa_2 \end{pmatrix}, \quad A_1 = \begin{pmatrix} \theta_1 & 0 \\ q_1 & 0 \end{pmatrix}, \quad A_i \sim \begin{pmatrix} \theta_i & \\ & 0 \end{pmatrix} \quad (1 \le i \le p),$$

where κ_1, κ_2 are the eigenvalues of A_0 and satisfy

$$\kappa_1 + \kappa_2 + \sum_{i=1}^{p} \theta_i = 0.$$

By using Lemma 7.10, we can parametrize $A_2, \ldots, A_p$ as

$$A_i = \begin{pmatrix} \theta_i - q_i\, p_i & (\theta_i - q_i\, p_i)p_i \\ q_i & q_i\, p_i \end{pmatrix} \quad (2 \le i \le p). \tag{8.21}$$

By a similar transformation with a diagonal matrix, we can send p_2 to 1, and then get

$$A_2 = \begin{pmatrix} \theta_2 - q_2 & \theta_2 - q_2 \\ q_2 & q_2 \end{pmatrix}.$$

From the condition $\sum_{i=0}^{p} A_i = O$ we obtain

$$\begin{cases} \kappa_1 + \theta_1 + (\theta_2 - q_2) + \displaystyle\sum_{i=3}^{p}(\theta_i - q_i\, p_i) = 0, \\[2mm] a + (\theta_2 - q_2) + \displaystyle\sum_{i=3}^{p}(\theta_i - q_i\, p_i)p_i = 0, \\[2mm] q_1 + \displaystyle\sum_{i=2}^{p} q_i = 0. \end{cases}$$

Then a, q_1, q_2 can be expressed as polynomials in q_i, p_i $(3 \le i \le p)$. In particular, we have

$$q_1 = \kappa_2 + \sum_{i=3}^{p} q_i(p_i - 1), \quad q_2 = -\kappa_2 - \sum_{i=3}^{p} q_i\, p_i. \tag{8.22}$$

Thus we can choose q_i, p_i $(3 \le i \le p)$ as $2(p-2)$ accessory parameters.

The deformation equation is obtained by putting $A_1, \ldots, A_p$, which are parametrized by the accessory parameters, into the Schlesinger system (8.13).

Proposition 8.9 *The deformation equation for the spectral type (8.20) is given by the following system of differential equations:*

$$
\left\{
\begin{aligned}
\frac{\partial q_i}{\partial t_i} &= -\frac{1}{t_i - t_1}(\theta_1 q_i - \theta_i q_1 + 2q_1 q_i p_i) \\
&\quad - \frac{1}{t_i - t_2}(\theta_2 q_i - \theta_i q_2 + 2q_2 q_i (p_i - 1)) \\
&\quad - \sum_{\substack{j=3 \\ j \neq i}}^{p} \frac{1}{t_i - t_j}(\theta_j q_i - \theta_i q_j + 2q_j q_i (p_i - p_j)), \\
\frac{\partial q_i}{\partial t_j} &= \frac{1}{t_i - t_j}(\theta_j q_i - \theta_i q_j + 2q_j q_i (p_i - p_j)), \\
\frac{\partial p_i}{\partial t_i} &= \frac{1}{t_i - t_1}(\theta_1 + q_1 p_i)p_i + \frac{1}{t_i - t_2}(\theta_2 + q_2(p_i - 1))(p_i - 1) \\
&\quad + \sum_{\substack{j=3 \\ j \neq i}}^{p} \frac{1}{t_i - t_j}(\theta_j + q_j(p_i - p_j))(p_i - p_j), \\
\frac{\partial p_i}{\partial t_j} &= -\frac{1}{t_i - t_j}(\theta_j + q_j(p_i - p_j))(p_i - p_j),
\end{aligned}
\right.
\tag{8.23}
$$

$(3 \leq i, j \leq p, i \neq j)$, *where* q_1, q_2 *are given by (8.22).*

The system (8.23) is one of the expressions of the Garnier system.

Remark 8.2 By a Möbius transformation, which is an automorphism of $\mathbb{P}^1$, we can send any three points of $\mathbb{P}^1$ to any prescribed three points. Then we may assume that, among the singular points of (8.5), t_1, t_2 are fixed besides $t_0 = \infty$. Usually we normalize $t_1 = 0, t_2 = 1$. Then in the deformation equation (8.23), we understand that t_1, t_2 are fixed, and that the remaining points $t_3, \ldots, t_p$ are the independent variables.

Corollary 8.10 *The deformation equation for the spectral type* $((11), (11),$ $(11), (11))$ *is given by the following system of differential equations:*

$$
\left\{
\begin{aligned}
\frac{dq}{dt} &= -\frac{1}{t}(\theta_1 q - \theta_3(\kappa_2 + q(p - 1)) + 2(\kappa_2 + q(p - 1))qp) \\
&\quad - \frac{1}{t - 1}(\theta_2 q + \theta_3(\kappa_2 + qp) - 2(\kappa_2 + qp)q(p - 1)), \\
\frac{dp}{dt} &= \frac{1}{t}(\theta_1 + (\kappa_2 + q(p - 1))p)p \\
&\quad + \frac{1}{t - 1}(\theta_2 - (\kappa_2 + qp)(p - 1))(p - 1).
\end{aligned}
\right.
\tag{8.24}
$$

This assertion is a particular case of Proposition 8.9 where $p = 3$ and $t_1 = 0, t_2 = 1$. We we set $t_3 = t, q_3 = q, p_3 = p$. (Note that the last p denotes the unknown function $p(t)$, which is different from the number p of the singular points in finite place.) The deformation equation (8.24) is one of the expressions of the Painlevé equation. Precisely speaking, there are six Painlevé equations, and the deformation equation (8.24) corresponds to the sixth Painlevé equation. The first to the fifth Painlevé equations are obtained as deformation equations of linear ordinary differential equations having irregular singular points.

Here we note that, since we have obtained the Garnier system by putting $(A_1, \ldots, A_p)$ into the Schlesinger system (8.13), for the solution of the Garnier system the local solution at $x = \infty$ of the Fuchsian system (8.5) gives isomonodromic fundamental system of solutions.

P. Painlevé tried to find new special functions defined by non-linear ordinary differential equations. He formulated the problem as an analysis of behaviors of solutions, and then he and his collaborators succeeded to find six Painlevé equations. Several years later, Painlevé equations were rediscovered as results of isomonodromic deformations [38, 83, 84, 86]. More recently, it has been shown that the Painlevé equations have remarkable properties from the viewpoint of the space of initial values and symmetry [129, 131, 132, 148]. These properties can be shown also from the viewpoint of the moduli spaces of connections and monodromy representations [78]. It is just a few years ago that the τ function of the sixth Painlevé equation is shown to coincide with some conformal block in the conformal field theory [17, 79, 80]. This coincidence suggests a new development of the study of the Painlevé equations. For the τ function, please refer to [130]. The Garnier system is obtained as a natural extension of the Painlevé equations from the viewpoint of isomonodromic deformations, and there are many analogous studies [82, 107]. We have, in this section, introduced these systems of equations in terms of basic spectral types.

As we noted in the beginning of this section, thanks to Theorem 8.7, it is enough to obtain the deformation equations for the basic spectral types. Here we note that, in order to obtain a deformation equation, we need to have at least four singular points in the linear differential equation. Then we are interested in the basic spectral types with at least four singular points. For the case of the index of rigidity 0, there is only one basic spectral type

$$((11), (11), (11), (11))$$

with at least four singular points. This corresponds to the Painlevé equation. This implies that the Painlevé equation has a canonical meaning among the deformation equations. Next we consider the basic spectral types with the index of rigidity -2 and with at least four singular points. We find that there are four such spectral types

$$((11), (11), (11), (11), (11)), \quad ((21), (21), (111), (111)),$$

$$((22), (22), (22), (211)), \quad ((31), (22), (22), (1111)).$$

The first one corresponds to the Garnier system. The deformation equations for the other three spectral types were obtained by Sasano [153], Fuji-Suzuki [39] and Sakai [149] in the form of Hamiltonian systems. For the Hamiltonian structure of deformation equations, we shall explain in the next section. Suzuki [170] obtained a similar result for the index of rigidity -4. Thus the viewpoint of spectral types works also in deformation theory.

8.5   Hamiltonian Structure of Deformation Equations

Explicit forms of deformation equations depend on the choice of the accessory parameters. It is important to choose accessory parameters so that the deformation equation takes some good form. For example, if we can take accessory parameters so that any entry of each matrix A_i is rational in the parameters, the deformation equation becomes rational in unknowns, which makes the analysis simpler. It is also important to choose accessory parameters so as to be compatible with some natural structure, if any.

It has been known that several deformation equations including Painlevé equations can be written in Hamiltonian systems. Jimbo-Miwa-Môri-Sato [85] generalized these facts, and showed that there exists a Hamiltonian structure for deformation equations in general. A Hamiltonian structure seems to be substantial for the study of deeper structure of deformation equations [66]. In this section we explain the result in [85] for the deformation of Fuchsian ordinary differential equations, and consider the relation to the choice of the accessory parameters.

In the following, we shall show that the system (8.8) describing isomonodromic deformation can be written in a Hamiltonian system. Note that, as described in Corollary 8.4, the system (8.8) reduces to the Schlesinger system (8.13) if there exists an isomonodromic fundamental system of solutions satisfying some condition. In deriving a Hamiltonian system, we start from the Schlesinger system (8.13) for a moment. As we noted in Remark 8.1, the Schlesinger system (8.13) is the integrability condition of the Pfaffian system

$$dY = \left(\sum_{j=1}^{p} A_j d\log(s - t_j) \right) Y, \tag{8.25}$$

and is given by the non-linear Pfaffian system

$$dA_i = - \sum_{j \neq i} [A_i, A_j] d\log(t_i - t_j) \quad (1 \leq i \leq p). \tag{8.26}$$

We denote the Jordan canonical form of A_i by Λ_i. Thanks to Theorem 8.2, Λ_i is a constant matrix which is also independent of $t_1, t_2, \ldots, t_p$. We further assume, for a moment, that the local fundamental system of solutions

$$\mathcal{Y}_i(x) = F_i(x)(x - t_i)^{\Lambda_i}, \quad F_i(x) = \sum_{m=0}^{\infty} F_{im}(x - t_i)^m$$

at $x = t_i$ satisfies the linear Pfaffian system (8.25). Set $F_{i0} = Q_i$. Then, as is shown in Corollary 4.6, Q_i is an invertible matrix satisfying

$$Q_i^{-1} A_i Q_i = \Lambda_i.$$

We set $P_i = Q_i^{-1} A_i$, so that we have

$$Q_i P_i = A_i, \quad P_i Q_i = \Lambda_i.$$

Put $\mathcal{Y}_i(x)$ into (8.25) to obtain

$$(dF_i + F_i \Lambda_i d\log(x - t_i))(x - t_i)^{\Lambda_i} = \left(\sum_{j=1}^{p} A_j d\log(x - t_j) \right) F_i(x - t_i)^{\Lambda_i}.$$

We can remove $(x - t_i)^{\Lambda_i}$ in both sides. Then we see that the coefficients of $(x - t_i)^{-1}$ in both sides are equal by $Q_i \Lambda_i = A_i Q_i$. Remove the terms in both sides, and put $x = t_i$ to obtain

$$dQ_i = \left(\sum_{j \neq i} A_j d\log(t_i - t_j) \right) Q_i.$$

On the other hand, by using the equation $d(P_i Q_i) = d\Lambda_i = O$, we get

$$dP_i = -P_i dQ_i \cdot Q_i^{-1}$$

$$= -P_i \left(\sum_{j \neq i} A_j d\log(t_i - t_j) \right).$$

Put $A_j = Q_j P_j$ into the above. Then we obtain the Pfaffian system

$$\begin{cases} dQ_i = \left(\displaystyle\sum_{j \neq i} Q_j P_j d\log(t_i - t_j) \right) Q_i, \\[4mm] dP_i = -P_i \left(\displaystyle\sum_{j \neq i} Q_j P_j d\log(t_i - t_j) \right) \end{cases} \quad (1 \leq i \leq p) \qquad (8.27)$$

with unknowns Q_i, P_i $(1 \leq i \leq p)$. In the above, we have assumed that $\mathcal{Y}_i(x)$ is isomonodromic. However, we can derive from (8.27) and $A_i = Q_i P_i$ the Schlesinger Pfaffian system (8.26) without using the assumption. Thus we can remove the assumption, and hence the Pfaffian system (8.27) describes the isomonodromic deformation.

Next we shall show that the Pfaffian system (8.27) is a Hamiltonian system. A bilinear form $\{\cdot, \cdot\}$ is called *Poisson bracket* if it satisfies

$$\{g, f\} = -\{f, g\},$$
$$\{fg, h\} = \{f, h\}g + f\{g, h\}, \quad \{f, gh\} = \{f, g\}h + g\{f, h\},$$
$$\{f, \{g, h\}\} + \{g, \{h, f\}\} + \{h, \{f, g\}\} = 0.$$

We denote the entries of the matrices Q_i, P_i $(1 \leq i \leq p)$ as

$$Q_i = (q_{iuv})_{1 \leq u,v \leq n}, \quad P_i = (p_{iuv})_{1 \leq u,v \leq n},$$

and define a Poisson bracket for functions in these entries by

$$\{q_{iuv}, p_{ivu}\} = 1,$$

for the other combinations $= 0$.

In other words, we define the bracket by

$$\{f, g\} = \sum_{i=1}^{p} \sum_{u,v=1}^{n} \left(\frac{\partial f}{\partial q_{iuv}} \frac{\partial g}{\partial p_{ivu}} - \frac{\partial f}{\partial p_{ivu}} \frac{\partial g}{\partial q_{iuv}} \right). \tag{8.28}$$

For a matrix function $F = (f_{uv})_{u,v}$ and a scalar function g, we generalize the Poisson bracket by

$$\{F, g\} = (\{f_{uv}, g\})_{u,v}.$$

We are going to rewrite the Pfaffian system (8.27) by using this Poisson bracket.

We write the first equation in the Pfaffian system (8.27) entry-wise as

$$dq_{iuv} = \sum_{j \neq i} \left(\sum_{r=1}^{n} (Q_j P_j)_{ur} q_{irv} \right) d\log(t_i - t_j).$$

Noting the equality

$$q_{irv} = -\left\{ \sum_{s=1}^{n} q_{irs} p_{isu}, q_{iuv} \right\} = \{(Q_i P_i)_{ru}, q_{iuv}\},$$

we get

$$dq_{iuv} = -\sum_{j\neq i}\left(\sum_{r=1}^{n} A_{jur}\{A_{iru}, q_{iuv}\}\right) d\log(t_i - t_j)$$

$$= -\sum_{j\neq i}\left\{\sum_{r=1}^{n} A_{jur}A_{iru}, q_{iuv}\right\} d\log(t_i - t_j)$$

$$= \sum_{j\neq i}\{q_{iuv}, \operatorname{tr}(A_i A_j)\}d\log(t_i - t_j).$$

We introduce the 1-form

$$\omega = \frac{1}{2}\sum_{\substack{k,l=1\\k\neq l}}^{p} \operatorname{tr}(A_k A_l)d\log(t_k - t_l). \tag{8.29}$$

Then the above equation can be expressed as

$$dq_{iuv} = \{q_{iuv}, \omega\}.$$

We have a similar expression for p_{iuv}, and then we obtain the system of equations

$$dQ_i = \{Q_i, \omega\}, \quad dP_i = \{P_i, \omega\} \quad (1 \le i \le p) \tag{8.30}$$

from (8.27). Since ω is defined by using $\operatorname{tr}(A_k A_l)$, it is an invariant under simultaneous similar transformations for $(A_1, A_2, \ldots, A_p)$. Thus, although we have started from the Schlesinger system (8.13), the resulting system (8.30) can be understood as derived from the general system (8.8), or equivalently from the Pfaffian system (8.10).

The system (8.30) is a Hamiltonian system. We shall give an explicit form of the Hamiltonian. If we write ω as

$$\omega = \sum_{i=1}^{p} H_i \, dt_i,$$

we have

$$H_i = \operatorname{tr}\left(A_i \sum_{j\neq i}\frac{A_j}{t_i - t_j}\right). \tag{8.31}$$

By using this expression and the definition of the exterior derivative, the Pfaffian system (8.30) can be written as

$$\frac{\partial Q_i}{\partial t_j} = \{Q_i, H_j\}, \quad \frac{\partial P_i}{\partial t_j} = \{P_i, H_j\} \quad (1 \leq i, j \leq p).$$

By the definition (8.28) of the Poisson bracket, we finally obtain

$$\frac{\partial q_{iuv}}{\partial t_j} = \frac{\partial H_j}{\partial p_{ivu}}, \quad \frac{\partial p_{ivu}}{\partial t_j} = -\frac{\partial H_j}{\partial q_{iuv}} \quad (1 \leq i, j \leq p, 1 \leq u, v \leq n). \tag{8.32}$$

This system is a Hamiltonian system with multi-time variables $(t_1, t_2, \ldots, t_p)$, and H_j is the Hamiltonian corresponding to the time variable t_j. Thus we obtain the following.

Theorem 8.11 *Isomonodromic deformation of non-resonant normalized Fuchsian system (8.5) is described by the Hamiltonian system (8.30) with the entries of the matrices Q_i, P_i ($1 \leq i \leq p$) as the canonical variables, where ω is the 1-form defined by (8.29). The system (8.30) is rewritten in the form (8.32) with Hamiltonians H_i given by (8.31).*

This theorem tells that the deformation equation intrinsically has a Hamiltonian structure. Moreover the Hamiltonians (8.31) are polynomials in the canonical variables, which makes the analysis simple. However, the number of the canonical variables is $2pn^2$, which is too big compared with the number $2 - \iota$ of the independent variables of the deformation equation, that is the number of the accessory parameters. It is even too big compared with the number pn^2 of the entries of the residue matrices A_i. Then we want to reduce the number of the canonical variables. If each A_i has lower rank, we can reduce the number. For the purpose, we use the following fact.

Lemma 8.12 *Let A be an $n \times n$-matrix of rank m which is less than n. Let*

$$\begin{pmatrix} \Lambda' & \\ & O \end{pmatrix}$$

be the Jordan canonical form of A, where Λ' is an $m \times m$-matrix. Then there exist $n \times m$-matrix Q' and $m \times n$-matrix P' such that

$$Q'P' = A, \quad P'Q' = \Lambda'$$

hold.

Proof Take a matrix Q such that

$$Q^{-1}AQ = \begin{pmatrix} \Lambda' & \\ & O \end{pmatrix}.$$

Partition Q and Q^{-1} into $(m, n - m) \times (m, n - m)$-blocks and set

$$Q = \begin{pmatrix} S & T \\ U & V \end{pmatrix}, \quad Q^{-1} = \begin{pmatrix} X & Y \\ Z & W \end{pmatrix}.$$

Then we have

$$A = Q \begin{pmatrix} \Lambda' & \\ & 0 \end{pmatrix} Q^{-1} = \begin{pmatrix} S \\ U \end{pmatrix} \Lambda' (X \ Y).$$

Now set

$$Q' = \begin{pmatrix} S \\ U \end{pmatrix}, \quad P' = \Lambda' (X \ Y).$$

Then we have the first equation of the lemma. From $Q^{-1} Q = I_n$, we obtain $XS + YU = I_m$, which implies the second equation of the lemma. $\square$

In the above derivation of the Hamiltonian system, we can replace Q_i, P_i by Q', P' of the lemma when the rank of A_i is less than n. Thus we can reduce the number of the canonical variables of the Hamiltonian system (8.30) to $2n \sum_{i=1}^{p} \mathrm{rank} A_i$.

In general, however, the number of the canonical variables still exceeds the number of the accessory parameters. It is an open problem whether the number of the canonical variables can be reduced to the number of the accessory parameters while keeping the polynomial structure of the Hamiltonians. For the knowns cases, we have affirmative results of the problem [149, 170].

In Proposition 8.9 in Sect. 8.4, we obtained the Garnier system (8.23) without referring to the Hamiltonian structure. Now we see that the dependent variables of the Garnier system (8.23) are the canonical variables, because the parametrization

$$A_i = \begin{pmatrix} \theta_i - q_i p_i \\ q_i \end{pmatrix} (1 \ p_i)$$

is just the one in Lemma 8.12. By using this parametrization, we can calculate the Hamiltonians, and get

$$H_i = \sum_{j \neq i} \frac{(\theta_i + q_i(p_i - p_j))(\theta_j + q_j(p_i - p_j))}{t_i - t_j}.$$

It is easily checked that the Hamiltonian system for these Hamiltonians coincides with the Garnier system (8.23).

Thus Lemmas 8.12, 8.8 and 7.10 may be useful to parametrize the tuple of the residue matrices A_i. The problem of the parametrization will also be connected to the character varieties.

8.6   Deformation of Rigid Differential Equations

The word rigid is used to express the impossibility of deformation, and then the deformation of rigid differential equation may sound contradictory. However, by the definition, we see that the rigid differential equations can be easily deformed.

We consider a rigid Fuchsian differential equation (8.1) with singular points $t_0, t_1, \ldots, t_p$, among which t_0, t_1, t_2 have been sent to $\infty, 0, 1$, respectively, by a Möbius transformation as normalization. We assume that $p \geq 3$. Since the monodromy is uniquely determined by the local monodromies, it does not depend on the positions of the singular points $t_3, \ldots, t_p$. This implies the existence of isomonodromic fundamental system of solutions, and hence the rigid differential equation (8.1) can be deformed. By Theorem 8.1, we see that there exists a system of differential equations

$$\frac{\partial Y}{\partial t_j} = B_j(x, t)Y \quad (3 \leq j \leq p)$$

in t_j which is compatible with the differential equation (8.1) in x. As we have shown in the proof of Theorem 8.3, differential equations in t_j for a normalized Fuchsian system (8.5) are given by

$$\frac{\partial Y}{\partial t_j} = \left(\frac{A_j}{t_j - x} + B_j^0(t) \right) Y,$$

where the Fuchsian system is not necessarily rigid. Since A_j in the right hand side does not depend on t, the problem is only to determine $B_j^0(t)$. If the normalized Fuchsian system (8.5) is rigid, we can explicitly construct $B_j^0(t)$ by using the middle convolution for completely integrable systems in several variables which will be defined in Part II, Chap. 14. Please refer to Sect. 14.2 for the construction. As a result, we see that $B_j^0(t)$ is rational not only in x but also in $t = (t_3, \ldots, t_p)$. Here we give one example.

Example 8.1 The tuple (A_0, A_1, A_2, A_3) of 3×3-matrices with the rigid spectral type $((21), (21), (21), (21))$ satisfying $A_0 + A_1 + A_2 + A_3 = O$ is given by

$$A_1 = \begin{pmatrix} \alpha_1 + \lambda & \alpha_2 & \alpha_3 \\ 0 & 0 & 0 \\ 0 & 0 & 0 \end{pmatrix}, \ A_2 = \begin{pmatrix} 0 & 0 & 0 \\ \alpha_1 & \alpha_2 + \lambda & \alpha_3 \\ 0 & 0 & 0 \end{pmatrix}, \ A_3 = \begin{pmatrix} 0 & 0 & 0 \\ 0 & 0 & 0 \\ \alpha_1 & \alpha_2 & \alpha_3 + \lambda \end{pmatrix},$$

where $\alpha_1, \alpha_2, \alpha_3, \lambda \in \mathbb{C}$. Let us consider the deformation of the normalized Fuchsian system with these matrices as the residues. We normalize two singular points other than ∞ to 0, 1, and denote the remaining singular point by t. Then we have the system

$$\frac{dY}{dx} = \left(\frac{A_1}{x} + \frac{A_2}{x-t} + \frac{A_3}{x-1} \right) Y.$$

We can construct the partial differential equation

$$\frac{\partial Y}{\partial t} = \left(\frac{B_1}{t} + \frac{A_2}{t-x} + \frac{B_3}{t-1} \right) Y,$$

in t compatible with the Fuchsian system, where

$$B_1 = \begin{pmatrix} \alpha_2 & -\alpha_2 & 0 \\ -\alpha_1 & \alpha_1 & 0 \\ 0 & 0 & 0 \end{pmatrix}, \quad B_3 = \begin{pmatrix} 0 & 0 & 0 \\ 0 & \alpha_3 & -\alpha_3 \\ 0 & -\alpha_2 & \alpha_2 \end{pmatrix}.$$

The compatibility can be shown by checking for $(A_1, A_2, A_3, B_1, B_3)$ the integrability condition (11.15) which will be given in Example 11.1 in Chap. 11.

Chapter 9
Integral Representations of Solutions of Euler Type

Some differential equations have integral representations of solutions of Euler type. A typical example is the Gauss hypergeometric differential equation. The study of such integrals is one of the main topics in differential equations in the complex domain, however, we do not develop the general theory of integrals of Euler type. Instead, we shall study several illustrative examples, which will make the readers understand the main notions and main ideas.

9.1 Integrals with Standard Loading

Let a, b, c be real numbers satisfying $a < b < c$, and λ, μ, ν complex numbers (Fig. 9.1). We consider the integral

$$I_{p,q} = \int_p^q (t-a)^\lambda (t-b)^\mu (t-c)^\nu \, dt, \qquad (9.1)$$

where p, q are two adjacent points in $\{-\infty, a, b, c, +\infty\}$ with $p < q$. Since p, q are singular points of the integrand, the integral (9.1) becomes an improper integral. Note that the beta function is defined by such improper integral

$$B(\alpha, \beta) = \int_0^1 t^{\alpha-1}(1-t)^{\beta-1} \, dt, \qquad (9.2)$$

from which we obtain many useful properties. On the other hand, of course improper integrals may diverge. Then we assume that the exponents λ, μ, ν are so taken that

© The Editor(s) (if applicable) and The Author(s), under exclusive
license to Springer Nature Switzerland AG 2020
Y. Haraoka, *Linear Differential Equations in the Complex Domain*, Lecture Notes
in Mathematics 2271, https://doi.org/10.1007/978-3-030-54663-2_9

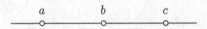

Fig. 9.1   Location of a, b, c

the integral (9.1) converges. For example, if one of p, q is a, we assume

$$\operatorname{Re}\lambda > -1.$$

Later we will relax this condition to the condition

$$\lambda \notin \mathbb{Z}_{<0}.$$

The integrand of (9.1) is a multi-valued function, so that we need to specify a branch. Here we use the standard loading, which is defined by Mimachi [124]. Let $f(t)$ be one of the linear functions $t - a, t - b, t - c$. Then on the domain of integration (i.e. on the interval (p, q)), either $f(t) > 0$ or $f(t) < 0$ holds. If $f(t) > 0$, we do not change it, while if $f(t) < 0$, we replace $f(t)$ by $-f(t)$. Then we assign 0 to the argument (as a complex number) of the linear function. We operate this to every linear function. The result is said to be an integral with standard loading. Explicitly, every integral (9.1) with standard loading is given by

$$I_{-\infty,a} = \int_{-\infty}^{a} (a - t)^{\lambda}(b - t)^{\mu}(c - t)^{\nu}\, dt,$$

$$I_{a,b} = \int_{a}^{b} (t - a)^{\lambda}(b - t)^{\mu}(c - t)^{\nu}\, dt,$$

$$\qquad\qquad (9.3)$$

$$I_{b,c} = \int_{b}^{c} (t - a)^{\lambda}(t - b)^{\mu}(c - t)^{\nu}\, dt,$$

$$I_{c,+\infty} = \int_{c}^{+\infty} (t - a)^{\lambda}(t - b)^{\mu}(t - c)^{\nu}\, dt,$$

where the arguments of the linear functions in the integrands are all assigned to be 0. We can understand that the beta function (9.2) is defined by the integral with standard loading.

9.2   Linear Relations Among the Integrals

In the following, we regard a, b, c as independent variables. Then the integrals $I_{p,q}$ become functions in a, b, c. We are going to obtain linear relations among these functions.

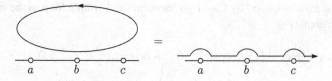

Fig. 9.2 Deformation of a loop

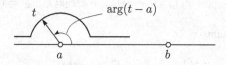

Fig. 9.3 Change of arg$(t - a)$

The branch of the integrand

$$U(t) = (a - t)^{\lambda}(b - t)^{\mu}(c - t)^{\nu}$$

of $I_{-\infty,a}$ is definite on $(-\infty, a)$ by the standard loading. We continue this branch to the upper half plane $\{t \in \mathbb{C} \mid \mathrm{Im}\, t > 0\}$ to obtain a single valued function there. Then, by Cauchy's theorem, the integral of this function over a simple closed curve in the upper half plane becomes 0. Again by Cauchy's theorem, this integral is equal to the sum of the integrals over $(-\infty, a)$, (a, b), (b, c) and $(c, +\infty)$, where the branch of each interval is continued from the upper half plane (Fig. 9.2).

When we continue $(a - t)^{\lambda}$ on $(-\infty, a)$ to (a, b) through upper half plane, $\arg(a - t)$ changes from 0 to $-\pi$, so that we have

$$(a - t)^{\lambda} \rightsquigarrow (e^{-\pi\sqrt{-1}}(t - a))^{\lambda} = e^{-\pi\sqrt{-1}\lambda}(t - a)^{\lambda}$$

with $\arg(t - a) = 0$ (Fig. 9.3).

Then, by this analytic continuation, we get

$$(a - t)^{\lambda}(b - t)^{\mu}(c - t)^{\nu} \rightsquigarrow e^{-\pi\sqrt{-1}\lambda}(t - a)^{\lambda}(b - t)^{\mu}(c - t)^{\nu}$$

as t moves from $(-\infty, a)$ to (a, b) (via upper half plane), where we use the standard loading. Similarly, we have

$$(t - a)^{\lambda}(b - t)^{\mu}(c - t)^{\nu} \rightsquigarrow e^{-\pi\sqrt{-1}\mu}(t - a)^{\lambda}(t - b)^{\mu}(c - t)^{\nu},$$

$$(t - a)^{\lambda}(t - b)^{\mu}(c - t)^{\nu} \rightsquigarrow e^{-\pi\sqrt{-1}\nu}(t - a)^{\lambda}(t - b)^{\mu}(t - c)^{\nu}$$

as t moves from (a, b) to (b, c), and from (b, c) to $(c, +\infty)$. Now we set

$$e^{\pi\sqrt{-1}\lambda} = e_1, \; e^{\pi\sqrt{-1}\mu} = e_2, \; e^{\pi\sqrt{-1}\nu} = e_3.$$

Then the relation obtained by Cauchy's theorem for a simple loop in the upper half plane is written as

$$I_{-\infty,a} + e_1^{-1}I_{a,b} + (e_1e_2)^{-1}I_{b,c} + (e_1e_2e_3)^{-1}I_{c,+\infty} = 0. \qquad (9.4)$$

Similarly, by considering a simple loop in the lower half plane, we get the relation

$$I_{-\infty,a} + e_1 I_{a,b} + e_1e_2 I_{b,c} + e_1e_2e_3 I_{c,+\infty} = 0. \qquad (9.5)$$

If the rank of the matrix

$$\begin{pmatrix} 1 & e_1^{-1} & (e_1e_2)^{-1} & (e_1e_2e_3)^{-1} \\ 1 & e_1 & e_1e_2 & e_1e_2e_3 \end{pmatrix} \qquad (9.6)$$

of coefficients in the relations (9.4), (9.5) is two, two integrals among four $I_{p,q}$ can be written as linear combinations of the remaining two integrals. This implies that the dimension of the linear space spanned by $I_{p,q}$ is at most two. In particular, if λ, μ, ν is generic, any two columns of (9.6) are linearly independent, and hence these can be a basis. Then, in this case, we can take the integrals $I_{a,b}$ and $I_{b,c}$ on bounded intervals as a basis of the linear space spanned by $I_{p,q}$.

9.3 Asymptotic Behaviors of the Integrals

We study the asymptotic behavior of the integral $I_{p,q}$ as a function of (a, b, c). First we fix b, c, and regard it as a function in a, which, for a moment, is assumed to be in the interval $-\infty < a < b$. We have the following asymptotic behavior as a tends to the boundaries $-\infty$ and b.

Proposition 9.1

$$I_{-\infty,a} \sim B(-\lambda - \mu - \nu - 1, \lambda + 1)(-a)^{\lambda+\mu+\nu+1} \quad (a \to -\infty), \qquad (9.7)$$

$$I_{b,c} \sim (c - b)^{\mu+\nu+1} B(\mu + 1, \nu + 1)(-a)^{\lambda} \quad (a \to -\infty). \qquad (9.8)$$

$$I_{a,b} \sim (c - b)^{\nu} B(\lambda + 1, \mu + 1)(b - a)^{\lambda+\mu+1} \quad (a \to b), \qquad (9.9)$$

$$I_{c,+\infty} \sim (c - b)^{\lambda+\mu+\nu+1} B(-\lambda - \mu - \nu - 1, \nu + 1) \quad (a \to b). \qquad (9.10)$$

Notational Remark Let A, γ be constants and ξ a local coordinate at $x = x_0$. Here we use the notation

$$f(x) \sim A\xi^{\gamma} \quad (x \to x_0)$$

if there exists a convergent power series $\sum_{n=1}^{\infty} A_n \xi^n$ at $\xi = 0$ such that

$$f(x) = \xi^\gamma \left(A + \sum_{n=1}^{\infty} A_n \xi^n \right)$$

holds. Of course, in this case

$$\lim_{x \to x_0} \frac{f(x)}{A \, \xi^\gamma} = 1$$

holds, however, the last equality does not mean $f(x) \sim A \, \xi^\gamma \ (x \to x_0)$.

Proof First we shall show (9.7). We change the variable of integration to

$$s = \frac{a}{t},$$

so that we have $0 < s < 1$. We assume that a is sufficiently close to $-\infty$, and set $\arg(-a) = 0$. Then, if we write

$$a - t = (-a)\frac{1 - s}{s},$$

$$b - t = (-a)\frac{1 - \frac{b}{a}s}{s},$$

$$c - t = (-a)\frac{1 - \frac{c}{a}s}{s},$$

each factor in the right hand sides has argument 0. Since a is sufficiently close to $-\infty$ and $|s| < 1$ holds, we have

$$\left| \frac{b}{a} s \right| < 1, \qquad \left| \frac{c}{a} s \right| < 1. \tag{9.11}$$

Then we get

$$\int_{-\infty}^{a} (a - t)^\lambda (b - t)^\mu (c - t)^\nu \, dt$$

$$= \int_0^1 \left((-a)\frac{1 - s}{s} \right)^\lambda \left((-a)\frac{1 - \frac{b}{a}s}{s} \right)^\mu \left((-a)\frac{1 - \frac{c}{a}s}{s} \right)^\nu \left(-\frac{a}{s^2} \right) ds$$

$$= (-a)^{\lambda+\mu+\nu+1} \int_0^1 s^{-\lambda-\mu-\nu-2}(1 - s)^\lambda \left(1 - \frac{b}{a}s \right)^\mu \left(1 - \frac{c}{a}s \right)^\nu ds$$

$$\sim B(-\lambda - \mu - \nu - 1, \lambda + 1)(-a)^{\lambda+\mu+\nu+1}.$$

Note that, thanks to (9.11), we have Taylor expansions of $\left(1 - \dfrac{b}{a}s\right)^{\mu}$ and $\left(1 - \dfrac{c}{a}s\right)^{\nu}$ in $(-a)^{-1}$ with the initial term 1, which induces the last $\sim$ in the above.

The asymptotic behavior (9.8) follows directly from

$$\int_b^c (t - a)^{\lambda} (t - b)^{\mu} (c - t)^{\nu} \, dt$$

$$= (-a)^{\lambda} \int_b^c \left(1 - \frac{t}{a}\right)^{\lambda} (t - b)^{\mu} (c - t)^{\nu} \, dt$$

$$\sim (-a)^{\lambda} \int_b^c (t - b)^{\mu} (c - t)^{\nu} \, dt.$$

The asymptotic behavior (9.9) (resp. (9.10)) can be derived in a similar manner as (9.7) (resp. (9.8)).   □

The asymptotic behaviors as $a \to -\infty$ of the other two integrals $I_{a,b}$, $I_{c,+\infty}$ are not simple as in (9.7), (9.8), but complex because of the linear relations (9.4), (9.5). Namely, for example we have

$$I_{a,b} \sim A(-a)^{\lambda+\mu+\nu+1} + B(-a)^{\lambda} \tag{9.12}$$

with some constants A, B. The asymptotic behavior as $a \to b$ of $I_{-\infty,a}$ and $I_{b,c}$ are similarly obtained.

How can we find intervals of integration which give simple asymptotic behaviors? Let us see the case $a \to -\infty$. We notice that the behaviors of (9.7) and (9.8) are obtained in different reasons. The behavior $(-a)^{\lambda+\mu+\nu+1}$ in (9.7) comes from the collapse of the interval $(-\infty, a)$ as $a \to -\infty$. On the other hand, the behavior (9.8) is obtained directly from

$$(t - a)^{\lambda} = (-a)^{\lambda} \left(1 - \frac{t}{a}\right)^{\lambda} \sim (-a)^{\lambda},$$

since the interval of integration (b, c) does not change by the movement $a \to -\infty$.

The case $a \to b$ is similar. The asymptotic behavior (9.9) comes from the collapse of the interval, and the asymptotic behavior (9.10) comes from the integrand because the interval does not change when $a \to b$.

This is the basic viewpoint to study the behavior of the integrals. Applying this, we see without explicit computation the following behaviors:

$$\begin{cases} I_{a,b} \sim C_1 (b - a)^{\lambda+\mu+1} & (b \to a), \\ I_{c,+\infty} \sim C_2 & (b \to a), \end{cases} \tag{9.13}$$

$$\begin{cases} I_{b,c} \sim C_3(c-b)^{\mu+\nu+1} & (b \to c), \\ I_{-\infty,a} \sim C_4 & (b \to c), \end{cases} \tag{9.14}$$

$$\begin{cases} I_{b,c} \sim C_3(c-b)^{\mu+\nu+1} & (c \to b), \\ I_{-\infty,a} \sim C_4 & (c \to b), \end{cases} \tag{9.15}$$

$$\begin{cases} I_{c,+\infty} \sim C_5 c^{\lambda+\mu+\nu+1} & (c \to +\infty), \\ I_{a,b} \sim C_6 c^{\nu} & (c \to +\infty), \end{cases} \tag{9.16}$$

where $C_1, C_2, \ldots, C_6$ are constants which can be evaluated by the integrals.

The Gauss-Kummer formula (6.18) introduced in Chap. 6 can be also shown by studying the behavior of the integral $I_{p,q}$. However, the behavior in this case is different from one which we have seen in the above. Namely we study the asymptotic behavior of the integral $I_{a,b}$ as $c \to b$ where $a < b < c$. (We should study the behavior like (9.12).) The Gauss-Kummer formula will be explained later in Sect. 9.5.

9.4 Regularization of Integral

The regularization of integral is a method to give a meaning to divergent integrals, and is equivalent to Hadamard's finite part of divergent integrals.

We explain by the integral $I_{a,b}$. This integral converges only if

$$\operatorname{Re} \lambda > -1, \quad \operatorname{Re} \mu > -1 \tag{9.17}$$

holds. If λ, μ satisfy this condition, the integral $I_{a,b}$ converges, and gives a holomorphic function in λ, μ. If the holomorphic function is analytically continued to a wider domain than (9.17), we can regard the integral $I_{a,b}$ as defined in the wider domain. This is the main idea of the regularization. Explicitly, the regularization is to realize the analytic continuation in the following manner.

For simplicity we assume $\operatorname{Re} \mu > -1$, and consider the analytic continuation with respect to λ. Suppose $\lambda \notin \mathbb{Z}$. Take sufficiently small $\varepsilon > 0$. We set S_ε to be the circle $|t - a| = \varepsilon$, L_ε to be the interval $[a + \varepsilon, b)$, and consider the integral

$$I'_{a,b} = \frac{1}{e^{2\pi\sqrt{-1}\lambda} - 1} \int_{S_\varepsilon} (t-a)^\lambda (b-t)^\mu (c-t)^\nu \, dt + \int_{L_\varepsilon} (t-a)^\lambda (b-t)^\mu (c-t)^\nu \, dt.$$

Here we understand that S_ε is a chain starting from $a + \varepsilon$ with positive direction, and the branch on S_ε is specified by the standard loading at the initial point $a + \varepsilon$. The branch on L_ε is also specified by the standard loading. Then we have the following assertion.

Proposition 9.2 *The equality*

$$I'_{a,b} = I_{a,b}$$

holds on $\{\lambda \in \mathbb{C} \mid \operatorname{Re}\lambda > -1\} \cap \{\lambda \in \mathbb{C} \mid \lambda \notin \mathbb{Z}\}.$

Proof Take $0 < \varepsilon_1 < \varepsilon_2$, and let $I'_{a,b,1}$ (resp. $I'_{a,b,2}$) be $I'_{a,b}$ with $\varepsilon = \varepsilon_1$ (resp. $\varepsilon = \varepsilon_2$). We consider the integral of

$$U(t) = (t - a)^{\lambda}(b - t)^{\mu}(c - t)^{\nu}$$

on the closed curve in Fig. 9.4. The branch of $U(t)$ is specified by the standard loading at the initial point P_1 of S_{ε_2}. Then we have

$$\arg(t - a) = 2\pi$$

at P_2 and also on the segment P_2Q_2, and, by encircling S_{ε_1} in the negative direction, have

$$\arg(t - a) = 0$$

at Q_1 and on the segment Q_1P_1. Thus we get

$$\int_{S_{\varepsilon_2}} U(t)\,dt + e^{2\pi\sqrt{-1}\lambda}\int_{P_2}^{Q_2} U(t)\,dt - \int_{S_{\varepsilon_1}} U(t)\,dt + \int_{Q_1}^{P_1} U(t)\,dt = 0.$$

Since the direction of the segment P_2Q_2 is reverse to one of the segment Q_1P_1, we obtain from the above

$$\frac{1}{e^{2\pi\sqrt{-1}\lambda} - 1}\int_{S_{\varepsilon_1}} U(t)\,dt + \int_{Q_1}^{P_1} U(t)\,dt = \frac{1}{e^{2\pi\sqrt{-1}\lambda} - 1}\int_{S_{\varepsilon_2}} U(t)\,dt.$$

Fig. 9.4 S_{ε_1}, S_{ε_2} and the closed curve

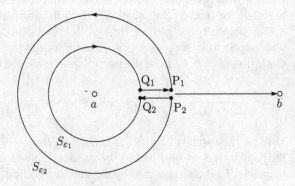

By adding $\int_{P_1}^b U(t)\,dt$ to both sides, we get

$$\frac{1}{e^{2\pi\sqrt{-1}\lambda}-1}\int_{S_{\varepsilon_1}} U(t)\,dt + \int_{a+\varepsilon_1}^b U(t)\,dt = \frac{1}{e^{2\pi\sqrt{-1}\lambda}-1}\int_{S_{\varepsilon_2}} U(t)\,dt + \int_{a+\varepsilon_2}^b U(t)\,dt.$$

Thus we have $I'_{a,b,1} = I'_{a,b,2}$, which implies that $I'_{a,b}$ does not depend on ε.

Now we assume $\operatorname{Re}\lambda > -1$. Since

$$\int_{S_\varepsilon} U(t)\,dt = \varepsilon^{\lambda+1}\int_0^{2\pi} (b-a-\varepsilon e^{\sqrt{-1}\theta})^\mu(c-a-\varepsilon e^{\sqrt{-1}\theta})^\nu \sqrt{-1}\, e^{\sqrt{-1}(\lambda+1)\theta}\,d\theta,$$

we have

$$\lim_{\varepsilon\to+0}\int_{S_\varepsilon} U(t)\,dt = 0.$$

Also we have

$$\lim_{\varepsilon\to+0}\int_{L_\varepsilon} U(t)\,dt = I_{a,b},$$

and hence we obtain

$$I'_{a,b} = \lim_{\varepsilon\to+0} I'_{a,b} = I_{a,b}.$$

Note that the first equality holds by the independence of $I'_{a,b}$ on ε. $\qquad\square$

It follows from this proposition that $I_{a,b}$ can be analytically continued to the domain $\lambda \in \mathbb{C}\setminus\mathbb{Z}_{<0}$. We denote the result of the analytic continuation by the same symbol $I_{a,b}$.

In a similar way, we can regularize the integral with respect to μ, and the result together with the regularization with respect to λ is given by

$$I_{a,b} = \frac{1}{e^{2\pi\sqrt{-1}\lambda}-1}\int_{|t-a|=\varepsilon} U(t)\,dt + \int_{a+\varepsilon}^{b-\varepsilon} U(t)\,dt - \frac{1}{e^{2\pi\sqrt{-1}\mu}-1}\int_{|t-b|=\varepsilon} U(t)\,dt.$$

$$(9.18)$$

Here $|t-b| = \varepsilon$ has the initial point $b-\varepsilon$, at which the branch is determined by the standard loading. In Fig. 9.5 we illustrate the path of integration of the right hand side of (9.18). The branch is connected where the path is connected in the figure.

Fig. 9.5 Path for
regularization

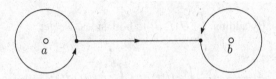

The right hand side of (9.18) is equivalent to the Pochhammer contour divided by
$(e^{2\pi\sqrt{-1}\lambda} - 1)(e^{2\pi\sqrt{-1}\mu} - 1)$. For the Pochhammer contour, please refer to [182].

For the integral $I_{a,b}$, the path of integration is an open interval (a, b), which can
be regarded as a locally finite infinite chain in the terminology of homology theory.
Precisely speaking, it is a locally finite cycle with coefficients in a local system, and
then is an element of the homology group

$$H_1^{lf}(X, \mathcal{L}).$$

On the other hand, the regularized path in the right hand side of (9.18) is a finite sum
of compact chains, and then can be regarded as an element of the homology group

$$H_1(X, \mathcal{L})$$

with coefficient in a local system. Then the regularization is regarded as a map

$$H_1^{lf}(X, \mathcal{L}) \to H_1(X, \mathcal{L}).$$

Homology groups with coefficient in a local system will be briefly explained in
Sect. 9.9.

9.5   Differential Equation Satisfied by the Integral

We are going to obtain a differential equation satisfied by $I_{p,q}$ as a function in
a, b, c. In order to avoid the inconvenience of changing integrands according to the
choice of p and q, we fix $I_{p,q}$ as in (9.1). Although it is different from (9.3) with
standard loading, the difference is a constant multiple, which does not influence the
differential equation.

We first fix b, c, and shall obtain a differential equation in the variable a. The
dimension of the linear space spanned by $I_{p,q}$ is known to be at most 2, and then
we see a priori that the rank of the differential equation is also at most 2.

We shall derive the differential equation by using cohomologies (differential
forms). We assume that the integrals are regularized, and then we can freely

interchange the order of differentiation and integration. Or we may understand that we assumed that Re λ etc. are sufficiently large.

Set

$$U(t) = (t - a)^{\lambda}(t - b)^{\mu+1}(t - c)^{\nu},$$

and define 1-forms φ_1, φ_2 by

$$\varphi_1 = \frac{dt}{t - b}, \quad \varphi_2 = \frac{dt}{t - c}.$$

By using them, we define

$$y_1(a) = \int_p^q U(t)\,\varphi_1, \quad y_2(a) = \int_p^q U(t)\,\varphi_2.$$

In particular, we have $y_1(a) = I_{p,q}$. Note that

$$\int_p^q d(U(t)) = [U(t)]_p^q = U(q) - U(p) = 0$$

holds. On the other hand we have

$$\int_p^q d(U(t)) = \int_p^q U(t) \left(\frac{\lambda}{t - a} + \frac{\mu + 1}{t - b} + \frac{\nu}{t - c} \right) dt.$$

Here we introduce the notation $\varphi \equiv \psi$ for 1-forms φ, ψ when

$$\int_p^q U(t)\,\varphi = \int_p^q U(t)\,\psi$$

holds. Then we have

$$\left(\frac{\lambda}{t - a} + \frac{\mu + 1}{t - b} + \frac{\nu}{t - c} \right) dt \equiv 0. \tag{9.19}$$

Now we compute $\partial y_1 / \partial a$. Temporarily we denote by φ_1' the 1-form ψ_1 such that

$$\frac{\partial y_1}{\partial a} = \int_p^q U(t)\,\psi_1.$$

Then we get

$$\varphi_1' = -\frac{\lambda}{t-a}\varphi_1$$

$$= -\frac{\lambda}{(t-a)(t-b)}\,dt$$

$$= \frac{1}{a-b}\left(\frac{\lambda}{t-b}-\frac{\lambda}{t-a}\right)dt$$

$$\equiv \frac{1}{a-b}\left(\frac{\lambda}{t-b}+\frac{\mu+1}{t-b}+\frac{\nu}{t-c}\right)dt$$

$$= \frac{1}{a-b}[(\lambda+\mu+1)\varphi_1 + \nu\varphi_2].$$

In a similar way, we also get

$$\varphi_2' \equiv \frac{1}{a-c}[(\mu+1)\varphi_1 + (\lambda+\nu)\varphi_2].$$

Hence we obtain the system of differential equations

$$\begin{cases} \dfrac{\partial y_1}{\partial a} = \dfrac{1}{a-b}[(\lambda+\mu+1)y_1 + \nu y_2], \\[2mm] \dfrac{\partial y_2}{\partial a} = \dfrac{1}{a-c}[(\mu+1)y_1 + (\lambda+\nu)y_2] \end{cases} \tag{9.20}$$

for y_1, y_2. By using the vector $Y = {}^t(y_1, y_2)$, it can be written as

$$\frac{\partial Y}{\partial a} = \left(\frac{B}{a-b}+\frac{C}{a-c}\right)Y,$$

$$B = \begin{pmatrix} \lambda+\mu+1 & \nu \\ 0 & 0 \end{pmatrix}, \quad C = \begin{pmatrix} 0 & 0 \\ \mu+1 & \lambda+\nu \end{pmatrix}.$$

From (9.20) we can derive a scalar differential equation for y_1, and get

$$(a-b)(a-c)\frac{\partial^2 y}{\partial a^2} - \{(\lambda+\mu)(a-c) + (\lambda+\nu)(a-b)\}\frac{\partial y}{\partial a}$$

$$+ \lambda(\lambda+\mu+\nu+1)y = 0. \tag{9.21}$$

This is the differential equation satisfied by $I_{p,q}$. The differential equation is nothing but the Gauss hypergeometric differential equation

$$x(1-x)\frac{d^2y}{dx^2} + (\gamma - (\alpha + \beta + 1)x)\frac{dy}{dx} - \alpha\beta y = 0. \qquad (9.22)$$

In order to relate them, we write the independent variable a as x, normalize the singular points b, c into $0, 1$ by a linear transformation of the independent variable, and define the parameters α, β, γ by

$$\gamma = -(\lambda + \mu), \ \alpha + \beta + 1 = -(2\lambda + \mu + \nu), \ \alpha\beta = \lambda(\lambda + \mu + \nu + 1).$$

To solve this relation in (α, β), we get two sets of solutions. One of them is given by

$$\alpha = -(\lambda + \mu + \nu + 1), \ \beta = -\lambda, \ \gamma = -(\lambda + \mu), \qquad (9.23)$$

which is solved conversely as

$$\lambda = -\beta, \ \mu = \beta - \gamma, \ \nu = \gamma - \alpha - 1.$$

The other is given by

$$\alpha = -\lambda, \ \beta = -(\lambda + \mu + \nu + 1), \ \gamma = -(\lambda + \mu), \qquad (9.24)$$

which gives

$$\lambda = -\alpha, \ \mu = \alpha - \gamma, \ \nu = \gamma - \beta - 1.$$

These two sets of solutions are transformed by interchanging α and β. Now, the result in this section can be regarded from the viewpoint of the hypergeometric differential equation as follows. We use the correspondence (9.23).

Theorem 9.3 *There is an integral representation of solutions of the hypergeometric differential equation (9.22) of the form*

$$y(x) = \int_p^q t^{\beta-\gamma}(t-1)^{\gamma-\alpha-1}(t-x)^{-\beta} \, dt, \qquad (9.25)$$

where p, q take values in $\{0, 1, x, \infty\}$.

Since we have established the relation between the integral $I_{p,q}$ and the hypergeometric differential equation, the results in this section can be interpreted to the assertions for the hypergeometric differential equations. For example, Gauss-Kummer identity, which we used in Chap. 6, can be derived from the result in

Sect. 9.3. The hypergeometric series $F(\alpha, \beta, \gamma; x)$ is a solution of the hypergeometric differential equation (9.22) of exponent 0 at $x = 0$, and is specified by the leading term 1 of the Taylor series at $x = 0$. By Proposition 9.1, we see that such a solution is a constant multiple of the solution $I_{c,+\infty} = I_{1,+\infty}$, whose asymptotic behavior is given by

$$(c - b)^{\lambda+\mu+\nu+1} B(-\lambda - \mu - \nu - 1, \nu + 1) = B(\alpha, \gamma - \alpha) = \frac{\Gamma(\alpha)\Gamma(\gamma - \alpha)}{\Gamma(\gamma)}.$$

Here we used the correspondence (9.23) among (λ, μ, ν) and (α, β, γ). Hence we get the integral representation

$$F(\alpha, \beta, \gamma; x) = \frac{\Gamma(\gamma)}{\Gamma(\alpha)\Gamma(\gamma - \alpha)} \int_1^\infty t^{\beta-\gamma}(t - 1)^{\gamma-\alpha-1}(t - x)^{-\beta} dt \qquad (9.26)$$

of the hypergeometric series. By the change $t = 1/s$ of the variable of integration, we can derive another integral representation

$$F(\alpha, \beta, \gamma; x) = \frac{\Gamma(\gamma)}{\Gamma(\alpha)\Gamma(\gamma - \alpha)} \int_0^1 s^{\alpha-1}(1 - s)^{\gamma-\alpha-1}(1 - xs)^{-\beta} ds. \qquad (9.27)$$

Note that we can obtain other representations by interchanging α and β in the right hand sides in (9.26), (9.27). Now take a limit $x \to 1$ in the representation (9.27) to get

$$\lim_{x \to 1} \int_0^1 s^{\alpha-1}(1 - s)^{\gamma-\alpha-1}(1 - xs)^{-\beta} ds = \int_0^1 s^{\alpha-1}(1 - s)^{\gamma-\alpha-\beta-1} ds$$

$$= B(\alpha, \gamma - \alpha - \beta).$$
$$(9.28)$$

Then we have the following assertion.

Theorem 9.4 (Gauss-Kummer Identity)   *Under the assumption $\gamma \notin \mathbb{Z}_{\leq 0}$, $\mathrm{Re}(\gamma - \alpha - \beta) > 0$, we have*

$$F(\alpha, \beta, \gamma; 1) = \frac{\Gamma(\gamma)\Gamma(\gamma - \alpha - \beta)}{\Gamma(\gamma - \alpha)\Gamma(\gamma - \beta)}. \qquad (9.29)$$

Proof Formally, by combining the limit formula (9.28) and the integral representation (9.27), we get

$$\frac{\Gamma(\gamma)}{\Gamma(\alpha)\Gamma(\gamma - \alpha)} \cdot B(\alpha, \gamma - \alpha - \beta) = \frac{\Gamma(\gamma)\Gamma(\gamma - \alpha - \beta)}{\Gamma(\gamma - \alpha)\Gamma(\gamma - \beta)},$$

which is the right hand side of (9.29). The assumption $\gamma \notin \mathbb{Z}_{\leq 0}$ is necessary for the existence of the series $F(\alpha, \beta, \gamma; x)$. The limit formula (9.28) holds under the assumption

$$\mathrm{Re}\,\alpha > 0,\ \mathrm{Re}(\gamma - \alpha) > 0,\ \mathrm{Re}(\gamma - \alpha - \beta) > 0.$$

Among these, first two conditions can be removed by the analytic continuation after obtaining (9.29).   □

In deriving the differential equation (9.21) in a, we do not use the positional relation $a < b < c$ among a, b, c. Hence the differential equations in b and in c can be obtained from (9.21) by interchanging (a, λ) with (b, μ) and with (c, ν), respectively.

9.6 Connection Problem

In the differential equation (9.21), we fix b, c, and normalize them as $b = 0, c = 1$ (Fig. 9.6). We denote the variable a by x, and then get the differential equation

$$x(x-1)\frac{d^2y}{dx^2} - \{(\lambda+\mu)(x-1) + (\lambda+\nu)x\}\frac{dy}{dx} + \lambda(\lambda+\mu+\nu+1)y = 0. \quad (9.30)$$

We consider the connection problem between two singular points $x = \infty, 0$.

By the asymptotic behaviors (9.7), (9.8) (resp. (9.9), (9.10)), we see that the differential equation (9.30) has exponents $-\lambda - \mu - \nu - 1, -\lambda$ (resp. $\lambda + \mu + 1, 0$) at $x \to -\infty$ (resp. $x = 0$). Then we define four solutions $y_{\infty1}, y_{\infty2}, y_{01}, y_{02}$ of the differential equation (9.30) by

$$
\begin{aligned}
y_{\infty1} &\sim (-x)^{\lambda+\mu+\nu+1} & (x \to -\infty), \\
y_{\infty2} &\sim (-x)^{\lambda} & (x \to -\infty), \\
y_{01} &\sim (-x)^{\lambda+\mu+1} & (x \to 0), \\
y_{02} &\sim 1 & (x \to 0).
\end{aligned}
\quad (9.31)
$$

The connection problem between $x = \infty$ and $x = 0$ is to obtain the linear relation among $(y_{\infty1}, y_{\infty2})$ and (y_{01}, y_{02}).

We can solve the connection problem in the following way. By comparing (9.7), (9.8), (9.9), (9.10) with (9.31), we obtain proportional relations among $y_{\bullet\bullet}$ and $I_{p,q}$. On the other hand, there are linear relations (9.4), (9.5) among the integrals $I_{p,q}$. Then we have only to combine these relations.

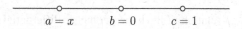

$$a = x \qquad b = 0 \qquad c = 1$$

Fig. 9.6 Normalization of a, b, c

Explicitly, we obtain from (9.7), (9.8), (9.9), (9.10) and (9.31) the proportional relations

$$y_{\infty 1} = \frac{1}{B(-\lambda - \mu - \nu - 1, \lambda + 1)} I_{-\infty, x},$$

$$y_{\infty 2} = \frac{1}{B(\mu + 1, \nu + 1)} I_{0,1},$$

$$y_{01} = \frac{1}{B(\lambda + 1, \mu + 1)} I_{x,0},$$ (9.32)

$$y_{02} = \frac{1}{B(-\lambda - \mu - \nu - 1, \nu + 1)} I_{1,+\infty}.$$

By solving the linear relations (9.4), (9.5) for $I_{a,b}$, $I_{c,+\infty}$, we get

$$I_{a,b} = -\frac{e_1 e_2 e_3 - (e_1 e_2 e_3)^{-1}}{e_2 e_3 - (e_2 e_3)^{-1}} I_{-\infty, a} - \frac{e_3 - e_3^{-1}}{e_2 e_3 - (e_2 e_3)^{-1}} I_{b,c},$$

$$I_{c,+\infty} = \frac{e_1 - e_1^{-1}}{e_2 e_3 - (e_2 e_3)^{-1}} I_{-\infty, a} - \frac{e_2 - e_2^{-1}}{e_2 e_3 - (e_2 e_3)^{-1}} I_{b,c},$$

where we read $(a, b, c) = (x, 0, 1)$. Thanks to (9.32), the above equalities can be translated to the relations among $y_{\bullet\bullet}$. We finally obtain the result

$$y_{01} = \frac{\Gamma(\lambda + \mu + 2)\Gamma(\mu + \nu + 1)}{\Gamma(\lambda + \mu + \nu + 2)\Gamma(\mu + 1)} y_{\infty 1} + \frac{\Gamma(\lambda + \mu + 2)\Gamma(-\mu - \nu - 1)}{\Gamma(\lambda + 1)\Gamma(-\nu)} y_{\infty 2},$$

$$y_{02} = \frac{\Gamma(-\lambda - \mu)\Gamma(\mu + \nu + 1)}{\Gamma(-\lambda)\Gamma(\nu + 1)} y_{\infty 1} + \frac{\Gamma(-\lambda - \mu)\Gamma(-\mu - \nu - 1)}{\Gamma(-\lambda - \mu - \nu - 1)\Gamma(-\mu)} y_{\infty 2}.$$

where we have used the identity

$$e^{\pi \sqrt{-1}\lambda} - e^{-\pi \sqrt{-1}\lambda} = 2\sqrt{-1} \sin \pi \lambda = \frac{2\pi \sqrt{-1}}{\Gamma(\lambda)\Gamma(1 - \lambda)}.$$

This result can be written in terms of the parameters α, β, γ by using (9.24), so that we have

$$y_{01} = \frac{\Gamma(2 - \gamma)\Gamma(\alpha - \beta)}{\Gamma(\alpha - \gamma + 1)\Gamma(1 - \beta)} y_{\infty 1} + \frac{\Gamma(2 - \gamma)\Gamma(\beta - \alpha)}{\Gamma(\beta - \gamma + 1)\Gamma(1 - \alpha)} y_{\infty 2},$$

$$y_{02} = \frac{\Gamma(\alpha - \beta)\Gamma(\gamma)}{\Gamma(\alpha)\Gamma(\gamma - \beta)} y_{\infty 1} + \frac{\Gamma(\beta - \alpha)\Gamma(\gamma)}{\Gamma(\beta)\Gamma(\gamma - \alpha)} y_{\infty 2}.$$

This is the connection relation of the hypergeometric differential equation between $x = 0$ and $x = \infty$. Note that we have derived this relation from the linear relations (9.4), (9.5) among the paths of integration, while, in Chap. 6, we derived the connection relation between $x = 0$ and $x = 1$ from Gauss-Kummer identity (Theorem 9.4).

The differential equation (9.30) has regular singular points at $x = 0, 1, \infty$. If you want to obtain the connection relation between $x = 0$ and $x = 1$, normalize $a = 0, b = x, c = 1$ in the differential equation (9.21), and operate an appropriate permutation for the parameters. Then you will get (9.30), and can apply the above method. Similarly you will solve the connection problem between $x = 1$ and $x = \infty$ by the normalization $a = 0, b = 1, c = x$.

9.7 Monodromy

By monodromy we mean transformations of a linear space of solutions of a differential equation caused by analytic continuations along loops. On the two dimensional linear space spanned by the integrals $I_{p,q}$, we can also define the monodromy by regarding a, b, c as variables. If you want to regard the linear space as a solution space of some ordinary differential equation, you can take, say, $x = b$ as an independent variable with a and c fixed. However, in this section we regard (a, b, c) as a variable in a three dimensional space.

Let a_0, b_0, c_0 be real numbers satisfying $a_0 < b_0 < c_0$, and a, b, c mutually distinct complex variables with the initial positions a_0, b_0, c_0, respectively. Set

$$\Delta = \{(x_1, x_2, x_3) \in \mathbb{C}^3 \mid (x_1 - x_2)(x_1 - x_3)(x_2 - x_3) = 0\}.$$

Then (a, b, c) is a variable in the space $\mathbb{C}^3 \setminus \Delta$. We denote by V the linear space over $\mathbb{C}$ spanned by the integrals $I_{p,q}$. The purpose of this section is to determine the monodromy representation

$$\pi_1(\mathbb{C}^3 \setminus \Delta, (a_0, b_0, c_0)) \rightarrow \mathrm{GL}(V). \tag{9.33}$$

First we explain that the fundamental group $\pi_1(\mathbb{C}^3 \setminus \Delta, (a_0, b_0, c_0))$ in (9.33) is the pure braid group. The *braid group* B_n is a group defined as follows. Take n distinct points in a plane $\mathbb{R}^2$. We put the names $1, 2, \ldots, n$ to the n points. We call the plane $\mathbb{R}^2$ with the n marked points Π_1, and put its copy Π_2 in a parallel position to Π_1 in $\mathbb{R}^3$. We take n strings with initial points $1, 2, \ldots, n$ on Π_1 tending monotonically to the n points $1, 2, \ldots, n$ on Π_2 without intersecting each other. For each string, the end point can be different from the initial point (Fig. 9.7). We call the tuple of n strings a braid. Precisely speaking, we identify two braids which are deformed continuously, and a braid means an equivalence class. For two braids, we define the product by connecting each strings as in Fig. 9.8. Then the set of braids makes a group. We denote the group by B_n, and call it the braid group.

For $i = 1, 2, \ldots, n - 1$, we denote by s_i the braid which interchanges i and $i + 1$ as in Fig. 9.9 while connecting the other points straightly to themselves. It is known that the braid group B_n is generated by $s_1, s_2, \ldots, s_{n-1}$ and has the following

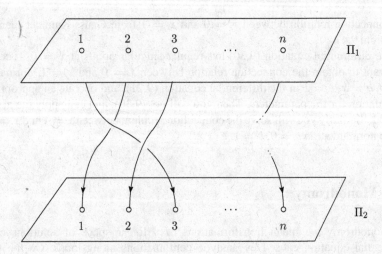

Fig. 9.7 B_n

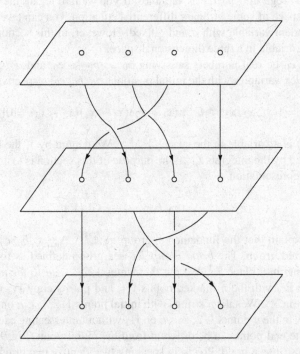

Fig. 9.8 Product in B_n

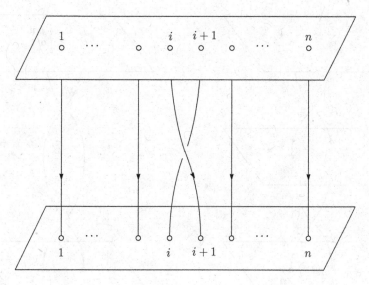

Fig. 9.9 s_i

presentation [7]:

$$B_n = \left\langle s_1, s_2, \ldots, s_{n-1} \;\middle|\; \begin{array}{l} s_i s_j = s_j s_i \ (|i - j| > 1), \\ s_i s_{i+1} s_i = s_{i+1} s_i s_{i+1} \ (1 \le i \le n - 2) \end{array} \right\rangle.$$

The relation $s_i s_{i+1} s_i = s_{i+1} s_i s_{i+1}$ is called the braid relation.

Each braid induces a permutation of n letters by associating the initial points to the end points. Then we have a homomorphism

$$B_n \to S_n$$

of groups, where S_n denotes the symmetric group of n letters. The *pure braid group* P_n is defined as the kernel of this homomorphism:

$$1 \to P_n \to B_n \to S_n \to 1.$$

The pure braid group P_n consists of braids with n strings each of which has the same initial and end point. By the definition, P_n is isomorphic to B_n/S_n. It is known that the pure braid group P_n is generated by s_i^2 $(1 \le i \le n - 1)$ [21, 46] (Fig. 9.10). Precisely speaking, we have

$$P_n = \langle \alpha s_i^2 \alpha^{-1} \ (1 \le i \le n - 1, \alpha \in B_n) \rangle.$$

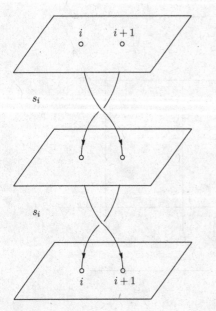
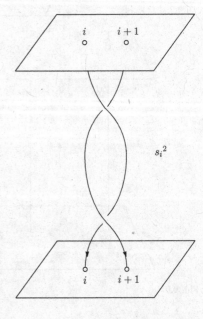

Fig. 9.10 s_i^2

Let Δ_n be a subset of $\mathbb{C}^n$ defined by

$$\Delta_n = \{(x_1, x_2, \ldots, x_n) \in \mathbb{C}^n \mid \prod_{i<j}(x_j - x_i) = 0\}.$$

Δ_n is called the diagonal of $\mathbb{C}^n$. Then we have the isomorphism

$$\pi_1(\mathbb{C}^n \setminus \Delta_n) \cong P_n.$$

The correspondence is given by regarding a point in $\mathbb{C}^n \setminus \Delta_n$ as a tuple of n distinct points in a complex line $\mathbb{C}$, which is regarded as the plane $\mathbb{R}^2$ in the definition of B_n.

Now we shall come back to the monodromy representation (9.33). As a basis of the linear space V, we take the integrals $I_{a,b}$, $I_{b,c}$ over the finite intervals (a, b), (b, c). Here we assume that (a, b, c) is on the initial position (a_0, b_0, c_0), and define the branches of $I_{a,b}$, $I_{b,c}$ by the standard loading. We are going to study the action of the generators s_1^2, s_2^2 of the fundamental group on this basis.

The action of s_1^2 is realized in $\mathbb{C}\setminus\{a_0, b_0, c_0\}$ by a curve where b encircles around a_0 once in the positive direction with base point b_0 (Fig. 9.11). The interval (a, b) is deformed during the movement of b, and finally coincides with the original interval

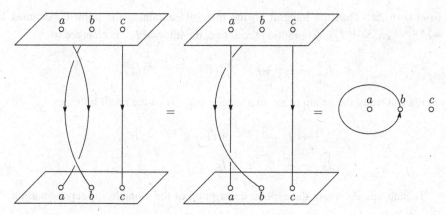

Fig. 9.11 Realization of $s_1{}^2$

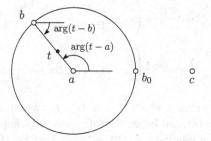

Fig. 9.12 Changes of $\arg(t-a)$ and $\arg(t-b)$

(a_0, b_0) when b comes back to b_0. By this movement, both of $\arg(t-a)$, $\arg(t-b)$ increase from 0 to 2π (Fig. 9.12). Thus the integral $I_{a,b}$ changes

$$I_{a,b} \rightsquigarrow e^{2\pi\sqrt{-1}(\lambda+\mu)} I_{a,b}.$$

The interval (b, c) is deformed, by the movement of b, as in Fig. 9.13. Thus the result is a union of the above three intervals. We look at the branch of each interval. The branch on (b, c) does not change. At the point Q, the branch is determined by the continuation of the branch on (b, c). Since $\arg(t-b)$ becomes 0 to π when t goes from (b, c) to Q via the upper half plane, the integral becomes $e^{\pi\sqrt{-1}\mu} I_{a,b}$. The branch at the point P is determined by $\arg(t-a)$, which is increased

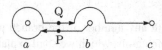

Fig. 9.13 Change of (b, c)

from 0 to 2π. Thus the integral on the interval containing the point P becomes $-e^{\pi\sqrt{-1}\mu}e^{2\pi\sqrt{-1}\lambda}I_{a,b}$. Combining the above, the integral $I_{b,c}$ is changed to

$$I_{b,c} \rightsquigarrow I_{b,c} + e^{\pi\sqrt{-1}\mu}(1 - e^{2\pi\sqrt{-1}\lambda})I_{a,b}.$$

We can study the action of $s_2{}^2$ in a similar way. Then the result becomes

$$I_{a,b} \rightsquigarrow I_{a,b} + e^{\pi\sqrt{-1}\mu}(1 - e^{2\pi\sqrt{-1}\nu})I_{b,c},$$
$$I_{b,c} \rightsquigarrow e^{2\pi\sqrt{-1}(\mu+\nu)}I_{b,c}.$$

To sum up, we obtain the explicit description of the monodromy representation (9.33):

$$(I_{a,b}, I_{b,c}) \overset{s_1{}^2}{\longmapsto} (I_{a,b}, I_{b,c})\begin{pmatrix} e^{2\pi\sqrt{-1}(\lambda+\mu)} & e^{\pi\sqrt{-1}\mu}(1 - e^{2\pi\sqrt{-1}\lambda}) \\ 0 & 1 \end{pmatrix},$$

$$(I_{a,b}, I_{b,c}) \overset{s_2{}^2}{\longmapsto} (I_{a,b}, I_{b,c})\begin{pmatrix} 1 & 0 \\ e^{\pi\sqrt{-1}\mu}(1 - e^{2\pi\sqrt{-1}\nu}) & e^{2\pi\sqrt{-1}(\mu+\nu)} \end{pmatrix}.$$

Here we explain the reason why we chose $I_{a,b}$, $I_{b,c}$ as a basis of V. The intervals (a, b), (b, c) are bounded, and hence their closures are compact. Then the images of these closures by a continuous map are also compact. Hence the images of (a, b), (b, c) by the action of π_1 become linear combinations of themselves. Namely this basis is closed under the action of π_1. If we use an unbounded interval as a member of a basis, the result of an action of π_1 may becomes a linear combination containing some bounded intervals. In such case, we have to use the linear relations (9.4), (9.5) in order to describe the result of the action in terms of the basis.

9.8   Multiple Integrals

In the above, we have explained several notions and techniques in the study of the integral representation by using a simple integral $I_{p,q}$. We can easily extend the results to the integrals

$$\int_p^q (t - a_1)^{\lambda_1}(t - a_2)^{\lambda_2}\cdots(t - a_m)^{\lambda_m}\, dt$$

of one integral variable with any number of the branch points. Moreover, we can also extend to the multiple integrals, which we shall explain in this section.

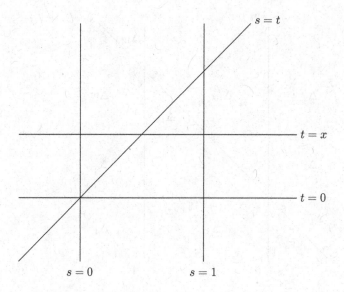

Fig. 9.14 Configuration of the branch locus

As an illustrative example, we consider the following integral

$$I_\Delta = \int_\Delta s^{\lambda_1}(s-1)^{\lambda_2}t^{\lambda_3}(t-x)^{\lambda_4}(s-t)^{\lambda_5}\,ds\,dt.$$

The variable x is a complex variable, however, for a moment we assume it is a real variable on the interval $0 < x < 1$. The exponents $\lambda_1, \lambda_2, \ldots, \lambda_5$ are complex constants. The set of the branch points of the integrand consists of five lines, which are illustrated in $\mathbb{R}^2$ as Fig. 9.14. We take a domain Δ of integration from the open regions in $\mathbb{R}^2$ surrounded by these lines.

On each Δ, the signature of each of the linear functions

$$s,\ s-1,\ t,\ t-x,\ s-t$$

in the integrand is definite, and hence we can define the branch on Δ by the standard loading. For example, when Δ is the region surrounded by $s = 0, t = x, s = t$, we have

$$s > 0,\ s-1 < 0,\ t > 0,\ t-x < 0,\ s-t < 0$$

on Δ, so that we define

$$I_\Delta = \int_\Delta s^{\lambda_1}(1-s)^{\lambda_2}t^{\lambda_3}(x-t)^{\lambda_4}(t-s)^{\lambda_5}\,ds\,dt,$$

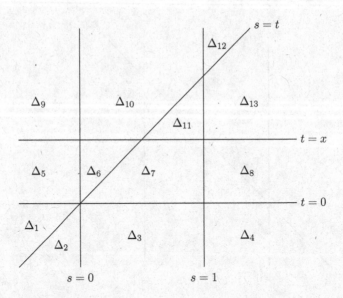

Fig. 9.15 Domains Δ_j

where we set $\arg s = \arg(1 - s) = \arg t = \arg(x - t) = \arg(t - s) = 0$. We set

$$U(s, t) = s^{\lambda_1}(s - 1)^{\lambda_2}t^{\lambda_3}(t - x)^{\lambda_4}(s - t)^{\lambda_5},$$

and denote by $U_\Delta(s, t)$ the result of the standard loading of U on each Δ.

We are going to obtain linear relations over $\mathbb{C}$ among the integrals I_Δ. We put a number on every domain as in Fig. 9.15. We consider the integral over Δ_1. Geometrically Δ_1 is given by

$$\Delta_1 = \{(s, t) \in \mathbb{R}^2 \mid s < t < 0\},$$

and hence the integral over this chain is given by the iterated integral

$$\int_{\Delta_1} U_{\Delta_1}(s, t)\, ds\, dt = \int_{-\infty}^{0} dt \int_{-\infty}^{t} U_{\Delta_1}(s, t)\, ds. \tag{9.34}$$

In a similar way, we have the iterated integrals for Δ_2, Δ_3, Δ_4:

$$\int_{\Delta_2} U_{\Delta_2}(s, t)\, ds\, dt = \int_{-\infty}^{0} dt \int_{t}^{0} U_{\Delta_2}(s, t)\, ds,$$

$$\int_{\Delta_3} U_{\Delta_3}(s, t)\, ds\, dt = \int_{-\infty}^{0} dt \int_{0}^{1} U_{\Delta_3}(s, t)\, ds, \tag{9.35}$$

$$\int_{\Delta_4} U_{\Delta_4}(s, t)\, ds\, dt = \int_{-\infty}^{0} dt \int_{1}^{+\infty} U_{\Delta_4}(s, t)\, ds.$$

We fix t satisfying $-\infty < t < 0$, and set

$$J_{p,q} = \int_p^q U(s,t)\,ds \quad (p, q \in \{-\infty, t, 0, 1, +\infty\}).$$

Then $J_{p,q}$ is essentially similar to the integral $I_{p,q}$ we studied so far. Therefore, if we assume that each integral becomes definite by the standard loading, we get two linear relations

$$J_{-\infty,t} + e_5 J_{t,0} + e_1 e_5 J_{0,1} + e_1 e_2 e_5 J_{1,+\infty} = 0,$$

$$J_{-\infty,t} + e_5^{-1} J_{t,0} + (e_1 e_5)^{-1} J_{0,1} + (e_1 e_2 e_5)^{-1} J_{1,+\infty} = 0$$

as a similar manner in Sect. 9.2, where

$$e_j = e^{\pi \sqrt{-1}\lambda_j} \quad (1 \le j \le 5).$$

Integrating these relations from $-\infty$ to 0 with respect to t, we get the linear relations for the integrals over $\Delta_1, \Delta_2, \Delta_3, \Delta_4$ by the help of (9.34), (9.35). Thus we get

$$\begin{cases} I_{\Delta_1} + e_5 I_{\Delta_2} + e_1 e_5 I_{\Delta_3} + e_1 e_2 e_5 I_{\Delta_4} = 0, \\ I_{\Delta_1} + e_5^{-1} I_{\Delta_2} + (e_1 e_5)^{-1} I_{\Delta_3} + (e_1 e_2 e_5)^{-1} I_{\Delta_4} = 0. \end{cases} \tag{9.36}$$

In a similar way, we have the following sets of linear relations:

$$\begin{cases} I_{\Delta_5} + e_1 I_{\Delta_6} + e_1 e_5 I_{\Delta_7} + e_1 e_2 e_5 I_{\Delta_8} = 0, \\ I_{\Delta_5} + e_1^{-1} I_{\Delta_6} + (e_1 e_5)^{-1} I_{\Delta_7} + (e_1 e_2 e_5)^{-1} I_{\Delta_8} = 0, \end{cases}$$

$$\begin{cases} I_{\Delta_9} + e_1 I_{\Delta_{10}} + e_1 e_5 I_{\Delta_{11}} + e_1 e_2 I_{\Delta_{12}} + e_1 e_2 e_5 I_{\Delta_{13}} = 0, \\ I_{\Delta_9} + e_1^{-1} I_{\Delta_{10}} + (e_1 e_5)^{-1} I_{\Delta_{11}} + (e_1 e_2)^{-1} I_{\Delta_{12}} + (e_1 e_2 e_5)^{-1} I_{\Delta_{13}} = 0, \end{cases}$$

$$\begin{cases} I_{\Delta_2} + e_5 I_{\Delta_1} + e_3 e_5 I_{\Delta_5} + e_3 e_4 e_5 I_{\Delta_9} = 0, \\ I_{\Delta_2} + e_5^{-1} I_{\Delta_1} + (e_3 e_5)^{-1} I_{\Delta_5} + (e_3 e_4 e_5)^{-1} I_{\Delta_9} = 0, \end{cases}$$

$$\begin{cases} I_{\Delta_3} + e_3 I_{\Delta_7} + e_3 e_5 I_{\Delta_6} + e_3 e_4 I_{\Delta_{11}} + e_3 e_4 e_5 I_{\Delta_{10}} = 0, \\ I_{\Delta_3} + e_3^{-1} I_{\Delta_7} + (e_3 e_5)^{-1} I_{\Delta_6} + (e_3 e_4)^{-1} I_{\Delta_{11}} + (e_3 e_4 e_5)^{-1} I_{\Delta_{10}} = 0, \end{cases}$$

$$\begin{cases} I_{\Delta_4} + e_3 I_{\Delta_8} + e_3 e_4 I_{\Delta_{13}} + e_3 e_4 e_5 I_{\Delta_{12}} = 0, \\ I_{\Delta_4} + e_3^{-1} I_{\Delta_8} + (e_3 e_4)^{-1} I_{\Delta_{13}} + (e_3 e_4 e_5)^{-1} I_{\Delta_{12}} = 0. \end{cases} \tag{9.37}$$

We obtained these relations by applying the method in Sect. 9.2 to the lines parallel to the coordinate axes. In general case, we can apply the method in Sect. 9.2 by taking an intersection of two lines in the set of the branch points. For example,

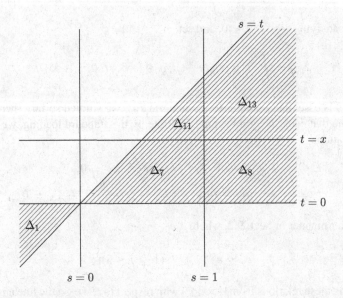

Fig. 9.16 Domains swept by a line passing through $(0, 0)$

we take the intersection $(0, 0)$. Then the integrals over $\Delta_1, \Delta_7, \Delta_8, \Delta_{11}, \Delta_{13}$ can be regarded as iterated integrals on a line passing through $(0, 0)$ and on the slope of the line (Fig. 9.16). From the integrals over a line passing through $(0, 0)$, we obtain linear relations of the integrals, and then we integrate these relations with respect to the slope to get the linear relations among I_{Δ_j}. Noting that the integral with respect to the slope has different signature on the two sides with respect to $(0, 0)$, we get the following relations:

$$
\begin{cases}
I_{\Delta_1} - e_1 e_3 e_5 I_{\Delta_7} - e_1 e_2 e_3 e_5 I_{\Delta_8} - e_1 e_3 e_4 e_5 I_{\Delta_{11}} - e_1 e_2 e_3 e_4 e_5 I_{\Delta_{13}} = 0, \\
I_{\Delta_1} - (e_1 e_3 e_5)^{-1} I_{\Delta_7} - (e_1 e_2 e_3 e_5)^{-1} I_{\Delta_8} - (e_1 e_3 e_4 e_5)^{-1} I_{\Delta_{11}} \\
\hspace{6cm} - (e_1 e_2 e_3 e_4 e_5)^{-1} I_{\Delta_{13}} = 0.
\end{cases}
$$

We can get such relations for each intersection of the lines in the set of the branch points. This method is given by Aomoto [4].

In this way, we can get many linear relations among the integrals I_{Δ_i} ($1 \leq i \leq 13$), and then these integrals become linearly dependent. It turns out that the dimension of the linear space spanned by these integrals becomes three, which coincides with the number of the bounded domains. The coincidence of the dimension and the number of the bounded domains is a consequence of the general assertion given by Kohno [110].

Next we study the asymptotic behavior of the integrals. We consider the behaviors for the limits $x \to 0$ and $x \to 1$. We start from finding domains of integration such that the integrals give simple asymptotic behavior. As in Sect. 9.3, we can find such domains among collapsing domains by the limit and among invariant domains by the limit. When we consider the limit $x \to 0$, the collapsing domains are $\Delta_5, \Delta_6, \Delta_7, \Delta_8$, and the invariant domain is Δ_{12}. The integrals over $\Delta_5, \Delta_6, \Delta_7, \Delta_8$ are linearly dependent owing to the relation (9.37), and two of them are linearly independent. We shall find the domains among these four such that the integrals give simple asymptotic behaviors for the limit $x \to 0$. Assume that the integral over a domain Δ_A gives a simple asymptotic behavior

$$ I_{\Delta_A} \sim C x^\mu \quad (x \to 0). $$

Then, by the analytic continuation along a path encircling $x = 0$ once in the positive direction, we get

$$ I_{\Delta_A} \rightsquigarrow e^{2\pi \sqrt{-1}\mu} I_{\Delta_A}. $$

Thus each integral which gives a simple asymptotic behavior is an eigenfunction for the analytic continuation. Now we explain how to obtain the analytic continuation for multiple integrals. As an example, we take the domain Δ_7, which seems relatively complicated.

Let r $(0 < r < 1)$ be the initial position of x. We study the change of I_{Δ_7} when x encircles 0 along $|x| = r$. The multiple integral I_{Δ_7} can be written as the iterated integral

$$ I_{\Delta_7} = \int_0^x dt \int_t^1 U_{\Delta_7}(s, t)\, ds. $$

If we understand the integral with respect to t as the integral on the segment $\overline{0x}$ from 0 to x in the complex t-plane, and also the integral with respect to s as the integral on the segment $\overline{t1}$ from t to 1 in the complex s-plane, this integral has a definite meaning. In this iterated integral, the singular points in the complex t-plane are $t = 0, x$, and the singular points in the complex s-plane are $s = 0, 1, t$. When x encircles 0 once in the x-plane, the singular point x in the t-plane encircles $t = 0$ once. Then the singular point t in the s-plane, which is on the segment $\overline{0x}$, encircles $s = 0$ once. Hence we have the change of the paths of integration as in Fig. 9.17. As a result, there appears a new integral over the segment $(0, t)$ in the s-plane. By considering the branches on each segments, we get the result of the analytic continuation

$$ I_{\Delta_7} \rightsquigarrow e^{2\pi \sqrt{-1}(\lambda_3 + \lambda_4)} \left(I_{\Delta_7} + e^{\pi \sqrt{-1}\lambda_5}(1 - e^{2\pi \sqrt{-1}\lambda_1}) I_{\Delta_6} \right). $$

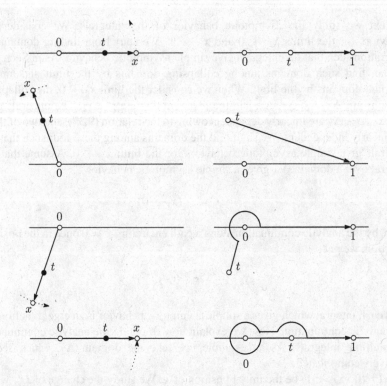

Fig. 9.17 Change of paths for Δ_7

In particular, the integral over I_{Δ_7} is not an eigenfunction for the analytic continuation around $x = 0$.

By analyzing in this way, we find that the integrals over Δ_6 and Δ_8 give eigenfunctions. Then we get the asymptotic behaviors

$$I_{\Delta_6} \sim C_1 x^{\lambda_1+\lambda_3+\lambda_4+\lambda_5+2} \quad (x \to 0),$$

$$I_{\Delta_8} \sim C_2 x^{\lambda_3+\lambda_4+1} \qquad (x \to 0)$$

by appropriate changes of the integral variables. On the other hand, for Δ_{12}, we have immediately

$$I_{\Delta_{12}} \sim C_3 \quad (x \to 0).$$

The asymptotic behaviors for the limit $x \to 1$ looks somewhat different. The collapsing domain is Δ_{11}, and invariant domains are $\Delta_1, \Delta_2, \Delta_3, \Delta_4$. By an appropriate change of the integral variables, we get the asymptotic behavior

$$I_{\Delta_{11}} \sim C_4 (1 - x)^{\lambda_2+\lambda_4+\lambda_5+2} \quad (x \to 1).$$

On the other hand, we immediately get the asymptotic behaviors

$$I_{\Delta_j} \sim D_j \quad (x \to 1), \quad (j = 1, 2, 3, 4),$$

where D_j $(1 \le j \le 4)$ are constants. Then every $\Delta_1, \ldots, \Delta_4$ is an eigenfunction for the analytic continuation around $x = 1$. Thanks to the relation (9.36), among $\Delta_1, \ldots, \Delta_4$ two are linearly independent. Thus the dimension of the linear space of holomorphic solutions at $x = 1$ is two, and all the four integrals are eigenfunctions. Therefore, we cannot choose a canonical basis from the four integrals on account of being eigenfunctions. (In general, there is no way to choose a canonical basis of a linear space of dimension greater than one.) This makes the analysis difficult for the case of higher dimensional eigenspaces. (We have already discussed in Sect. 6.3.3 on the connection problem for such case.)

Next we shall derive the differential equation satisfied by the integral I_Δ. Note that we can define the regularization also for the double integral I_Δ. This will be explained in the next Sect. 9.9. Then we assume that the integrals are already regularized. Or, we take $\mathrm{Re}\,\lambda_j$ etc. sufficiently large so that all improper integrals are convergent.

In order to derive the differential equation, we apply the method explained in Sect. 9.5 to the double integral. Let Δ be a domain surrounded by some of the lines $s = 0, s = 1, t = 0, t = x, s = t$ and ∞. Namely, Δ is one of $\Delta_1, \ldots, \Delta_{13}$ in Fig. 9.15 or a union of some of these domains. By the Stokes theorem, we have for any 1-form φ

$$\int_\Delta d(U\varphi) = \int_{\partial\Delta} U\varphi.$$

Here the right hand side is the integral of $U(s, t)$ over the boundary $\partial\Delta$, and, since we assume that the integral is regularized, we may understand $U(s, t)|_{\partial\Delta} = 0$. Hence the right hand side becomes 0. On the other hand, if we set

$$d \log U = \frac{dU}{U} = \omega,$$

we have

$$\int_\Delta d(U\varphi) = \int_\Delta U(\omega \wedge + d)\varphi.$$

Since the left hand side is equal to 0, we have the relation

$$(\omega \wedge + d)\varphi \equiv 0 \tag{9.38}$$

according to the notation in Sect. 9.5. By taking various φ, we get many relations. For example, if we take $\varphi = f(t)\,ds$ with any function $f(t)$ in t, the relation (9.38) induces

$$f(t)\left(\frac{\lambda_1}{s} + \frac{\lambda_2}{s-1} + \frac{\lambda_5}{s-t}\right) ds \wedge dt \equiv 0. \tag{9.39}$$

If we take $\varphi = g(s)\,dt$ with any function $g(s)$ in s, we get

$$g(s)\left(\frac{\lambda_3}{t} + \frac{\lambda_4}{t-x} + \frac{\lambda_5}{t-s}\right) ds \wedge dt \equiv 0. \tag{9.40}$$

We take particular 2-forms φ_1, φ_2, φ_3 given by

$$\varphi_1 = \frac{ds \wedge dt}{st}, \quad \varphi_2 = \frac{ds \wedge dt}{(s-1)t}, \quad \varphi_3 = \frac{ds \wedge dt}{(s-1)(t-s)}. \tag{9.41}$$

Set

$$y_j(x) = \int_\Delta U\varphi_j \quad (j = 1, 2, 3),$$

and denote the 2-form ψ_j satisfying

$$y_j'(x) = \int_\Delta U\psi_j$$

by φ_j'. Then, by using the relations (9.39), (9.40) and the relation obtained from $\varphi = d\log(t-s)$, we get

$$\begin{cases} \varphi_1' = \dfrac{1}{x}[(\lambda_1 + \lambda_3 + \lambda_4 + \lambda_5)\varphi_1 + \lambda_2\varphi_2], \\[2mm] \varphi_2' = \dfrac{1}{x}[(\lambda_3 + \lambda_4)\varphi_2 + \lambda_5\varphi_3], \\[2mm] \varphi_3' = \dfrac{1}{x-1}\left[\dfrac{\lambda_1(\lambda_1 + \lambda_3 + \lambda_5)}{\lambda_5}\varphi_1 + \dfrac{\lambda_1\lambda_2 + \lambda_2\lambda_3 + \lambda_3\lambda_5}{\lambda_5}\varphi_2 + (\lambda_2 + \lambda_4 + \lambda_5)\varphi_3\right]. \end{cases}$$

From this, we readily obtain the differential equation satisfied by $Y(x) = {}^t(y_1(x), y_2(x), y_3(x))$:

$$\frac{dY}{dx} = \left(\frac{A}{x} + \frac{B}{x-1} \right) Y, \qquad (9.42)$$

$$A = \begin{pmatrix} \lambda_1 + \lambda_3 + \lambda_4 + \lambda_5 & \lambda_2 & 0 \\ 0 & \lambda_3 + \lambda_4 & \lambda_5 \\ 0 & 0 & 0 \end{pmatrix},$$

$$B = \begin{pmatrix} 0 & 0 & 0 \\ 0 & 0 & 0 \\ \frac{\lambda_1(\lambda_1+\lambda_3+\lambda_5)}{\lambda_5} & \frac{\lambda_1\lambda_2+\lambda_2\lambda_3+\lambda_3\lambda_5}{\lambda_5} & \lambda_2 + \lambda_4 + \lambda_5 \end{pmatrix}.$$

If you want to obtain the differential equation satisfied by I_Δ, you only need to change the parameters λ_1, λ_3 to $\lambda_1 + 1, \lambda_3 + 1$, respectively, and to derive the differential equation satisfied by y_1 from (9.42). We note that the differential equation (9.42) can be reduced to the differential equation ($_3E_2$) satisfied by the generalized hypergeometric series ${}_3F_2 \begin{pmatrix} \alpha_1, \alpha_2, \alpha_3 \\ \beta_1, \beta_2 \end{pmatrix} ; x \end{pmatrix}$ by an appropriate change of parameters. The explicit form of ($_3E_2$) is given by (7.12) in Sect. 7.2.

We can also obtain the connection coefficients and monodromy representations in a similar manner as in Sects. 9.6 and 9.7. The connection coefficients are obtained by combining the linear relations (9.36), (9.37) among the integrals I_Δ and the asymptotic behaviors of the integrals. In order to obtain the monodromy, we take the integrals over the bounded domains $I_{\Delta_6}, I_{\Delta_7}, I_{\Delta_{11}}$ as a basis of the three dimensional linear space spanned by the integrals I_{Δ_j} ($1 \le j \le 13$), and study the change of the integrals under the action of the fundamental group. We have already calculated the change of Δ_7 by the analytic continuation around $x = 0$. The other cases are similarly calculated.

At the end of this section, we remark the reason why we took the 2-forms φ_j in (9.41). These 2-forms becomes a basis of the cohomology group, which will be explained in the next section, and can be written as follows:

$$\varphi_1 = d \log s \wedge d \log t, \quad \varphi_2 = d \log(s-1) \wedge d \log t, \quad \varphi_3 = d \log(s-1) \wedge d \log(t-s).$$

Thus each φ_j is an exterior product of logarithmic forms of the linear functions defining the branch points of the integrand $U(s, t)$. In general, we can take a basis of the cohomology group from such exterior products of logarithmic forms. Please refer to [37] for this property.

9.9 Theory of Twisted (Co)homology

The theory of integral representations explained so far is systematized as the theory of (co)homology with coefficients in a local system. The theory is also called the theory of twisted (co)homology. Here a *local system* means a locally constant sheaf. Namely, a sheaf $\mathcal{F}$ on a topological space X is a local system if, for any $x \in X$, there exists a neighborhood V of x such that $\mathcal{F}|_V$ is a constant sheaf. The theory of (co)homology with coefficients in a local system is established by Eilenberg and Steenrod [169]. Aomoto [2–4] noticed that this theory becomes a powerful tool for the study of hypergeometric integrals, and developed the theory.

In this section we briefly explain the outline of the theory. If you want to study precisely, we recommend the books [5, 136] and [67].

Let M be a complex manifold of dimension n, and U a multi-valued function on M such that its logarithmic derivative

$$d \log U = \omega$$

is a single-valued holomorphic 1-form on M. We can regard the multi-valued function U as a section of some local system on M. Namely, we can define a local system $\mathcal{K}$ over M whose local section is given by a branch of U at each point of M. We denote by $\mathcal{L}$ the dual local system $\mathcal{K}^\vee$ of $\mathcal{K}$. In the following we mainly use $\mathcal{L}$, and denote $\mathcal{K}$ by $\mathcal{L}^\vee$.

Let Δ be a p-chain on M, and φ a holomorphic p-form on M. Consider the integral

$$\int_\Delta U\varphi. \tag{9.43}$$

In order to define the integral, we should determine the branch of U on the chain Δ. Then we understand that Δ is not only a chain but a chain with an assigned branch of U to the chain. We call such chains *twisted chains*. The boundary operator for twisted chains is defined as a usual boundary operator for topological chains where the branch of the boundary is determined by the continuation of the branch of the twisted chain. We denote this boundary operator by ∂_ω. Then we can define the homology group for twisted chains and the boundary operator ∂_ω. We denote the homology group by $H_p(M, \mathcal{L}^\vee)$, and call it the homology group with coefficients in a local system, or the twisted homology group. We call an element of the twisted homology group a twisted cycle.

Now we go to the definition of twisted cocycles. By the Stokes theorem, we have

$$\int_\Delta d(U\varphi) = \int_{\partial\Delta} U\varphi, \tag{9.44}$$

whose left hand side can be written as

$$\int_\Delta d(U\varphi) = \int_\Delta U\left(\frac{dU}{U} \wedge \varphi + d\varphi\right) = \int_\Delta U(\omega \wedge + d)\varphi.$$

Recall that we set $\omega = d\log U = dU/U$. Then, introducing the operator

$$\nabla_\omega = d + \omega\wedge,$$

we can write (9.44) as

$$\int_\Delta U\nabla_\omega\varphi = \int_{\partial\Delta} U\varphi.$$

It can be shown that $\nabla_\omega \circ \nabla_\omega = 0$ holds. Then we can define the cohomology group $H^p(\Omega^\bullet, \nabla_\omega)$ by the coboundary operator ∇_ω, where Ω^p denotes the sheaf of holomorphic p-forms on M. This cohomology group is called the cohomology group with coefficient in a local system, or the twisted cohomology group. An element of the twisted cohomology group is called a twisted cocycle. It can be shown that the twisted cohomology group is isomorphic to $H^p(M, \mathcal{L})$, and the integral (9.43) can be understood as the paring

$$\begin{aligned} H_p(M, \mathcal{L}^\vee) \times H^p(M, \mathcal{L}) &\to \quad \mathbb{C} \\ (\Delta, \varphi) \quad &\mapsto \int_\Delta U\varphi. \end{aligned}$$

We often consider integrals over an open interval and over an open polyhedron. Such domains can be understood as locally finite chains. For such infinite chains, we have the twisted homology group $H_p^{lf}(M, \mathcal{L}^\vee)$ of locally finite chains. There is a natural inclusion map

$$H_p(M, \mathcal{L}^\vee) \to H_p^{lf}(M, \mathcal{L}^\vee).$$

Then the regularization can be understood as the inverse operation of this map. If we consider an integral (9.43) over a locally finite chain Δ, we should take a differential form φ with compact support. We denote by $H_c^p(M, \mathcal{L})$ the set of twisted cocycles with compact support. Then the integral (9.43) in this case can be understood as the pairing

$$\begin{aligned} H_p^{lf}(M, \mathcal{L}^\vee) \times H_c^p(M, \mathcal{L}) &\to \quad \mathbb{C} \\ (\Delta, \varphi) \quad &\mapsto \int_\Delta U\varphi. \end{aligned}$$

The above formalization not only gives a definite meaning for the integral (9.43) but also brought a discovery of the intersection theory of twisted homology and of twisted cohomology. The intersection number of twisted cycles is defined as a paring

$$H_p(M, \mathcal{L}^\vee) \times H_p^{lf}(M, \mathcal{L}) \to \quad \mathbb{C}$$
$$(\sigma, \tau) \qquad\qquad \mapsto \sigma \cdot \tau,$$

which is determined by the intersection number as the topological chains together with the branches of U on the chains. The intersection number of twisted cocycles is defined as a pairing

$$H_c^{2n-p}(M, \mathcal{L}) \times H^p(M, \mathcal{L}^\vee) \to \quad \mathbb{C}$$
$$(\varphi, \psi) \qquad\qquad \mapsto \int_M \varphi \wedge \psi.$$

It is shown that these pairings are non-degenerate in general.

These intersection theories are interesting as theories, and also have many applications. For the Gauss hypergeometric differential equation and some generalizations, it is known the existence of the monodromy invariant Hermitian form when the characteristic exponents are real numbers. Kita-Yoshida [108] clarified that the invariant Hermitian form can be regarded as the inverse of the intersection matrix of twisted homology associated with the integral representation of solutions. Cho-Matsumoto [26] combined the intersection theory of twisted homology and one of twisted cohomology, and obtained the twisted version of Riemann's period relation. Applying the relation, we get systematically quadratic relations such as

$$F(\alpha, \beta, \gamma; x) F(1 - \alpha, 1 - \beta, 2 - \gamma; x)$$
$$= F(\alpha + 1 - \gamma, \beta + 1 - \gamma, 2 - \gamma; x) F(\gamma - \alpha, \gamma - \beta, \gamma; x) \tag{9.45}$$

for hypergeometric functions. So far, relations like (9.45) were found sporadically. Now we get a systematic way to obtain such relations, not only for the Gauss hypergeometric functions but also hypergeometric functions in several variables such as Appell-Lauricella.

Moreover, the intersection theory of twisted homology brings a powerful tool to the connection problem. For solutions represented by integrals, we can directly obtain the coefficients of linear relations of these solutions by calculating the intersection numbers of twisted cycles in the integral. Then we do not need to obtain linear relations in a way as in Sects. 9.2 and 9.8. This tool is particularly useful for the connection problem of multiple integrals. For more detail, please refer to [124–126].

9.10 Integral Representation of Solutions of Legendre's Equation

Although we have not mentioned explicitly, we assumed that the exponents λ, μ, ν and $\lambda_1, \lambda_2, \ldots$ in the integrals I_{pq} and I_Δ in Sections from 9.1 to 9.7 are generic. Here generic means that there is no linear relation among the exponents and 1 with coefficients in integers. Namely, λ, μ, ν are generic if

$$k + l\lambda + m\mu + n\nu = 0 \quad (k, l, m, n \in \mathbb{Z})$$

implies $k = l = m = n = 0$. Under this assumption, many operations become well-defined and many results hold, however, there are many important integrals with non-generic exponents. Such integrals should be studied individually.

In this section, as an example of such integrals, we study the integral representation of solutions of Legendre's differential equation. Legendre's equation has already appeared in Chap. 6, and the explicit form is given by

$$(1 - t^2)\frac{d^2 y}{dt^2} - 2t\frac{dy}{dt} + \lambda y = 0. \tag{9.46}$$

This is a Fuchsian differential equation of the second order. Here $\lambda \in \mathbb{C}$ is a parameter. The Riemann scheme is given by (6.13). By the Riemann scheme, we see that there exists logarithmic solutions at $t = \pm 1$.

Legender's equation (9.46) can be obtained from the hypergeometric differential equation (9.22) by specializing the parameters as

$$\alpha + \beta = 1, \ \gamma = 1, \tag{9.47}$$

and by changing the independent variable as

$$x = \frac{1 - t}{2}. \tag{9.48}$$

Then we have

$$\lambda = -\alpha\beta = \alpha(\alpha - 1). \tag{9.49}$$

As we have seen in Sect. 9.5, the hypergeometric differential equation (9.22) has the integral I_{pq} as a solution. In Theorem 9.3 we gave an explicit integral representation of solutions of (9.22), however, here we use the following integral

$$y(x) = \int_p^q s^{\alpha - \gamma}(1 - s)^{\gamma - \beta - 1}(s - x)^{-\alpha}\, ds,$$

which is obtained by exchanging α and β. We can derive the integral representation of solutions of Legendre's equation by the specialization (9.47) of the parameters and the change (9.48) of the independent variable. Since the change (9.48) makes the description a little longer, we continue to use x. We recall that x is the right hand side of (9.48) if necessary. In particular, $x = 0$ means $t = 1$, and $x = 1$ means $t = -1$. Thus we have the integral representation

$$y(t) = \int_\Delta s^{\alpha-1}(1-s)^{\alpha-1}(s-x)^{-\alpha}\,ds \qquad (9.50)$$

of solutions of Legendre's equation. Note that the integral is a function in t, and the parameter α is related to the parameter λ of Legendre's equation by the relation (9.49). The path of integration is regarded as a twisted cycle, and is denoted by Δ. We see that, in the integral (9.50), the exponents of the integrand is not generic. For a moment we assume

$$\alpha \notin \mathbb{Z}.$$

We shall solve the connection problem between two singular points $t = 1$ and $t = -1$ of Legendre's equation by using the above integral. Then we shall specify the paths of integration which correspond to solutions with pure asymptotic behaviors, and shall obtain linear relations among these paths.

The characteristic exponents at the regular singular point $t = 1$ is $0, 0$, and then there are linearly independent solutions of (9.46) of the form

$$\begin{cases} y_1^+(t) = 1 + \displaystyle\sum_{n=1}^{\infty} a_n(t-1)^n, \\[2mm] y_2^+(t) = y_1^+(t)\log(t-1) + \displaystyle\sum_{n=0}^{\infty} b_n(t-1)^n. \end{cases} \qquad (9.51)$$

Note that $y_1^+(t)$ is uniquely determined, while $y_2^+(t)$ has an ambiguity of adding a constant multiple of $y_1^+(t)$. This ambiguity is fixed if we specify the value of b_0. Similarly, we have linearly independent solutions at the regular singular point $t = -1$:

$$\begin{cases} y_1^-(t) = 1 + \displaystyle\sum_{n=1}^{\infty} c_n(t+1)^n, \\[2mm] y_2^-(t) = y_1^-(t)\log(t+1) + \displaystyle\sum_{n=0}^{\infty} d_n(t+1)^n. \end{cases} \qquad (9.52)$$

Also $y_2^-(t)$ has an ambiguity, which is fixed by specifying the value of d_0.

Fig. 9.18 Paths $\Delta_0, \Delta_1, \Delta_x$

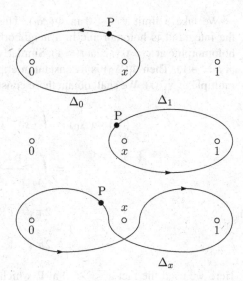

We want to obtain the paths Δ of integration corresponding to these solutions. However in this non-generic case, we cannot apply the method in Sect. 9.3. Nevertheless, by the help of the relations among the exponents, we can find clever paths of integration. Namely we take the paths $\Delta_0, \Delta_1, \Delta_x$ as in Fig. 9.18. We assume that x is a real number on $0 < x < 1$, and take a common initial point P for $\Delta_0, \Delta_1, \Delta_x$ which is located in the upper half plane with its real part between 0 and x. We fix the branch of the integrand

$$s^{\alpha-1}(1-s)^{\alpha-1}(s-x)^{-\alpha}$$

on each path by specifying the arguments at P as

$$\arg s \approx 0, \ \arg(1-s) \approx 0, \ \arg(s-x) \approx \pi. \tag{9.53}$$

We set

$$\Phi(\Delta) = \int_\Delta s^{\alpha-1}(1-s)^{\alpha-1}(s-x)^{-\alpha}\,ds.$$

When s encircles along Δ_0 once, both $\arg s$ and $\arg(s-x)$ increase by 2π, and these change are cancelled in branch because the exponents $\alpha - 1$, $-\alpha$ are negative each other modulo integers. Thus the path Δ_0 becomes a twisted cycle in the sense of Sect. 9.9. Namely, we have $\partial_\omega \Delta_0 = 0$. This implies that the integral $\Phi(\Delta_0)$ becomes a solution of the differential equation (9.46). Similarly $\Phi(\Delta_1)$ becomes a solution. When s moves along Δ_x, $\arg s$ increases by 2π, while $\arg(1-s)$ decreases by 2π. Then also these changes are cancelled in branch, and hence Δ_x becomes a twisted cycle. Thus $\Phi(\Delta_x)$ is also a solution.

We take a limit $x \to 0$ in $\Phi(\Delta_0)$. This causes no change to the path, and the integrand is holomorphic in a neighborhood of $x = 0$. Therefore $\Phi(\Delta_0)$ is holomorphic at $x = 0$ (i.e. at $t = 1$). Similarly, $\Phi(\Delta_1)$ is holomorphic at $x = 1$ (i.e. at $t = -1$). Then $\Phi(\Delta_0)$ is a constant multiple of $y_1^+(t)$, and $\Phi(\Delta_1)$ is a constant multiple of $y_1^-(t)$. We shall obtain these constants. We have

$$
\begin{aligned}
\lim_{x \to 0} \Phi(\Delta_0) &= \int_{\Delta_0} s^{\alpha-1}(1-s)^{\alpha-1} s^{-\alpha} \, ds \\
&= \int_{\Delta_0} \frac{(1-s)^{\alpha-1}}{s} \, ds \\
&= 2\pi\sqrt{-1} \operatorname*{Res}_{s=0} \frac{(1-s)^{\alpha-1}}{s} \, ds \\
&= 2\pi\sqrt{-1}.
\end{aligned}
$$

Here we used the fact $\arg s \approx 0$ at P, which comes from the specification (9.53). Also we used $\arg(1-s) \approx 0$, which comes from (9.53), to show

$$
\lim_{s \to 0} (1-s)^{\alpha-1} = 1.
$$

In a similar way, we have

$$
\begin{aligned}
\lim_{x \to 1} \Phi(\Delta_1) &= \int_{\Delta_1} s^{\alpha-1}(1-s)^{\alpha-1} (e^{\pi\sqrt{-1}}(1-s))^{-\alpha} \, ds \\
&= e^{-\pi\sqrt{-1}\alpha} \int_{\Delta_1} \frac{s^{\alpha-1}}{1-s} \, ds \\
&= e^{-\pi\sqrt{-1}\alpha} 2\pi\sqrt{-1} \operatorname*{Res}_{s=1} \frac{s^{\alpha-1}}{1-s} \, ds \\
&= e^{\pi\sqrt{-1}(1-\alpha)} 2\pi\sqrt{-1}.
\end{aligned}
$$

Then we get

$$
\begin{aligned}
\Phi(\Delta_0) &= 2\pi\sqrt{-1}\, y_1^+(t), \\
\Phi(\Delta_1) &= e^{\pi\sqrt{-1}(1-\alpha)} 2\pi\sqrt{-1}\, y_1^-(t).
\end{aligned} \tag{9.54}
$$

Then we may think that $\Phi(\Delta_x)$ is a solution containing a logarithmic term. To show this, we may study the asymptotic behavior of $\Phi(\Delta_x)$ as $x \to 0$ as we have done in Sect. 9.3. However, here we will go another way. We study the local monodromy as in Sect. 9.8. Namely we study the change of Δ_x when x encircles $s = 0$ once.

Fig. 9.19 Dividing Δ_x into Γ_1 and Γ_2

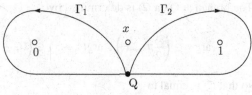

Fig. 9.20 Change of Γ_1

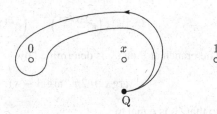

Fig. 9.21 Rewriting the change of Γ_1

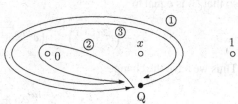

We divide Δ_x into the left part Γ_1 and the right part Γ_2 as in Fig. 9.19, and let Q be the common initial point. The branch on Γ_1, Γ_2 are determined by (9.53). Explicitly, the branch at the initial point Q of Γ_1 is determined by

$$\arg s \in \left(-\frac{\pi}{2}, 0\right), \ \arg(1 - s) \in \left(0, \frac{\pi}{2}\right), \ \arg(s - x) = \frac{3}{2}\pi,$$

and the branch at the initial point Q of Γ_2 is determined by

$$\arg s \in \left(\frac{3}{2}\pi, 2\pi\right), \ \arg(1 - s) \in \left(0, \frac{\pi}{2}\right), \ \arg(s - x) = \frac{3}{2}\pi.$$

When x encircles $s = 0$ once in the positive direction, Γ_2 does not change, and Γ_1 changes as in Fig. 9.20. We rewrite the result as in Fig. 9.21. Apart from the difference of the branches, ①, ②, ③ are equal to Δ_0, Γ_1, $-\Delta_0$, respectively. The branch at P on ① is determined by

$$\arg s \approx 0, \ \arg(1 - s) \approx 0, \ \arg(s - x) \approx 3\pi,$$

so that ① is equal to

$$\left(e^{2\pi\sqrt{-1}}\right)^{-\alpha} \Delta_0.$$

The branch at Q on ② is determined by

$$\arg s \in \left(\frac{3}{2}\pi, 2\pi\right), \quad \arg(1 - s) \in \left(0, \frac{\pi}{2}\right), \quad \arg(s - x) = \frac{3}{2} + 2\pi,$$

so that ② is equal to

$$\left(e^{2\pi\sqrt{-1}}\right)^{\alpha-1} \left(e^{2\pi\sqrt{-1}}\right)^{-\alpha} \Gamma_1 = \Gamma_1.$$

The branch at P on ③ is determined by

$$\arg s \approx 2\pi, \ \arg(1 - s) \approx 0, \ \arg(s - x) \approx 3\pi,$$

so that ③ is equal to

$$-\left(e^{2\pi\sqrt{-1}}\right)^{\alpha-1} \left(e^{2\pi\sqrt{-1}}\right)^{-\alpha} \Delta_0 = -\Delta_0.$$

Thus we have the change

$$\Delta_x \rightsquigarrow e^{-2\pi\sqrt{-1}\alpha} \Delta_0 + \Gamma_1 - \Delta_0 + \Gamma_2$$

$$= \Delta_x + (e^{-2\pi\sqrt{-1}\alpha} - 1)\Delta_0$$

when x encircles $x = 0$ once in the positive direction. Namely, γ_0 being a loop in x space where x encircles 0 once in the positive direction, we get

$$\gamma_{0*}\Phi(\Delta_x) = \Phi(\Delta_x) + (e^{-2\pi\sqrt{-1}\alpha} - 1)\Phi(\Delta_0).$$

Now we set

$$f(x) = \frac{e^{-2\pi\sqrt{-1}\alpha} - 1}{2\pi\sqrt{-1}} \Phi(\Delta_0) \log x.$$

Since $\Phi(\Delta_0)$ is holomorphic at $x = 0$, the analytic continuation along γ_0 becomes

$$\gamma_{0*}f(x) = f(x) + (e^{-2\pi\sqrt{-1}\alpha} - 1)\Phi(\Delta_0).$$

Then, if we set

$$\Phi(\Delta_x) - f(x) = g(x),$$

$g(x)$ is single-valued around $x = 0$. By the relation

$$\Phi(\Delta_x) = \frac{e^{-2\pi\sqrt{-1}\alpha} - 1}{2\pi\sqrt{-1}} \Phi(\Delta_0) \log x + g(x),$$

we see that $\Phi(\Delta_x)$ is a solution at $x = 0$ containing a logarithmic term. Comparing this with (9.51) and (9.54), we get

$$\Phi(\Delta_x) = (e^{-2\pi\sqrt{-1}\alpha} - 1)y_2^+(t). \tag{9.55}$$

The ambiguity of $y_2^+(t)$ is also determined by the equality.

Similarly we study Δ_x at $x = 1$. Then we see that, when x encircles 1 once in the positive direction, we have

$$\Delta_x \rightsquigarrow \Delta_x + (1 - e^{2\pi\sqrt{-1}\alpha})\Delta_1.$$

Therefore we obtain

$$\Phi(\Delta_x) = e^{\pi\sqrt{-1}(1-\alpha)}(1 - e^{2\pi\sqrt{-1}\alpha})\, y_2^-(t). \tag{9.56}$$

Finally we shall obtain the relation among the twisted cycles. By deforming the paths as in Fig. 9.22, we get

$$\Delta_0 - \Delta_1 = \Delta_x.$$

Fig. 9.22 Relation among Δ_1, Δ_1 and Δ_x

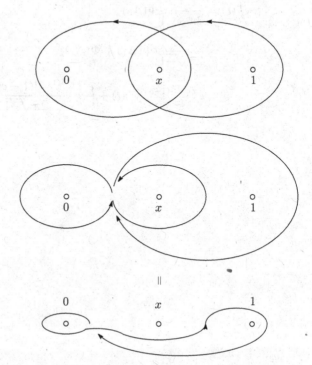

Thus we have

$$\Phi(\Delta_0) = \Phi(\Delta_1) + \Phi(\Delta_x). \tag{9.57}$$

Combining the above results, we solve the connection problem for Legendre's equation.

Theorem 9.5 *In Legendre's equation (9.46), we set $\lambda = \alpha(\alpha - 1)$, and assume $\alpha \notin \mathbb{Z}$. We define a fundamental system of solutions $y_1^+(t)$, $y_2^+(t)$ at $t = 1$ by (9.51), and a fundamental system of solutions $y_1^-(t)$, $y_2^-(t)$ at $t = -1$ by (9.52). Here the ambiguities in $y_2^+(t)$, $y_2^-(t)$ are specified by the integral representations (9.55), (9.56). Then, in the domain $\{|t - 1| < 2\} \cap \{|t + 1| < 2\}$, the connection relation*

$$y_1^+(t) = Ay_1^-(t) + By_2^-(t)$$

holds, where

$$A = e^{\pi\sqrt{-1}(1-\alpha)}, \quad B = e^{\pi\sqrt{-1}(1-\alpha)}\frac{1 - e^{2\pi\sqrt{-1}\alpha}}{2\pi\sqrt{-1}}.$$

Proof We have only to use (9.54), (9.56) and (9.57) to get

$$y_1^+(t) = \frac{1}{2\pi\sqrt{-1}}\Phi(\Delta_0)$$

$$= \frac{1}{2\pi\sqrt{-1}}(\Phi(\Delta_1) + \Phi(\Delta_x))$$

$$= e^{\pi\sqrt{-1}(1-\alpha)}\, y_1^-(t) + \frac{e^{\pi\sqrt{-1}(1-\alpha)}(1 - e^{2\pi\sqrt{-1}\alpha})}{2\pi\sqrt{-1}}\, y_2^-(t). \qquad \square$$

Chapter 10
Irregular Singular Points

For a linear differential equation, a pole of the coefficient of the equation is called an irregular singular point if it is not a regular singular point. In this chapter, we explain the outline of the theory of irregular singular point. Several assertions are proved.

First we consider how to distinguish irregular singular points from regular singular points. For a scalar differential equation

$$y^{(n)} + p_1(x)y^{(n-1)} + \cdots + p_n(x)y = 0, \tag{10.1}$$

let $x = a$ be a pole of at least one of the coefficients $p_1(x), p_2(x), \ldots, p_n(x)$. Let $k_1, k_2, \ldots, k_n$ be the orders of the pole for $p_1(x), p_2(x), \ldots, p_n(x)$, respectively. Then $x = a$ is a regular singular point of (10.1) if and only if

$$k_j \leq j \tag{10.2}$$

holds for every $j = 1, 2, \ldots, n$ (see Theorem 4.3 (Fuchs' theorem)). Therefore, if there exists j such that the inequality (10.2) does not hold, then $x = a$ is an irregular singular point of (10.1).

On the other hand, for a system of differential equations

$$\frac{dY}{dx} = A(x)Y, \tag{10.3}$$

there is no such simple criterion. Nevertheless, in many cases as follows, we can see whether a singular point is regular or irregular. Let $x = a$ be a pole of $A(x)$ in

© The Editor(s) (if applicable) and The Author(s), under exclusive license to Springer Nature Switzerland AG 2020
Y. Haraoka, *Linear Differential Equations in the Complex Domain*, Lecture Notes in Mathematics 2271, https://doi.org/10.1007/978-3-030-54663-2_10

(10.3), and q the order. We write the Laurent series expansion of $A(x)$ at $x = a$ as

$$A(x) = \frac{1}{(x-a)^q} \sum_{m=0}^{\infty} A_m (x-a)^m.$$

If $q = 1$, $x = a$ is a regular singular point of (10.3) as shown in Theorem 4.4. If $q > 1$ and A_0 is not nilpotent, then $x = a$ is an irregular singular point of (10.3) [165, Theorem 5.5.4]. If $q > 1$ and A_0 is nilpotent, we have both possibilities. A criterion for such case is obtained by Jurkat and Lutz [87], which is also explained in [165].

The basis of the theory of irregular singular point are the construction of formal solutions, the construction of analytic solutions asymptotic to the formal solutions, and the description of Stokes phenomenon. A good review of the history of the study of irregular singular points is found in [120, Section II.1]. In the present book, we only explain basic notions and basic results on irregular singularities. For the constructions of formal and analytic solutions, the description is based on the series of papers by Hukuhara [74–76] and a paper by Turrittin [176]. For the Stokes phenomenon, the description is based on the paper by Birkhoff [19]. Also the books by Balser [8] and by Wasow [180] are good references.

10.1   Formal Solutions

For the construction of a formal solution at an irregular singular point, the following theorem is decisive.

Theorem 10.1 *Let $x = a$ be an irregular singular point of the scalar differential equation (10.1) or the system (10.3) of differential equations. Then there exists an integer $d \geq 1$ such that, in the new variable t defined by*

$$t^d = x - a,$$

the Eq. (10.1) or the system (10.3) has a formal solution of the form

$$e^{h(t^{-1})} t^\rho \sum_{j=0}^{k} (\log t)^j \sum_{m=0}^{\infty} y_{jm} t^m, \tag{10.4}$$

where $h(t^{-1})$ is a polynomial in t^{-1}, $\rho \in \mathbb{C}$, and each y_{jm} is a scalar or a vector according to the case of the scalar equation or the system.

The polynomial $h(t^{-1})$ in the formal series (10.4) is called the *dominant term*, and ρ is called the *characteristic exponent*. If $d > 1$, we call the irregular singular point *ramified*, and if $d = 1$, we call it *unramified*. The series $\sum_{m=0}^{\infty} y_{jm} t^m$ in the formal solution (10.4) does not converge in general, which is a big difference from the case of regular singular points. An algorithm to obtain the integer d, the dominant term, the characteristic exponent and the coefficients y_{jm} of the series is known. In the following, we sketch the algorithm. For details, please refer to the papers [19, 75, 176] or the book [180].

We consider a system of differential equations

$$\frac{dY}{dx} = A(x)Y, \tag{10.5}$$

and, for simplicity, assume that $x = 0$ is an irregular singular point. Let the Laurent expansion of $A(x)$ at $x = 0$ be given by

$$A(x) = \frac{1}{x^q} \sum_{m=0}^{\infty} A_m x^m, \tag{10.6}$$

where $q > 1$. Let $\alpha_1, \alpha_2, \ldots, \alpha_l$ be the distinct eigenvalues of A_0 ($\alpha_i \neq \alpha_j$ for $i \neq j$). Then A_0 is similar to the direct sum

$$A_0 \sim \bigoplus_{i=1}^{l} A_0^i,$$

where, for each i, A_0^i has the only eigenvalue α_i. Operating the gauge transformation of (10.5) by the constant matrix which realizes this similarity, we can take

$$A_0 = \bigoplus_{i=1}^{l} A_0^i.$$

If we transform the system (10.5) by a matrix function $P(x)$:

$$Y = P(x)Z, \tag{10.7}$$

we get a transformed system

$$\frac{dZ}{dx} = B(x)Z, \tag{10.8}$$

where

$$A(x)P(x) = P(x)B(x) + P'(x) \tag{10.9}$$

holds. We assume that $P(x)$ is a formal Taylor series at $x = 0$, and $B(x)$ is a Laurent series at $x = 0$ with a pole of the same order q as $A(x)$. We set

$$P(x) = \sum_{m=0}^{\infty} P_m x^m, \quad B(x) = \frac{1}{x^q} \sum_{m=0}^{\infty} B_m x^m.$$

We put them together with the Laurent expansion of $A(x)$ into (10.9), and compare the coefficients of the both sides. By noting that there is no term of negative power of x in $P'(x)$, we get

$$A_0 P_0 = P_0 B_0, \tag{10.10}$$

$$\sum_{i=0}^{j} A_i P_{j-i} = \sum_{i=0}^{j} P_i B_{j-i} \quad (1 \le j \le q - 1), \tag{10.11}$$

$$\sum_{i=0}^{j} A_i P_{j-i} = \sum_{i=0}^{j} P_i B_{j-i} + (j - q + 1) P_{j-q+1} \quad (j \ge q). \tag{10.12}$$

We set $P_0 = I$, $B_0 = A_0$, and then (10.10) is satisfied. Next we rewrite (10.11) with $j = 1$ to get

$$B_1 = A_1 + [A_0, P_1]. \tag{10.13}$$

We partition P_1 and B_1 into blocks as the direct sum decomposition A_0:

$$P_1 = \begin{pmatrix} P_1^{11} & \cdots & P_1^{1l} \\ \vdots & & \vdots \\ P_1^{l1} & \cdots & P_1^{ll} \end{pmatrix}, \quad B_1 = \begin{pmatrix} B_1^{11} & \cdots & B_1^{1l} \\ \vdots & & \vdots \\ B_1^{l1} & \cdots & B_1^{ll} \end{pmatrix}.$$

Then the (i, j)-block of $[A_0, P_1]$ in the right hand side of (10.13) is written as

$$A_0^i P_1^{ij} - P_1^{ij} A_0^j.$$

For $i \ne j$, A_0^i and A_0^j have no common eigenvalue, and then, thanks to Lemma 4.7, the linear map

$$X \mapsto A_0^i X - X A_0^j$$

is an isomorphism, in particular is surjective. Then we can choose P_1^{ij} so that $B_1^{ij} = O$ holds. In this way, we can make B_1 block diagonal. We see that the equations in (10.11), (10.12) can be written as

$$B_j = C_j + [A_0, P_j],$$

where C_j is a matrix determined by A_i, B_i, P_i $(1 \leq i \leq j - 1)$. Therefore, by the same reason, we can choose P_j so that B_j becomes block diagonal. Thus we can make B_j for all j block diagonal as A_0. In the above process, we used only non-diagonal blocks of P_j $(j \geq 1)$. Then we can choose the diagonal blocks of P_j arbitrary. In particular, we may choose them as zero matrices. We note that the resulting series of matrices $P(x), B(x)$ not necessarily converge. As we have mentioned, this is the big difference from the case of regular singular points.

Thus, by the above operations, the system (10.5) is reduced to the system (10.8) with block diagonal matrix $B(x)$ as coefficient. Then we can reduce to the system of differential equations

$$\frac{dZ_i}{dx} = B^i(x)Z_i, \quad B^i(x) = \frac{1}{x^q} \sum_{m=0}^{\infty} B_m^i x^m \quad (1 \leq i \leq l), \tag{10.14}$$

where $B^i(x)$ is the i-th diagonal block of $B(x)$, and then $B_0^i = A_0^i$ has only one eigenvalue.

If the eigenvalues of A_0 are all distinct, all the systems (10.14) become of rank 1. Here we consider the formal solution of the rank 1 equation

$$\frac{dz}{dx} = \left(\frac{1}{x^q} \sum_{m=0}^{\infty} b_m x^m \right) z.$$

Set

$$h(x) = h_{-q} \frac{x^{1-q}}{1-q} + h_{-q+1} \frac{x^{2-q}}{2-q} + \cdots + h_{-2} \frac{x^{-1}}{-1} + h_{-1} \log x,$$

and look for a formal solution of the form

$$z = e^{h(x)} \sum_{m=0}^{\infty} z_m x^m. \tag{10.15}$$

Input this expression into the differential equation, and then we get

$$\left(\sum_{m=-q}^{-1} h_m x^m \right)\left(\sum_{m=0}^{\infty} z_m x^m \right) + \sum_{m=1}^{\infty} m z_m x^{m-1} = \left(\sum_{m=0}^{\infty} b_m x^{m-q} \right)\left(\sum_{m=0}^{\infty} z_m x^m \right).$$

Comparing the coefficients of both sides of negative powers of x, we get

$$h_m = b_{m+q} \quad (-q \leq m \leq -1).$$

Then we have

$$\sum_{m=0}^{\infty}(m+1)z_{m+1}x^m = \Big(\sum_{m=q}^{\infty} b_m x^{m-q}\Big)\Big(\sum_{m=0}^{\infty} z_m x^m\Big).$$

Hence we can uniquely determine the sequence $\{z_m\}$ by specifying z_0. Namely, we can construct the formal solution of the form (10.15). Then, by combining with the gauge transformation (10.7), we get the formal solution (10.4) in Theorem 10.1. Note that $e^{h-1\log x} = x^{h-1}$, and that the logarithmic terms do not appear in this case.

We consider the case where A_0 has multiple eigenvalues. In this case, there appears a system of rank at least 2 among the systems (10.14). We study such system. We assume that $B_0^i = A_0^i$ has the only eigenvalue α_i. If B_0^i is equal to the scalar matrix $\alpha_i I$, by the gauge transformation

$$Z_i = e^{\frac{\alpha_i}{1-q}x^{1-q}}W_i,$$

we get a system in W_i where q is reduced by 1. Thus we obtained an algorithm to reduce q by 1. If we can continue to apply this algorithm, finally we come to a system (10.14) with $q = 1$, which is a system of regular singular point treated in Theorem 4.5. Then applying the conclusion of the theorem, we get a formal solution of the form (4.17). Therefore in this case we obtain a formal solution of the form (10.4).

The remaining is the case where $B_0^i \neq \alpha_i I$. In this case, we introduce a new transformation which reduces a system to a rank one equation. The new transformation is the sharing transformation, which is a gauge transformation by the matrix of the form

$$S(x) = \begin{pmatrix} 1 & & & \\ & x^p & & \\ & & x^{2p} & \\ & & & \ddots \end{pmatrix}.$$

If we take an appropriate rational number p, we can bring some entries to the initial term B_0^i from other terms in the Laurent expansion so that the initial term has at least two distinct eigenvalues. We do not give a proof of the existence of such rational number p.

In conclusion, we can reduce the system (10.14) to a rank one equation or a system with $q = 1$, and then, in both cases, we can get a formal solution (10.4) in Theorem 10.1.

10.2 Asymptotic Expansion

When the formal series (10.4) diverges, it defines no analytic function, however, we can give it a meaning by the idea of asymptotic expansions. In this section, we explain basic notions and fundamental properties of asymptotic expansions.

Let $r > 0$ and $\theta_1 < \theta_2$. We call the domain S in t-plane given by

$$S = \{t \in \mathbb{C};\ 0 < |t| < r,\ \theta_1 < \arg t < \theta_2\} \tag{10.16}$$

a *sector* centered at $t = 0$. If we want to indicate r, θ_1, θ_2, we use the notation $S = S(\theta_1, \theta_2; r)$. We call $\theta_2 - \theta_1$ the *opening* of the sector S. We understand that a sector is a domain in the universal covering of $\mathbb{C} \setminus \{0\}$, and then it does not coincide with the annular domain $0 < |t| < r$ if $\theta_2 - \theta_1 > 2\pi$. The set defined by

$$S' = \{t \in \mathbb{C};\ 0 < |t| \leq r',\ \theta_1' \leq \arg t \leq \theta_2'\} \tag{10.17}$$

with $0 < r' < r, \theta_1 < \theta_1' < \theta_2' < \theta_2$ is called a *closed subsector* of the sector S (Fig. 10.1).

In the following we always consider sectors centered at 0. Remark that, in several books and articles, sectors centered at ∞ are often considered.

Definition 10.1 Let S be a sector. A holomorphic function $f(t)$ on S is said to be *asymptotically developable* on S if there exists a sequence $\{f_n\}_{n=0}^{\infty}$ such that, for any closed subsector S' of S and for any $n \geq 1$,

$$\left| f(t) - \sum_{m=0}^{n-1} f_m t^m \right| \leq C_{S',n} |t|^n \tag{10.18}$$

Fig. 10.1 Sector and closed subsector

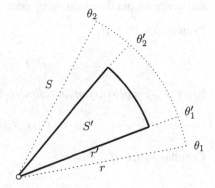

holds with some constant $C_{S',n} > 0$ and for any $t \in S'$. The formal series

$$\hat{f}(t) = \sum_{n=0}^{\infty} f_n t^n$$

with the sequence $\{f_n\}_{n=0}^{\infty}$ as coefficients is called the *asymptotic expansion* of $f(t)$ on S. We also say that $f(t)$ is asymptotically developable to $\hat{f}(t)$ on S, and denote as

$$f(t) \sim \hat{f}(t) \quad (t \in S).$$

We study basic properties of asymptotically developable functions and asymptotic expansions.

Proposition 10.2 *If*

$$f(t) \sim \hat{f}(t) = \sum_{n=0}^{\infty} f_n t^n \quad (t \in S)$$

holds, we have

$$\lim_{\substack{t \to 0 \\ t \in S'}} f(t) = f_0$$

for any closed subsector S' of S.

Proof Put $n = 1$ in (10.18). Then we have

$$|f(t) - f_0| \le C_{S',0}|t| \quad (t \in S'),$$

and hence we get the assertion by taking $t \to 0$ $(t \in S')$. $\square$

Proposition 10.3 *If*

$$f(t) \sim \hat{f}(t) \quad (t \in S)$$

holds, $f'(t)$ is also asymptotically developable on S, and we have

$$f'(t) \sim \hat{f}'(t) \quad (t \in S).$$

Corollary 10.4 *If*

$$f(t) \sim \hat{f}(t) = \sum_{n=0}^{\infty} f_n t^n \quad (t \in S)$$

holds, we have

$$\lim_{\substack{t \to 0 \\ t \in S'}} f^{(n)}(t) = n! f_n$$

for any subsector S' of S and any n ≥ 0.

Corollary 10.5 *For an asymptotically developable function on a sector S, its asymptotic expansion on S is uniquely determined.*

Proof of Proposition 10.3 Let the sector S and its closed subsector S' be given by (10.16) and (10.17), respectively. Take ε satisfying $0 < \varepsilon < \min\{\theta'_1 - \theta_1, \theta_2 - \theta'_2, \pi/2\}$, and set $\delta = \sin\varepsilon$. If $r' + r'\delta \geq r$, we retake ε smaller so that $r' + r'\delta < r$ holds. Then for any $t \in S'$, we have

$$f'(t) = \frac{1}{2\pi i} \int_{|\zeta - t| = |t|\delta} \frac{f(\zeta)}{(\zeta - t)^2} \, d\zeta. \tag{10.19}$$

We take another closed subsector S'' between S and S' by

$$S'' = \{t \in \mathbb{C}; \ 0 < |t| \leq r' + r'\delta, \ \theta'_1 - \varepsilon \leq \arg t \leq \theta'_2 + \varepsilon\}.$$

We use the estimate (10.18) for S''. For any $n \geq 1$, there exists a constant $C_n > 0$ such that, by setting

$$e_n(t) = f(t) - \sum_{m=0}^{n-1} f_m t^m, \tag{10.20}$$

$$|e_n(t)| \leq C_n |t|^n \tag{10.21}$$

holds for any $t \in S''$. We rewrite the right hand side of (10.19) by using (10.20):

$$f'(t) = \frac{1}{2\pi i} \int_{|\zeta - t| = |t|\delta} \frac{\sum_{m=0}^{n-1} f_m \zeta^m + e_n(\zeta)}{(\zeta - t)^2} \, d\zeta$$

$$= \sum_{m=0}^{n-1} \frac{1}{2\pi i} \int_{|\zeta - t| = |t|\delta} \frac{f_m \zeta^m}{(\zeta - t)^2} \, d\zeta + \frac{1}{2\pi i} \int_{|\zeta - t| = |t|\delta} \frac{e_n(\zeta)}{(\zeta - t)^2} \, d\zeta$$

$$= \sum_{m=1}^{n-1} m f_m t^{m-1} + \frac{1}{2\pi i} \int_{|\zeta - t| = |t|\delta} \frac{e_n(\zeta)}{(\zeta - t)^2} \, d\zeta.$$

Applying (10.21) to the last integral, we get

$$\left| \frac{1}{2\pi i} \int_{|\zeta - t| = |t|\delta} \frac{e_n(\zeta)}{(\zeta - t)^2} d\zeta \right| \leq \frac{1}{2\pi} \int_0^{2\pi} \frac{C_n |t| + |t|\delta e^{i\varphi}|^n}{(|t|\delta)^2} |t|\delta \, d\varphi$$

$$\leq \frac{1}{2\pi} \int_0^{2\pi} \frac{C_n (1+\delta)^n |t|^n}{|t|\delta} d\varphi$$

$$= C_n \frac{(1+\delta)^n}{\delta} |t|^{n-1}.$$

Therefore we have

$$\left| f'(t) - \sum_{m=0}^{n-2} (m+1) f_{m+1} t^m \right| \leq C_n \frac{(1+\delta)^n}{\delta} |t|^{n-1}$$

for any $n \geq 1$. This implies $f'(t) \sim \hat{f}'(t)$ $(t \in S)$.   □

Proof of Corollary 10.4 Applying Proposition 10.3 repeatedly, we get

$$f^{(n)}(t) \sim \hat{f}^{(n)}(t) = \sum_{m=n}^{\infty} \frac{m!}{(m-n)!} f_m t^{m-n} \quad (t \in S)$$

for any n. Then we apply Theorem 10.2 to $f^{(n)}(t)$ to get the assertion.   □

Proof of Corollary 10.5 Thanks to Corollary 10.4, the limit $\lim_{t \to 0, t \in S'} f^{(n)}(t)$ exists for any closed subsector S', and hence the limit does not depend on the choice of S'. Therefore f_n is uniquely determined by $f(t)$.   □

As we have seen above, a function asymptotically developable on a sector S has the limits of all derivatives as $t \to 0$ in S, and the coefficient of x^n in its asymptotic expansion is given by the limit of its n-th order derivative divided by $n!$. We shall show that the converse assertion holds.

Theorem 10.6 *Let $f(t)$ be a holomorphic function on a sector S. The followings are equivalent.*

(i) *$f(t)$ is asymptotically developable on S.*
(ii) *For any closed subsector S' of S and for any integer $n \geq 0$, the limit*

$$\lim_{\substack{t \to 0 \\ t \in S'}} f^{(n)}(t)$$

exists.

Proof We have only to show (ii)$\Rightarrow$(i). Let S' be any closed subsector of S, and take a point a in S'. By taking the Taylor expansion of $f(t)$ at $t = a$ up to degree $n - 1$ terms, we get

$$f(t) = \sum_{m=0}^{n-1} \frac{f^{(m)}(a)}{m!} (t - a)^m + \frac{1}{(n-1)!} \int_a^t (t - u)^{n-1} f^{(n)}(u)\, du.$$

We shall take the limit $a \to 0$ in S'. By the assumption, $|f^{(n)}(u)|$ is bounded on S', so that there is a constant $C_n > 0$ such that $|f^{(n)}(u)| \le C_n$ for $u \in S'$. Then the error term expressed by the integral is evaluated as

$$\left| \frac{1}{(n-1)!} \int_a^t (t - u)^{n-1} f^{(n)}(u)\, du \right|$$

$$= \left| \frac{1}{(n-1)!} \int_0^1 ((t-a)(1-v))^{n-1} f^{(n)}(a + (t-a)v)(t-a)\, dv \right|$$

$$\le \frac{C_n}{(n-1)!} |t - a|^n \int_0^1 (1 - v)^{n-1}\, dv.$$

Thus we have

$$\left| f(t) - \sum_{m=0}^{n-1} \frac{f^{(m)}(a)}{m!} (t - a)^m \right| \le C_n' |t - a|^n$$

for some constant C_n' which is independent of a. Then, by taking the limit $a \to 0$, we get (10.18). $\qquad\square$

Theorem 10.6 implies that asymptotically developable functions behave similarly as holomorphic functions at $t = 0$ in some respect. In considering the similarity and difference between holomorphic functions at $t = 0$ and asymptotically developable functions on a sector centered at $t = 0$, the following theorems will be useful.

Theorem 10.7 *A function holomorphic at $t = 0$ is asymptotically developable on any sector centered at $t = 0$, and the asymptotic expansion is given by the Taylor series a $t = 0$. If a function asymptotically developable on a sector centered at $t = 0$ with opening greater than 2π is single-valued on the sector, the function is holomorphic at $t = 0$.*

Theorem 10.8 *If*

$$f(t) \sim \hat{f}(t), \quad g(t) \sim \hat{g}(t) \quad (t \in S),$$

then we have

$$af(t) + bg(t) \sim a\hat{f}(t) + b\hat{g}(t),$$
$$f(t)g(t) \sim \hat{f}(t)\hat{g}(t) \qquad (t \in S),$$

where a, b are constants.

The above two theorems can be shown by applying the definition directly.

Theorem 10.9 (Borel-Ritt Theorem) *For any power series $\hat{f}(t)$ and for any sector S, there exists a holomorphic function $f(t)$ on S satisfying*

$$f(t) \sim \hat{f}(t) \quad (t \in S).$$

For the proof of Theorem 10.9, please refer to Wasow [180]. Note that the function $f(t)$ in Theorem 10.9 is not unique. The reason will be seen by the following example, which is also useful to get an image of asymptotically developable functions.

Example 10.1 The function

$$f(t) = e^{-\frac{1}{t}}$$

is asymptotically developable to 0 in the sector

$$S = \left\{ t \in \mathbb{C};\ 0 < |t| < r,\ -\frac{\pi}{2} < \arg t < \frac{\pi}{2} \right\}.$$

In order to show this, we first remark the identity

$$\left| e^{-\frac{1}{t}} \right| = \left| e^{-\frac{1}{|t|}(\cos\theta - i\sin\theta)} \right|$$
$$= e^{-\frac{\cos\theta}{|t|}},$$

where we set $\arg t = \theta$. If t is in a closed subsector of S, there exists a $\delta > 0$ such that $\cos\theta \geq \delta$, and then the limit of $f(t)$ as $t \to 0$ becomes 0. The derivatives of $f(t)$ have the following forms:

$$f^{(n)}(t) = \frac{h_n(t)}{t^{2n}} e^{-\frac{1}{t}},$$

where $h_n(t)$ is a polynomial in t. Then these functions converges to 0 as $t \to 0$ in any closed subsector of S. Hence, by Theorem 10.6, $f(t)$ is asymptotically developable on S, and all the coefficients of the asymptotic expansion become 0.

As will be suggested by this example, we can construct various functions asymptotically developable to 0 by using the exponential function. If we add such function to an asymptotically developable function, thanks to Theorem 10.8, the asymptotic expansion does not change. Borel-Ritt theorem can be proved by using such mechanism. Moreover, as a direct application of Example 10.1, for any sector, we can construct a function asymptotically developable to 0 in the sector.

Theorem 10.9 and Example 10.1 illustrate some basic properties of the asymptotic expansion. Here we give remarks on the properties.

Remark 10.1 Let S be a sector.

(i) For any power series $\hat{f}(t)$, the function $f(t)$ satisfying

$$f(t) \sim \hat{f}(t) \quad (t \in S)$$

is not unique.

(ii) For any divergent power series $\hat{f}(t)$, there exists a function $f(t)$ satisfying

$$f(t) \sim \hat{f}(t) \quad (t \in S).$$

(iii) Even if the power series $\hat{f}(t)$ is convergent, the function satisfying

$$f(t) \sim \hat{f}(t) \quad (t \in S)$$

is not necessarily holomorphic at $t = 0$.

Remark 10.2 We defined asymptotic developable functions on an open sector, and studied their properties. There is also a definition of asymptotic developable functions on a half line. Such functions possess similar properties, however, the analogue of Proposition 10.3 does not hold in general.

10.3 Existence of Asymptotic Solutions

We have studied the definition and basic properties of asymptotic expansion. Now we are going to put a meaning to a formal solution of a differential equation at an irregular singular point by using the asymptotic expansion. For the purpose, we slightly extend the notion of asymptotic expansion.

Definition 10.2 Let S be a sector, and $\phi(t)$ a holomorphic function on S which does not vanish on S. For a holomorphic function $f(t)$ on S and for a power series $\hat{f}(t)$, we denote

$$f(t) \sim \phi(t)\hat{f}(t) \quad (t \in S)$$

if

$$\frac{f(t)}{\phi(t)} \sim \hat{f}(t) \quad (t \in S)$$

holds. In this case, we call $\phi(t)\hat{f}(t)$ the *asymptotic behavior* of $f(t)$ on S.

The above notation has a definite meaning. By generalizing it, we often use the following notation:

$$h(t) \sim \phi(t)\hat{f}(t) + \psi(t)\hat{g}(t) \quad (t \in S), \tag{10.22}$$

where $\phi(t)$, $\psi(t)$ are holomorphic and non-vanishing function on S, and $\hat{f}(t)$, $\hat{g}(t)$ are power series in t. The notation (10.22) means that, there exist functions $f(t)$, $g(t)$ asymptotically developable as

$$\begin{aligned} f(t) &\sim \phi(t)\hat{f}(t) \\ g(t) &\sim \psi(t)\hat{g}(t) \end{aligned} \quad (t \in S),$$

such that

$$h(t) = f(t) + g(t).$$

In order to read the asymptotic behavior of $h(t)$ from (10.22), we need the following consideration. Suppose that, for a subsector S_1 of S,

$$\frac{\psi(t)}{\phi(t)} \sim \hat{u}(t) \quad (t \in S_1)$$

holds. Then we have

$$\frac{h(t)}{\phi(t)} = \frac{f(t)}{\phi(t)} + \frac{\psi(t)}{\phi(t)} \cdot \frac{g(t)}{\psi(t)} \sim \hat{f}(t) + \hat{u}(t)\hat{g}(t),$$

so that we get the asymptotic behavior

$$h(t) \sim \phi(t)(\hat{f}(t) + \hat{u}(t)\hat{g}(t)) \quad (t \in S_1)$$

of $h(t)$ on S_1. In the analysis of irregular singular point, we often encounter with the case

$$\frac{\psi(t)}{\phi(t)} \sim 0 \quad (t \in S_1).$$

In this case, we have $\hat{u}(t) = 0$, and then the asymptotic behavior

$$h(t) \sim \phi(t)\hat{f}(t) \quad (t \in S_1)$$

holds. On the other hand, by (10.22), the asymptotic behavior

$$h(t) \sim \phi(t)\hat{f}(t) + \psi(t)\hat{g}(t) \quad (t \in S_1)$$

also holds. Thus, in this generalized notation, the expression of the asymptotic behavior is not unique.

Now we consider the differential equation (10.5) with an irregular singular point at $x = 0$, where the matrix function $A(x)$ is given by (10.6) and has a pole of order q with $q > 1$ at $x = 0$. We assume that the differential equation (10.5) has a formal solution of the form

$$\hat{y}(x) = \hat{F}(x)x^{R}e^{H(x^{-1})}, \tag{10.23}$$

where

$$\hat{F}(x) = \sum_{m=0}^{\infty} F_m x^m$$

is a formal power series with F_0 a non-singular matrix, R a constant diagonal matrix, and $H(T)$ a diagonal matrix with diagonal entries in the polynomials in T. One may feel that the assumption to have the formal fundamental matrix solution of the form (10.23) is too strong compared with the general form (10.4) of the formal solution in Theorem 10.1, however, it is not so strong by the following reasons. First, without loss of generality, we may assume that the irregular singular point is at $x = 0$. Second, in the formal solution (10.4) in Theorem 10.1, we take the independent variable t defined by $t^d = x$, however, if we rewrite the differential equation (10.5) in the variable t, we can obtain a differential equation of the form (10.5) (with a possibly different value of q). Then we may regard the differential equation (10.5) as the transformed one in the variable t. Third, as explained in the previous section, the logarithmic functions in the formal solution (10.4) appear only when the differential equation (10.5) is reduced to a differential equation of regular singular type (i.e. the case $q = 1$). Since we are interested in the nature of irregular singular points, it may be natural to restrict to the case where logarithmic functions do not appear in the formal series, and then our assumption is not so strong.

In the formal matrix solution (10.23), the behaviors of x^R and $e^{H(x^{-1})}$ are explicitly known, and then what we should study is the formal power series $\hat{F}(x)$. In the analysis of $\hat{F}(x)$, it turns out that the exponential part $e^{H(x^{-1})}$ plays a decisive role. We set

$$H(T) = \begin{pmatrix} h_1(T) & & \\ & \ddots & \\ & & h_n(T) \end{pmatrix}, \tag{10.24}$$

$$h_i(T) = h_{i,m_i} T^{m_i} + h_{i,m_i-1} T^{m_i-1} + \cdots + h_{i,0} \quad (1 \le i \le n),$$

and denote the i-th column of $\hat{F}(x)$ by $\hat{f}_i(x)$. Then, for each i, the i-th column

$$\hat{Y}_i(x) = \hat{f}_i(x) x^{\rho_i} e^{h_i(x^{-1})} \tag{10.25}$$

of $\hat{\mathcal{Y}}(x)$ is the formal column solution, where we denote by ρ_i the (i, i) entry of the diagonal matrix R. In this section, we explain the existence of a solution $Y_i(x)$ with the formal solution (10.25) as the asymptotic behavior

$$Y_i(x) \sim \hat{Y}_i(x) \quad (x \in S)$$

on an appropriate sector S.

In order to study the behavior of the exponential factor $e^{h_i(x^{-1})}$ at the essential singular point $x = 0$, we consider in general a polynomial

$$a(T) = \alpha T^m + \alpha_1 T^{m-1} + \cdots + \alpha_m \quad (\alpha \ne 0)$$

and the behavior of its exponential $e^{a(T)}$ at $T \to \infty$. Let ω be the argument of α. If the argument of T is fixed as θ, we have

$$\left| e^{a(T)} \right| = e^{\mathrm{Re}(\alpha T^m + \alpha_1 T^{m-1} + \cdots)}$$

$$= e^{\mathrm{Re}\left(|\alpha||T|^m e^{\sqrt{-1}(\omega+m\theta)} + \alpha_1 |T|^{m-1} e^{\sqrt{-1}(m-1)\theta} + \cdots \right)}$$

$$= e^{|T|^m (|\alpha|\cos(\omega+m\theta) + O(1/|T|))}$$

as $T \to \infty$. Hence, if $\cos(\omega + m\theta) \ne 0$, the initial term of $a(T)$ determines the behavior of $e^{a(T)}$. Namely, if $\cos(\omega+m\theta) > 0$, $e^{a(T)}$ diverges, and if $\cos(\omega+m\theta) < 0$, it converges to 0. Note that $\cos(\omega + m\theta) = 0$ if the degree of the polynomial $\mathrm{Re}\, a(T)$ in $|T|$ becomes less than m. There are $2m$ such θ modulo 2π, and the $2m$ sectors bounded by these θ are arranged so that a sector in which $e^{a(T)}$ diverges and a sector in which it converges to 0 appear one after the other. Based on this observation, we give the following definition.

Definition 10.3 A direction θ is said to be a *singular direction* with respect to h_i if there exists j such that the degree of the polynomial ·

$$\text{Re}\left(h_j((re^{\sqrt{-1}\theta})^{-1}) - h_i((re^{\sqrt{-1}\theta})^{-1})\right)$$

in r^{-1} is smaller than the degree of the polynomial $h_j(T) - h_i(T)$ in T. A direction is said to be a singular direction if it is a singular direction with respect to some h_i.

Definition 10.4 A sector S centered at $x = 0$ is said to be a *proper sector* with respect to h_i if, for each j, the domain where $e^{h_j(x^{-1})-h_i(x^{-1})}$ diverges is connected in S (Fig. 10.2).

The following theorem, which is obtained by Hukuhara, is the basis of the analysis at an irregular singular point.

Theorem 10.10 *Assume that the differential equation (10.5) with an irregular singular point at $x = 0$ has a formal fundamental matrix solution (10.23). If a sector S is a proper sector with respect to h_i, then there exists an analytic solution which is asymptotic to the formal solution (10.25) in S.*

For the proof, please refer to [76] or [77]. Note that in several literature containing these references, an irregular singular point is located at ∞, and then, in reading these literature, we need to translate the statements in this book.

If a sector contains only one singular direction, it becomes a proper sector with respect to all h_i. Then we immediately obtain the following.

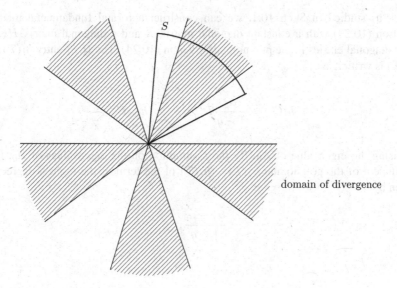

Fig. 10.2 Proper sector S

Corollary 10.11 *Let S be a sector containing only one singular direction. Then there exists a fundamental matrix solution which is asymptotic to the formal fundamental matrix solution (10.23) in S.*

10.4   Stokes Phenomenon

We have seen that there exists a formal solution at an irregular singular point, and an analytic solution which is asymptotic to the formal solution in an appropriate sector. The analytic solutions which are asymptotic to a formal solution is different by the sectors, in general. This is called the Stokes phenomenon, and the quantity which describes the difference is called the Stokes multipliers. In this section, we explain the Stokes phenomenon according to Birkhoff [19].

We consider the system (10.5) of differential equations of the first order. Let the coefficient $A(x)$ be expanded at $x = 0$ in the Laurent series (10.6). For the sake of simplicity, we assume that the eigenvalues $\alpha_1, \alpha_2, \ldots, \alpha_n$ of A_0 are mutually distinct. By operating a gauge transformation if necessary, we may assume that A_0 is already of the diagonal form

$$
A_0 = \begin{pmatrix} \alpha_1 & & & \\ & \alpha_2 & & \\ & & \ddots & \\ & & & \alpha_n \end{pmatrix}.
$$

Then, as studied in Sect. 10.1, we can construct a formal fundamental matrix solution (10.23) with a constant diagonal matrix R and a diagonal matrix $H(T)$ with diagonal entries in the polynomials given in (10.24). The (i, i)-entry $h_i(T)$ of $H(T)$ is written as

$$
h_i(T) = \frac{\alpha_i}{1-q} T^{q-1} + \sum_{m=0}^{q-2} h_{im} T^m
$$

by using the eigenvalue of A_0. By the assumption on the eigenvalues of A_0, the coefficient of the polynomial $h_j(T) - h_i(T)$ of the term of the highest degree is given by

$$
\frac{\alpha_j - \alpha_i}{1-q},
$$

so that the singular directions θ are determined by

$$\text{Re}\left(\frac{\alpha_j - \alpha_i}{1 - q}\left(re^{\sqrt{-1}\theta}\right)^{1-q}\right) = 0.$$

For each pair (i, j), there exist $2(q - 1)$ singular directions θ. For simplicity, we further assume that these θ's thus obtained are mutually distinct. Then the number of the singular directions becomes

$$\binom{n}{2} \times 2(q - 1) = n(n - 1)(q - 1) =: N.$$

We set these singular directions

$$\tau_1 < \tau_2 < \cdots < \tau_N < \tau_1 + 2\pi =: \tau_{N+1}.$$

For $i, j \in \{1, 2, \ldots, n\}$ and for a sector S, we denote $j >_S i$ if

$$\text{Re}\left(\frac{\alpha_j}{1 - q}x^{1-q}\right) > \text{Re}\left(\frac{\alpha_i}{1 - q}x^{1-q}\right) \quad (x \in S)$$

holds. Let a system of solutions $Y_1(x), Y_2(x), \ldots, Y_n(x)$ satisfy

$$Y_j(x) \sim \hat{Y}_j(x) \quad (1 \le j \le n)$$

in a sector S. Then, if $j >_S i$, for any $c \in \mathbb{C}$

$$Y_j(x) + cY_i(x) \sim \hat{Y}_j(x)$$

holds in S. Namely the asymptotic behavior does not change if we add a solution of relatively weak asymptotic behavior. By using this, we obtain the following important fact.

Proposition 10.12 *For a sector S containing no singular direction, $>_S$ becomes a total order. Let i_0 the minimal with respect to the order $>_S$. Then the solution $Y(x)$ satisfying the asymptotic behavior*

$$Y(x) \sim \hat{Y}_{i_0}(x) \quad (x \in S) \tag{10.26}$$

is uniquely determined.

Proof The first assertion is evident. Since S is a proper sector, there exists a fundamental matrix solution $\mathcal{Y}(x) = (Y_1(x), Y_2(x) \ldots, Y_n(x))$ such that $\mathcal{Y}(x) \sim \hat{\mathcal{Y}}(x)$. The solution $Y(x)$ satisfying (10.26) is written as a linear combination of the columns of $\mathcal{Y}(x)$ as

$$Y(x) = \sum_{i=1}^{n} c_i Y_i(x).$$

Let j_0 be the maximal with respect to $>_S$ among the indices j with $c_j \neq 0$. Then we have

$$Y(x) \sim c_{j_0} \hat{Y}_{j_0}(x),$$

since we can delete relatively weak asymptotic behaviors. Comparing this with (10.26), we get $j_0 = i_0$ and $c_{i_0} = 1$. Moreover, from $j_0 = i_0$ we obtain $c_j = 0$ ($j \neq i_0$), and hence $Y(x) = Y_{i_0}(x)$. This proves the uniqueness. $\square$

For a singular direction τ_m, we set $S(\tau_m, \tau_{m+1}; r) = S'$, $S(\tau_{m+1}, \tau_{m+2}; r) = S''$. Then by the definition of the singular directions and by the assumption that the singular directions are all distinct, there exists a unique pair (i_m, j_m) such that

$$j_m >_{S'} i_m, \quad j_m <_{S''} i_m. \tag{10.27}$$

For the other pairs, if an entry is greater than the other by the order $>_{S'}$, then it is also greater by the order $>_{S''}$. Moreover, there is no index k such that $i_m <_{S'} k <_{S'} j_m$.

Take sufficiently small $r > 0$ and sectors $S_1, S_2, \ldots, S_N$ so that their union covers the domain $0 < |x| < r$ and each S_m contains only one singular direction τ_m (Fig. 10.3). Then the following fundamental theorem due to Birkhoff holds.

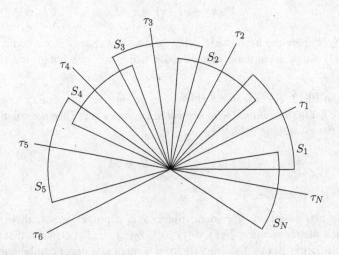

Fig. 10.3   Sectors $S_1, S_2, \ldots, S_N$

Theorem 10.13 *Notation being as above. Consider the differential equation (10.5) with the coefficient $A(x)$ given by (10.6). We assume that the eigenvalues of A_0 are mutually distinct, and that A_0 is diagonal. We also assume that N singular directions are all distinct. Then, for each sector S_m, there exists a fundamental matrix solution $\mathcal{Y}^{(m)}(x)$ such that*

$$\mathcal{Y}^{(m)}(x) \sim \hat{\mathcal{Y}}(x) \quad (x \in S_m),$$
$$\mathcal{Y}^{(m+1)}(x) = \mathcal{Y}^{(m)}\left(I + \gamma_m E_{i_m j_m}\right) \quad (1 \leq m \leq N), \tag{10.28}$$

where γ_m is a constant, and $\mathcal{Y}^{(N+1)}(x)$ is defined by

$$\mathcal{Y}^{(N+1)}(x e^{2\pi\sqrt{-1}}) = \mathcal{Y}^{(1)}(x) e^{2\pi\sqrt{-1}R} \quad (x \in S_1). \tag{10.29}$$

Proof Since each S_m is a proper sector, as we have seen in Corollary 10.11, there exists a fundamental matrix solution asymptotic to $\hat{\mathcal{Y}}(x)$ on S_m. We first take a fundamental matrix solution $\mathcal{Y}^{(1)}(x)$ asymptotic to $\hat{\mathcal{Y}}(x)$ on S_1. If we set $\mathcal{Y}^{(1)}(x) = (Y_1^{(1)}(x), Y_2^{(1)}(x), \ldots, Y_n^{(1)}(x))$, we have, on S_1,

$$Y_j^{(1)}(x) \sim \hat{Y}_j(x) \tag{10.30}$$

for $1 \leq j \leq n$. Set $S' = S(\tau_1, \tau_2; r)$, $S'' = S(\tau_2, \tau_3; r)$, and let (i_1, j_1) be defined by (10.27). Then the behavior (10.30) holds on S' for any j, and on S'' for any $j \neq j_1$. In particular, it holds on S_2 for $j \neq j_1$.

Next we take a fundamental matrix solution $\tilde{\mathcal{Y}}^{(2)}(x) = (\tilde{Y}_1^{(2)}(x), \tilde{Y}_2^{(2)}(x), \ldots, \tilde{Y}_n^{(2)}(x))$ satisfying

$$\tilde{\mathcal{Y}}^{(2)}(x) \sim \hat{\mathcal{Y}}(x) \quad (x \in S_2).$$

The solution $Y_{j_1}^{(1)}(x)$ can be written as a linear combination of the columns of $\tilde{\mathcal{Y}}^{(2)}(x)$. If a term $\tilde{Y}_i^{(2)}$ with $i >_{S'} j_1$ appeared in the linear combination, the behavior (10.30) would not hold on $S' \cap S_2$ for $j = j_1$. Hence the linear combination becomes

$$Y_{j_1}^{(1)} = \tilde{Y}_{j_1}^{(2)} + \sum_{i <_{S'} j_1} c_i \tilde{Y}_i^{(2)}.$$

In a similar reason, we have

$$Y_{i_1}^{(1)} = \tilde{Y}_{i_1}^{(2)} + \sum_{i <_{S'} i_1} d_i \tilde{Y}_i^{(2)}.$$

Since $i_1 <_{S'} j_1$ holds, and since the order with respect to $>_{S'}$ is kept if we replace it by $>_{S''}$ for any pair other than (i_1, j_1), we get

$$i <_{S'} i_1 \Rightarrow i <_{S''} j_1.$$

Also there is no k such that $i_1 <_{S'} k <_{S'} j_1$. Then we have

$$Y_{j_1}^{(1)} - c_{i_1} Y_{i_1}^{(1)} = \tilde{Y}_{j_1}^{(2)} + c_{i_1} \tilde{Y}_{i_1}^{(2)} - c_{i_1} \tilde{Y}_{i_1}^{(2)} + \sum_{i <_{S'} i_1} (c_i - c_{i_1} d_i) \tilde{Y}_i^{(2)}$$

$$= \tilde{Y}_{j_1}^{(2)} + \sum_{i <_{S'} i_1} (c_i - c_{i_1} d_i) \tilde{Y}_i^{(2)}$$

$$= \tilde{Y}_{j_1}^{(2)} + \sum_{i <_{S''} j_1} (c_i - c_{i_1} d_i) \tilde{Y}_i^{(2)},$$

and hence

$$Y_{j_1}^{(1)}(x) - c_{i_1} Y_{i_1}^{(1)}(x) \sim \hat{Y}_{j_1}(x) \quad (x \in S_2)$$

follows. Therefore, by setting

$$Y_j^{(2)}(x) = \begin{cases} Y_j^{(1)}(x) & (j \neq j_1), \\ Y_{j_1}^{(1)}(x) - c_{i_1} Y_{i_1}^{(1)}(x) & (j = j_1), \end{cases}$$

$$\mathcal{Y}^{(2)}(x) = (Y_1^{(2)}(x), Y_2^{(2)}(x), \ldots, Y_n^{(2)}(x)),$$

we have

$$\mathcal{Y}^{(2)}(x) \sim \hat{y}(x) \quad (x \in S_2),$$

$$\mathcal{Y}^{(2)}(x) = \mathcal{Y}^{(1)}(x) \left(I - c_{i_1} E_{i_1 j_1} \right).$$

These are the relations (10.28) with $\gamma_1 = -c_{i_1}$. Now we can start from this $\mathcal{Y}^{(2)}(x)$ to get $\mathcal{Y}^{(3)}(x)$ in a similar manner. Continuing this process recursively, we obtain $\mathcal{Y}^{(m)}(x)$ for each m $(1 \leq m \leq N + 1)$ such that (10.28) holds. Note that, for the fundamental matrix solution $\mathcal{Y}^{(N+1)}$, we consider the asymptotic behavior on the sector S_{N+1} which is obtained from S_1 by shifting the argument by 2π. Then the variable in S_{N+1} can be written as $\tilde{x} = x e^{2\pi \sqrt{-1}}$ in terms of the variable $x \in S_1$, and the asymptotic behavior becomes

$$\mathcal{Y}^{(N+1)}(\tilde{x}) \sim \hat{y}(x) e^{2\pi \sqrt{-1} R} \quad (x \in S_1).$$

We constructed $\mathcal{Y}^{(m)}$ recursively, and finally got $\mathcal{Y}^{(N+1)}$, however, the relation (10.29) does not necessarily hold. Then we will retake $\mathcal{Y}^{(1)}(x)$ appropriately so that (10.29) holds. In the following, we shall show that $\mathcal{Y}^{(N+1)}(x)$ is determined uniquely, and is independent of the choice of $\mathcal{Y}^{(1)}(x)$. If it is shown, we can take $\mathcal{Y}^{(1)}(x)$ as

$$\mathcal{Y}^{(N+1)}(xe^{2\pi\sqrt{-1}})e^{-2\pi\sqrt{-1}R},$$

which makes the relation (10.29) hold.

For each m, we set $S'_m = S(\tau_m, \tau_{m+1}; r)$. Let k_1 be the minimum with respect to the order $>_{S'_1}$. Then by Proposition 10.12, among the columns of $\mathcal{Y}^{(1)}(x)$ which we have taken at the first step, the k_1-th column $Y^{(1)}_{k_1}$ is uniquely determined. Set $N' = n(n-1)/2$. In constructing $\mathcal{Y}^{(2)}(x), \mathcal{Y}^{(3)}(x), \dots, \mathcal{Y}^{(N')}(x)$ recursively, the order of one pair (i, j) with respect to one $>_S$ becomes reversed at each step, and this occurs for all pairs (i, j) until we arrive at the N'-th step. Therefore, for the minimal element k_1 with respect to $>_{S'_1}$, the only possible direction of the reversion is the direction where k_1 becomes bigger. In such case, we have

$$Y^{(m+1)}_{k_1}(x) = Y^{(m)}_{k_1}(x)$$

by the definition, so that we obtain

$$Y^{(N')}_{k_1}(x) = Y^{(1)}_{k_1}(x).$$

Thus $Y^{(N')}_{k_1}(x)$ is uniquely determined independently of the choice of $\mathcal{Y}^{(1)}(x)$.

Next we take the minimal element k_2 among the indices except k_1 with respect to $>_{S'_1}$. If we take another fundamental matrix solution $\mathcal{Z}^{(1)}(x) = (Z^{(1)}_1(x), Z^{(1)}_2(x), \dots, Z^{(1)}_n(x))$ on S_1, we have a relation

$$Z^{(1)}_{k_2}(x) = Y^{(1)}_{k_2}(x) + aY^{(1)}_{k_1}(x),$$

because the uniqueness for the k_2-th column does not hold. When a reversion occurs, k_2 becomes bigger for any i except k_1, and becomes smaller for k_1. Assume that the reversion with k_1 occurs at the step from S'_m to S'_{m+1}. Before the step, we have

$$Y^{(1)}_{k_2}(x) = Y^{(2)}_{k_2}(x) = \cdots = Y^{(m)}_{k_2}(x),$$
$$Z^{(1)}_{k_2}(x) = Z^{(2)}_{k_2}(x) = \cdots = Z^{(m)}_{k_2}(x),$$

and at the step we have

$$Y_{k_2}^{(m+1)}(x) = Y_{k_2}^{(m)}(x) + bY_{k_1}^{(m)}(x),$$
$$Z_{k_2}^{(m+1)}(x) = Z_{k_2}^{(m)}(x) + cZ_{k_1}^{(m)}(x).$$

On the other hand, as we have seen for k_1, the relations

$$Y_{k_1}^{(m+1)}(x) = Y_{k_1}^{(m)}(x) = Z_{k_1}^{(m)}(x) = Z_{k_1}^{(m+1)}(x) \qquad (10.31)$$

hold. Therefore we get

$$Z_{k_2}^{(m+1)}(x) - cZ_{k_1}^{(m+1)}(x) = Y_{k_2}^{(m+1)}(x) - (b-a)Y_{k_1}^{(m+1)}(x).$$

Since $k_1 >_{S'_{m+1}} k_2$, the asymptotic behaviors on S'_{m+1} of both sides become

$$-c\hat{Y}_{k_1}(x) = -(b-a)\hat{Y}_{k_1}(x),$$

from which we obtain $c = b - a$. Then, by the help of (10.31), we get

$$Z_{k_2}^{(m+1)}(x) = Y_{k_2}^{(m+1)}(x).$$

Thereafter only the reversions in the direction where k_2 becomes bigger occur, and hence we obtain

$$Z_{k_2}^{(N')}(x) = Y_{k_2}^{(N')}(x).$$

Thus $Y_{k_2}^{(N')}(x)$ is also uniquely determined independently of the choice of $\mathcal{Y}^{(1)}(x)$.

The general case will be understood if we consider the case of three indices. Let k_3 be the minimal element among the indices except k_1, k_2 with respect to $>_{S'_1}$. Then we have

$$Z_{k_3}^{(1)}(x) = Y_{k_3}^{(1)}(x) + aY_{k_1}^{(1)}(x) + bY_{k_2}^{(1)}(x).$$

The inequality $k_1 <_{S'_1} k_2 <_{S'_1} k_3$ changes to $k_3 <_{S'_{N'}} k_2 <_{S'_{N'}} k_1$ by the three reversions between two contiguous indices. Before the first reversion among k_1, k_2, k_3, we have

$$Z_{k_3}^{(m)}(x) = Y_{k_3}^{(m)}(x) + aY_{k_1}^{(m)}(x) + bY_{k_2}^{(m)}(x). \qquad (10.32)$$

Assume that the reversion between k_1 and k_2 occurs first. Then $Y_{k_2}^{(m)}(x)$ changes to a linear combination of $Y_{k_1}^{(m+1)}(x)$ and $Y_{k_2}^{(m+1)}(x)$, where the relation (10.32) remains to hold with possibly different value of a. Consider the case where the reversion between k_3 and contiguous k_1 or k_2 occurs. For example let

$$k_1 <_{S'_m} k_2 <_{Sm'} k_3, \ k_1 <_{S'_{m+1}} k_3 <_{S'_{m+1}} k_2$$

hold. In this case, by the same reason as in the previous argument for k_2, we see that the coefficient of $Y_{k_2}^{(m+1)}(x)$ vanishes. Namely we have

$$Z_{k_3}^{(m+1)}(x) = Y_{k_3}^{(m+1)}(x) + a' Y_{k_1}^{(m+1)}(x).$$

Thus the reversions of indices smaller than k_3 only give rise to changes of coefficients, and the reversion where k_3 becomes smaller makes the coefficient of the solution with the reversed index zero. Thus finally we have

$$Z_{k_3}^{(N')}(x) = Y_{k_3}^{(N')}(x),$$

which implies that $Y_{k_3}^{(N')}(x)$ is also independent of the choice of $\mathcal{Y}^{(1)}(x)$.

In a similar way we see that, for all i, $Y_i^{(N')}(x)$ is uniquely determined independently of the choice of $\mathcal{Y}^{(1)}(x)$. Hence $\mathcal{Y}^{(N+1)}(x)$, which is defined by using $\mathcal{Y}^{(N')}(x)$, is independent of the choice of $\mathcal{Y}^{(1)}(x)$. $\square$

The coefficient γ_m in the relation (10.28) is called the *Stokes multiplier* ([180, Section 15], [164, Chapter 5]), and the matrix $I + \gamma_m E_{i_m j_m}$ is called the *Stokes matrix*.[1] The Stokes multipliers are quantities of global analytic nature, and very difficult to obtain; actually we have no general method to compute the Stokes multipliers. On the other hand, for several particular differential equations such as the Bessel equation, the Whittaker (Weber) equation and the Airy equation the Stokes multipliers are explicitly obtained [180–182]. Also please refer to [164] where a deep consideration for the Stokes multipliers is found.

For the recent research, we give several comments in "Notes and references to this book" at the end of this book.

---

[1]In [11–13], this matrix is called a *Stokes factor*.

Part II
Completely Integrable Systems

In Part II we study systems of partial differential equations called linear Pfaffian systems. A linear Pfaffian system is a system of linear partial differential equations of the first order, and has a finite dimensional solution space if it satisfies the integrability condition. Therefore linear Pfaffian systems behave like linear ordinary differential equations. On the other hand, there appear distinguished features arising from topological and complex analytic natures of the space in several variables. Therefore we are interested in linear Pfaffian systems.

There is a notion of *holonomic systems*, which is a generalization of completely integrable systems. The theory of holonomic systems is sophisticated, and provides many intrinsic results. In this book, however, we do not get into the theory of holonomic systems, while study completely integrable systems in a naïve way. For explicit computations, a naïve way may work well, and then we may make an image of completely integrable systems via such computations. This will be one of natural ways to study the theory of holonomic systems. For the theory of holonomic systems, please refer to [73, 92].

Chapter 11
Linear Pfaffian Systems and Integrability Condition

We denote by $x = (x_1, x_2, \ldots, x_n)$ the coordinate of $\mathbb{C}^n$. Let N be an integer. Let $X \subset \mathbb{C}^n$ be a domain, and $a_{ij}^k(x)$ $(1 \leq i, j \leq N, 1 \leq k \leq n)$ be holomorphic functions on X. The system

$$\frac{\partial u_i}{\partial x_k} = \sum_{j=1}^{N} a_{ij}^k(x) u_j \quad (1 \leq i \leq N, \ 1 \leq k \leq n) \tag{11.1}$$

of partial differential equations of the first order with unknown functions $u_1(x), u_2(x), \ldots, u_N(x)$ is called a *linear Pfaffian system*.

If we use the unknown vector

$$u = {}^t(u_1, u_2, \ldots, u_N)$$

and the matrix function

$$A_k(x) = \left(a_{ij}^k(x) \right)_{1 \leq i, j \leq N} \quad (1 \leq k \leq n),$$

the system (11.1) can be written in the form

$$\frac{\partial u}{\partial x_k} = A_k(x) u \quad (1 \leq k \leq n). \tag{11.2}$$

For each k, we call this system the differentia equation in x_k direction. Moreover, if we use the exterior derivative d in x and the 1-form

$$\Omega = \sum_{k=1}^{n} A_k(x) \, dx_k,$$

© The Editor(s) (if applicable) and The Author(s), under exclusive license to Springer Nature Switzerland AG 2020
Y. Haraoka, *Linear Differential Equations in the Complex Domain*, Lecture Notes in Mathematics 2271, https://doi.org/10.1007/978-3-030-54663-2_11

the system (11.2) can be written as

$$du = \Omega u. \tag{11.3}$$

Then a Pfaffian system is sometimes said to be a total differential equation.

The linear Pfaffian system (11.2) is determined by the matrix functions $A_1(x), A_2(x), \ldots, A_n(x)$, while, for arbitrarily given matrix functions, there may exist no solution in general. A solution $u(x)$ is differentiable in a domain in $\mathbb{C}^n$, and hence holomorphic, in particular lies in C^2-class. Thus, for any $k, l \in \{1, 2, \ldots, n\}$, we have

$$\frac{\partial^2 u}{\partial x_l \partial x_k} = \frac{\partial^2 u}{\partial x_k \partial x_l}.$$

By using (11.2), we get

$$\frac{\partial^2 u}{\partial x_l \partial x_k} = \frac{\partial}{\partial x_l}(A_k u)$$

$$= \frac{\partial A_k}{\partial x_l} u + A_k \frac{\partial u}{\partial x_l}$$

$$= \left(\frac{\partial A_k}{\partial x_l} + A_k A_l\right) u,$$

and we also get the relation with k and l exchanged. Hence we have

$$\left(\frac{\partial A_k}{\partial x_l} + A_k A_l\right) u = \left(\frac{\partial A_l}{\partial x_k} + A_l A_k\right) u,$$

which is necessary for the existence of a solution. Then we come to the following definition.

Definition 11.1 For a linear Pfaffian system (11.2), the condition

$$\frac{\partial A_k}{\partial x_l} + A_k A_l = \frac{\partial A_l}{\partial x_k} + A_l A_k \quad (1 \le k, l \le n) \tag{11.4}$$

is called the *complete integrability condition*.

The complete integrability condition (11.4) can be written in terms of the 1-form Ω as

$$d\Omega = \Omega \wedge \Omega. \tag{11.5}$$

This equality can be derived directly from (11.4), or obtained from $d^2u = 0$ as in the following way:

$$0 = d^2u = d(\Omega u) = d\Omega u - \Omega \wedge du = (d\Omega - \Omega \wedge \Omega)u.$$

The complete integrability condition (11.5) is a necessary condition for the existence of a solution, and also a sufficient condition. Namely we have the following theorem.

Theorem 11.1 *Assume that the linear Pfaffian system (11.3) satisfies the complete integrability condition (11.5). Then, for any $a \in X$ and any $u_0 \in \mathbb{C}^N$, there exists a unique solution $u(x)$ of (11.3) satisfying*

$$u(a) = u_0.$$

The solution $u(x)$ converges in any polydisc centered at a and contained in X.

Proof Without loss of generality, we may assume $a = 0 = (0, 0, \ldots, 0)$. First we show that there exists a unique formal power series solution

$$u(x) = \sum_{i_1, i_2, \ldots, i_n = 0}^{\infty} u_{i_1 i_2 \ldots i_n} x_1^{i_1} x_2^{i_2} \cdots x_n^{i_n} = \sum_I u_I x^I. \tag{11.6}$$

Here the right hand side is written in the standard notation with the multi-index $I = (i_1, i_2, \ldots, i_n)$.

We take the Taylor expansion of each $A_k(x)$ $(1 \leq k \leq n)$ at $x = 0$:

$$A_k(x) = \sum_{j_1, j_2, \ldots, j_n = 0}^{\infty} A_{j_1 j_2 \ldots j_n}^k x_1^{j_1} x_2^{j_2} \cdots x_n^{j_n} = \sum_J A_J^k x^J.$$

From the initial condition, we obtain

$$u_{00 \ldots 0} = u_0.$$

Input the formal series (11.6) into Eq. (11.2) to get

$$\sum_I i_k u_I x_1^{i_1} \cdots x_k^{i_k-1} \cdots x_n^{i_n} = \left(\sum_J A_J^k x^J\right)\left(\sum_K u_K x^K\right).$$

Comparing the coefficients of the terms $x^I = x_1^{i_1} x_2^{i_2} \cdots x_n^{i_n}$ in both sides, we get

$$(i_k + 1)u_{i_1 \ldots i_k+1 \ldots i_n} = \sum_{J+K=I} A_J^k u_K. \tag{11.7}$$

First we look at the relation (11.7) with $k = 1$ and $i_2 = \cdots = i_n = 0$:

$$(i_1 + 1)u_{i_1+1,0\ldots0} = \sum_{j_1+k_1=i_1} A^1_{j_1 0\ldots0} u_{k_1 0\ldots0}. \tag{11.8}$$

By this recurrence relation, the coefficients $\{u_{i_1 0\ldots0}\}_{i_1=0}^{\infty}$ are uniquely determined by $u_{00\ldots0}$. Next we consider (11.7) with $k = 2$ and $i_3 = \cdots = i_n = 0$:

$$(i_2 + 1)u_{i_1 i_2+1,0\ldots0} = \sum_{j_1+k_1=i_1} \sum_{j_2+k_2=i_2} A^2_{j_1 k_1 0\ldots0} u_{j_2 k_2 0\ldots0}. \tag{11.9}$$

If we put $i_1 = 0$ in this relation, we see that the coefficients $\{u_{0 i_2 0\ldots0}\}_{i_2=0}^{\infty}$ are uniquely determined by $u_{00\ldots0}$. If we put $i_1 = 1$ in (11.9), we see that the coefficients $\{u_{1 i_2 0\ldots0}\}_{i_2=0}^{\infty}$ are uniquely determined by $u_{00\ldots0}, u_{10\ldots0}$ and $\{u_{0 i_2 0\ldots0}\}_{i_2=0}^{\infty}$. In a similar way, by (11.9) with each i_1, we see that the coefficients $\{u_{i_1 i_2 0\ldots0}\}_{i_2=0}^{\infty}$ are uniquely determined by the data

$$u_{j_1 0\ldots0} \quad (0 \le j_1 \le i_1),$$

$$\{u_{j_1 i_2 0\ldots0}\}_{i_2=0}^{\infty} \quad (0 \le j_1 < i_1).$$

Thus by the relations (11.8) and (11.9), the coefficients $\{u_{i_1 i_2 0\ldots0}\}_{i_1,i_2=0}^{\infty}$ are uniquely determined by $u_{00\ldots0}$.

Similarly, by the relation (11.7) with $i_{k+1} = \cdots i_n = 0$, the coefficients $\{u_{i_1\ldots i_{k-1} i_k 0\ldots0}\}_{i_k=0}^{\infty}$ are uniquely determined by the data

$$u_{j_1\ldots j_{k-1} 0 0\ldots0} \quad (0 \le j_1 \le i_1, \ldots, 0 \le j_{k-1} \le i_{k-1}),$$

$$\{u_{j_1\ldots j_{k-1} i_k 0\ldots0}\}_{i_k=0}^{\infty} \quad (0 \le j_1 < i_1, \ldots, 0 \le j_{k-1} < i_{k-1})$$

with $i_1, \ldots, i_{k-1}$. By continuing up to $k = n$, we find that the coefficients $\{u_{i_1 i_2\ldots i_n}\}_{i_1,i_2,\ldots,i_n=0}^{\infty}$ are uniquely determined by $u_{00\ldots0}$. In the final step, we use the recurrence relation (11.7) with $k = n$, which implies the formal power series $u(x)$ given by (11.6) satisfies the differential equation in x_n direction

$$\frac{\partial u}{\partial x_n} = A_n(x)u. \tag{11.10}$$

On the other hand, we cannot derive from the above construction of $\{u_{i_1 i_2\ldots i_n}\}_{i_1,i_2,\ldots,i_n=0}^{\infty}$ that $u(x)$ satisfies the differential equations in the other $n - 1$ directions. This will be shown in the following.

Let $u(x)$ be the formal power series (11.6) determined by the above process, and define a formal power series $v_{n-1}(x)$ by

$$v_{n-1}(x) = \frac{\partial u}{\partial x_{n-1}}(x) - A_{n-1}(x)u(x).$$

Note that $u(x)$ satisfies the differential equation (11.10) as a formal power series. By using the complete integrability condition (11.4), we get

$$\frac{\partial v_{n-1}}{\partial x_n} = \frac{\partial}{\partial x_{n-1}}(A_n u) - \frac{\partial A_{n-1}}{\partial x_n}u - A_{n-1}A_n u$$

$$= A_n\frac{\partial u}{\partial x_{n-1}} + \left(\frac{\partial A_n}{\partial x_{n-1}} - \frac{\partial A_{n-1}}{\partial x_n} - A_{n-1}A_n\right)u$$

$$= A_n\left(\frac{\partial u}{\partial x_{n-1}} - A_{n-1}u\right)$$

$$= A_n v_{n-1}.$$

Thus $v_{n-1}(x)$ is a formal power series solution of (11.10). Moreover the power series

$$v_{n-1}(x_1, \ldots, x_{n-1}, 0)$$

$$= \frac{\partial u}{\partial x_{n-1}}(x_1, \ldots, x_{n-1}, 0) - A_{n-1}(x_1, \ldots, x_{n-1}, 0)u(x_1, \ldots, x_{n-1}, 0)$$

becomes 0 by the construction of the series $\{u_{i_1 \ldots i_{n-1} 0}\}_{i_1, \ldots, i_{n-1}=0}^\infty$. Therefore, owing to the uniqueness of formal power series solution of (11.10), we have $v_{n-1}(x) = 0$. Thus $u(x)$ is also a solution of the differential equation

$$\frac{\partial u}{\partial x_{n-1}} = A_{n-1}(x)u.$$

In a similar way, we can show that $u(x)$ becomes a formal power series solution of the differential equation (11.2) for any k. This completes the proof of the existence and uniqueness of the formal solution.

Next we shall show that the convergence of the formal power series solution thus obtained. Take any closed polydisc

$$|x_1| \le r_1, \ |x_2| \le r_2, \ldots, |x_n| \le r_n$$

contained in X. Since $A_k(x)$ $(1 \le k \le n)$ are holomorphic on X, by Cauchy's inequality, there exists $M > 0$ such that

$$\|A^k_{i_1 i_2 \ldots i_n}\| \le \frac{M}{r_1{}^{i_1} r_2{}^{i_2} \cdots r_n{}^{i_n}}$$

holds for any $(i_1, i_2, \ldots, i_n)$ and k. Define the series $\{U_{i_1 0 \ldots 0}\}_{i_1=0}^{\infty}$ by the recurrence relation

$$
\begin{cases}
U_{00 \ldots 0} = ||u_{00 \ldots 0}||, \\
(i_1 + 1)U_{i_1+1,0 \ldots 0} = \displaystyle\sum_{j_1+k_1=i_1} \frac{M}{r_1^{j_1}} U_{k_1 0 \ldots 0}.
\end{cases}
$$

Then, by induction, we can show that

$$
||u_{i_1 0 \ldots 0}|| \le U_{i_1 0 \ldots 0} \tag{11.11}
$$

holds for any i_1. Set

$$
U_1(x_1) = \sum_{i_1=0}^{\infty} U_{i_1 0 \ldots 0} x_1^{i_1}.
$$

Then, by the definition of $\{U_{i_1 0 \ldots 0}\}_{i_1=0}^{\infty}$, we have

$$
\frac{dU_1}{dx_1} = \frac{M}{1 - \frac{x_1}{r_1}} U_1, \ U_1(0) = ||u_{00 \ldots 0}||,
$$

so that we get

$$
U_1(x_1) = ||u_{00 \ldots 0}|| \left(1 - \frac{x_1}{r_1}\right)^{-Mr_1}.
$$

Next we define the series $\{U_{i_1 i_2 0 \ldots 0}\}_{i_1,i_2=0}^{\infty}$ by the recurrence relation

$$
(i_2 + 1)U_{i_1 i_2+1,0 \ldots 0} = \sum_{j_1+k_1=i_1} \sum_{j_2+k_2=i_2} \frac{M}{r_1^{j_1} r_2^{k_1}} U_{j_2 k_2 0 \ldots 0},
$$

where $\{U_{i_1 00 \ldots 0}\}_{i_1=0}^{\infty}$ are the ones already determined above. Then by the same reason for the uniqueness of $\{u_{i_1 i_2 0 \ldots 0}\}$, $\{U_{i_1 i_2 0 \ldots 0}\}_{i_1,i_2=0}^{\infty}$ are uniquely determined. Moreover, by induction with (11.11), we can show that

$$
||u_{i_1 i_2 0 \ldots 0}|| \le U_{i_1 i_2 0 \ldots 0}
$$

holds for any (i_1, i_2). Now we set

$$
U_2(x_1, x_2) = \sum_{i_1,i_2=0}^{\infty} U_{i_1 i_2 0 \ldots 0} x_1^{i_1} x_2^{i_2}.
$$

Then from the definition of $\{U_{i_1 i_2 0 \ldots 0}\}_{i_1, i_2 = 0}^{\infty}$ we derive

$$\frac{\partial U_2}{\partial x_2} = \frac{M}{\left(1 - \frac{x_1}{r_1}\right)\left(1 - \frac{x_2}{r_2}\right)} U_2, \quad U_2(x_1, 0) = U_1(x_1).$$

Thus we obtain

$$U(x_1, x_2) = U_1(x_1) \left(1 - \frac{x_2}{r_2}\right)^{-\frac{Mr_2}{1 - \frac{x_1}{r_1}}}$$

$$= \|u_{00\ldots 0}\| \left(1 - \frac{x_1}{r_1}\right)^{-Mr_1} \left(1 - \frac{x_2}{r_2}\right)^{-\frac{Mr_2}{1 - \frac{x_1}{r_1}}}.$$

Continuing these processes, we can show that the series $\{U_{i_1 i_2 \ldots i_n}\}_{i_1, i_2, \ldots, i_n = 0}^{\infty}$ is determined,

$$\|u_{i_1 i_2 \ldots i_n}\| \le U_{i_1 i_2 \ldots i_n}$$

holds for any $(i_1, i_2, \ldots, i_n)$, and, by setting

$$U(x_1, x_2, \ldots, x_n) = \sum_{i_1, i_2, \ldots, i_n = 0}^{\infty} U_{i_1 i_2 \ldots i_n} x_1^{i_1} x_2^{i_2} \cdots x_n^{i_n},$$

we have

$$U(x_1, x_2, \ldots, x_n) = \|u_{00\ldots 0}\| \prod_{k=1}^{n} \left(1 - \frac{x_k}{r_k}\right)^{-\frac{Mr_k}{\left(1 - \frac{x_1}{r_1}\right)\cdots\left(1 - \frac{x_{k-1}}{r_{k-1}}\right)}}.$$

This function $U(x_1, x_2, \ldots, x_n)$ is holomorphic in the open polydisc

$$|x_1| < r_1, \ |x_2| < r_2, \ldots, |x_n| < r_n,$$

and hence the formal power series $u(x_1, x_2, \ldots, x_n)$ with the majorant series $U(x_1, x_2, \ldots, x_n)$ converges in the same polydisc. □

In Chap. 3, we derived Theorems 3.2 and 3.3 from Theorem 3.1. By the same argument, we can derive the following two theorems.

Theorem 11.2 *Assume that the linear Pfaffian system (11.3) satisfies the complete integrable condition (11.5). Then any solution can be analytically continued along any curve in X, and the result becomes a solution.*

Theorem 11.3 *Let a, b be any two points in X. If two curves C_1, C_2 in X with the initial point a and the end point b are homotopic in X with the initial and end points fixed, then the analytic continuations of a solution of the Pfaffian system (11.3) at a along C_1 and along C_2 coincide.*

Therefore the domain of definition of any solution of the linear Pfaffian system (11.3) becomes the universal covering $\tilde{X}$ of the domain of definition X of the coefficients. In other words, any solution becomes a multi-valued function on X. It is a remarkable property for completely integrable systems that any solution can be analytically continued to the domain where the coefficients are holomorphic.

In the same reason as we have derived Theorem 3.4 from Theorem 3.1, we get the following result.

Proposition 11.4 *If a linear Pfaffian system (11.3) satisfies the complete integrability condition (11.5), for any point $a \in X$, the set of all solutions on a polydisc centered at a and contained in X becomes a linear space of dimension N.*

Next we study the linear independence of solutions. The following result corresponds to Theorem 3.6.

Theorem 11.5 *Let $u^1(x), u^2(x), \ldots, u^N(x)$ be solutions of the linear Pfaffian system (11.3) defined in a common domain. If the determinant of the matrix $(u^1(x), u^2(x), \ldots, u^N(x))$ becomes 0 at a point of the domain, then it becomes 0 at any point of the domain.*

Proof Let X_1 the common domain of definition. We regard X_1 as a subdomain of $\tilde{X}$. Assume that, at a point $a \in X_1$, we have

$$\left| u^1(a), u^2(a), \ldots, u^N(a) \right| = 0.$$

Take any point $b \in X_1$. Since X_1 is path-connected, we can take a path in X_1 connecting a and b, and moreover the path can be taken as a finite chain of paths each of which is parallel to an coordinate axis. Here a path ℓ is said to be parallel to an coordinate axis if there exist an index i $(1 \le i \le n)$ and constants $c_1, \ldots, c_{i-1}, c_{i+1}, \ldots, c_n \in \mathbb{C}$ such that the path can be parametrized as

$$t \mapsto (c_1, \ldots, c_{i-1}, \gamma(t), c_{i+1}, \ldots, c_n).$$

Assume that the determinant of $(u^1(x), u^2(x), \ldots, u^N(x))$ vanishes at the initial point of the path ℓ. Since $u^1(x), u^2(x), \ldots, u^N(x)$ are functions in one variable x_i in a neighborhood of ℓ with $x_j = c_j$ $(j \ne i)$ and are solutions of the linear ordinary differential equation

$$\frac{du}{dx_i} = A_i(c_1, \ldots, c_{i-1}, x_i, c_{i+1}, \ldots, c_n)u,$$

thanks to Theorem 3.6, the determinant vanishes on ℓ. In particular it vanishes at the end point of ℓ. Iterating this argument, we conclude that the determinant vanishes at b. $\qquad\square$

It follows that a set of N solutions linearly independent at a point of the domain of definition is linearly independent at any point of the domain. This assertion can also be shown as follows. If a set of solutions is linearly dependent over $\mathbb{C}$ at a point, the linear dependence relation remains hold under any analytic continuation, which implies that the set is linearly dependent at any point.

Anyway, we have the following assertion.

Theorem 11.6 *For a completely integrable linear Pfaffian system (11.3) defined on a domain X with the unknown vector of size N, the followings hold.*

  (i) *The set of solutions on the universal covering $\tilde{X}$ of X makes a linear space over $\mathbb{C}$ of dimension N.*
 (ii) *For any simply connected subdomain X_1 of X, the set of solutions on X_1 makes a linear space over $\mathbb{C}$ of dimension N.*
(iii) *If the determinant of N solutions with a common domain of definition does not vanish at a point of the domain, the set makes a basis of the linear space of the solutions on the domain.*

This theorem asserts that the rank of the linear Pfaffian system (11.3) is N. The set of N solutions in Theorem 11.6 (iii) is called a *fundamental system of solutions* of the linear Pfaffian system (11.3). The $N \times N$ matrix given by the N solutions is called a *fundamental matrix solution*.

Next we consider an inhomogeneous completely integrable linear Pfaffian system. For $1 \le k \le n$, let $b_k(x)$ be an N vector of holomorphic functions on X, and consider the inhomogeneous Pfaffian system

$$\frac{\partial u}{\partial x_k} = A_k(x)u + b_k(x) \quad (1 \le k \le n). \tag{11.12}$$

Theorem 11.7 *In the inhomogeneous Pfaffian system (11.12), we assume that $A_1(x), A_2(x), \ldots, A_n(x)$ satisfy the complete integrability condition (11.4), and that*

$$\frac{\partial b_k}{\partial x_l} - A_l b_k = \frac{\partial b_l}{\partial x_k} - A_k b_l \quad (1 \le k, l \le n) \tag{11.13}$$

hold. Then, for any $a \in X$ and any $u_0 \in \mathbb{C}^N$, there exists a unique solution of (11.12) satisfying

$$u(a) = u_0.$$

Any solution of (11.12) can be represented as a sum of a particular solution of (11.12) and a solution of the associated homogeneous linear Pfaffian system (11.2).

Proof The theorem can be proved by the method of variation of constants. Let $u_H(x)$ be a fundamental matrix solution of the associated homogeneous Pfaffian system (11.2). We are going to obtain a solution $u(x)$ of the inhomogeneous linear Pfaffian system (11.12) in the form

$$u(x) = u_H(x)c(x),$$

where $c(x)$ is an N vector. Put this into the equation (11.12) to get the differential equation

$$\frac{\partial c}{\partial x_k} = u_H(x)^{-1}b_k(x) \quad (1 \le k \le n)$$

in c. By the condition (11.13), we get

$$\frac{\partial}{\partial x_l}\left(u_H(x)^{-1}b_k\right) = \frac{\partial}{\partial x_k}\left(u_H(x)^{-1}b_l(x)\right),$$

which implies the existence of a solution c. The remaining assertions can be shown in a similar way as in the case of ordinary differential equations, and we omit them.
□

Example 11.1 We consider the case $n = 2$, and write the coordinates of $\mathbb{C}^2$ by (x, y). Let $A_{x,0}, A_{x,1}, A_{y,0}, A_{y,1}, A_{x,y}$ be $N \times N$ constant matrices, and consider the linear Pfaffian system

$$du = \left(A_{x,0}\frac{dx}{x} + A_{x,1}\frac{dx}{x-1} + A_{y,0}\frac{dy}{y} + A_{y,1}\frac{dy}{y-1} + A_{x,y}\frac{d(x-y)}{x-y}\right)u.$$
(11.14)

This system can be written in the form of (11.2) as

$$\begin{cases} \dfrac{\partial u}{\partial x} = \left(\dfrac{A_{x,0}}{x} + \dfrac{A_{x,1}}{x-1} + \dfrac{A_{x,y}}{x-y}\right)u, \\[3mm] \dfrac{\partial u}{\partial y} = \left(\dfrac{A_{y,0}}{y} + \dfrac{A_{y,1}}{y-1} + \dfrac{A_{x,y}}{y-x}\right)u. \end{cases}$$

We set the right hand sides as

$$A(x, y) = \frac{A_{x,0}}{x} + \frac{A_{x,1}}{x-1} + \frac{A_{x,y}}{x-y}, \quad B(x, y) = \frac{A_{y,0}}{y} + \frac{A_{y,1}}{y-1} + \frac{A_{x,y}}{y-x}.$$

Then the complete integrability condition (11.4) is given by

$$\frac{\partial A}{\partial y} + AB = \frac{\partial B}{\partial x} + BA.$$

Since we have $\partial A/\partial y = \partial B/\partial x$, the integrability condition becomes $[A, B] = O$. Note that $[A, B]$ is a rational function in (x, y). Then the condition $[A, B] = O$ can be written as relations in the coefficient matrices. Explicitly, we have

$$[A_{x,0}, A_{y,1}] = O,$$

$$[A_{x,1}, A_{y,0}] = O,$$

$$[A_{x,0}, A_{x,y} + A_{y,0}] = [A_{x,y}, A_{y,0} + A_{x,0}] = [A_{y,0}, A_{x,0} + A_{x,y}] = O,$$

$$[A_{x,1}, A_{x,y} + A_{y,1}] = [A_{x,y}, A_{y,1} + A_{x,1}] = [A_{y,1}, A_{x,1} + A_{x,y}] = O.$$

$$(11.15)$$

as the complete integrability condition. We will derive these relations in the next example, where more general case is considered. The linear Pfaffian system (11.14) appears as completely integrable systems satisfied by many special functions in several variables such as Appell's hypergeometric series in two variables.

Example 11.2 We extend Example 11.1 to obtain a linear Pfaffian system called of KZ type (KZ stands for Knizhnik-Zamolodchikov). Let $a_1, a_2, \ldots, a_p \in \mathbb{C}$ be mutually distinct points, and $A_{i,k}$ $(1 \le i \le n, 1 \le j \le p)$, $B_{i,j}$ $(1 \le i < j \le n)$ be $N \times N$ constant matrices. We call the linear Pfaffian system

$$du = \left(\sum_{i=1}^{n} \sum_{k=1}^{p} A_{i,k} \frac{dx_i}{x_i - a_k} + \sum_{1 \le i < j \le n} B_{i,j} \frac{d(x_i - x_j)}{x_i - x_j} \right) u \qquad (11.16)$$

of KZ type. We shall write the complete integrability condition in terms of the matrices $A_{i,k}$ and $B_{i,j}$. For $1 \le i \le n$, we set

$$A_i(x) = \sum_{k=1}^{p} \frac{A_{i,k}}{x_i - a_k} + \sum_{j \ne i} \frac{B_{i,j}}{x_i - x_j},$$

where we understand $B_{i,j} = B_{j,i}$ for $i > j$. As Example 11.1, the complete integrability condition (11.4) becomes

$$[A_i, A_j] = O.$$

Namely we have

$$\left(\sum_{k=1}^{p} \frac{A_{i,k}}{x_i - a_k} + \sum_{l \ne i} \frac{B_{i,l}}{x_i - x_l} \right) \left(\sum_{k=1}^{p} \frac{A_{j,k}}{x_j - a_k} + \sum_{l \ne j} \frac{B_{j,l}}{x_j - x_l} \right)$$

$$= \left(\sum_{k=1}^{p} \frac{A_{j,k}}{x_j - a_k} + \sum_{l \ne j} \frac{B_{j,l}}{x_j - x_l} \right) \left(\sum_{k=1}^{p} \frac{A_{i,k}}{x_i - a_k} + \sum_{l \ne i} \frac{B_{i,l}}{x_i - x_l} \right).$$

$$(11.17)$$

We multiply the both sides by $x_i - a_k$ and take the limit $x_i \to a_k$. Then we get

$$A_{i,k} \left(\sum_{m=1}^{p} \frac{A_{j,m}}{x_j - a_m} + \sum_{l \neq j,i} \frac{B_{j,l}}{x_j - x_l} + \frac{B_{j,i}}{x_j - a_k} \right)$$

$$= \left(\sum_{m=1}^{p} \frac{A_{j,m}}{x_j - a_m} + \sum_{l \neq j,i} \frac{B_{j,l}}{x_j - x_l} + \frac{B_{j,i}}{x_j - a_k} \right) A_{i,k}.$$

By comparing the residues at $x_j = a_m$ of both sides, we get

$$\begin{cases} A_{i,k} A_{j,m} = A_{j,m} A_{i,k} \quad (m \neq k), \\ A_{i,k}(A_{j,k} + B_{j,i}) = (A_{j,k} + B_{j,i}) A_{i,k}, \end{cases} \tag{11.18}$$

and by comparing the residues at $x_j = x_l$, we get

$$A_{i,k} B_{j,l} = B_{j,l} A_{i,k} \quad (l \neq i, j). \tag{11.19}$$

For $l \neq i, j$, we multiply the both sides of (11.17) by $x_i - x_l$, and take the limit $x_i \to x_l$ to obtain

$$B_{i,l} \left(\sum_{k=1}^{p} \frac{A_{j,k}}{x_j - a_k} + \sum_{m \neq j,i} \frac{B_{j,m}}{x_j - x_m} + \frac{B_{j,i}}{x_j - x_l} \right)$$

$$= \left(\sum_{k=1}^{p} \frac{A_{j,k}}{x_j - a_k} + \sum_{m \neq j,i} \frac{B_{j,m}}{x_j - x_m} + \frac{B_{j,i}}{x_j - x_l} \right) B_{i,l}.$$

The residues at $x_j = a_k$ of both sides bring the relation (11.19). Comparing the residues at $x_j = x_m$ of both sides, we get

$$\begin{cases} B_{i,l} B_{j,m} = B_{j,m} B_{i,l} \quad (m \neq l), \\ B_{i,l}(B_{j,l} + B_{i,j}) = (B_{j,l} + B_{i,j}) B_{i,l}. \end{cases} \tag{11.20}$$

Thus we have expressed the complete integrability condition in terms of the residue matrices. We can write the second equation in (11.18) and the second equation (11.20) in a cyclic form by combining several relations. Consequently, we obtain the complete list of the complete integrability condition:

$$[A_{i,k}, A_{j,m}] = O \quad (1 \le i, j \le n, i \ne j; 1 \le k, m \le p, k \ne m),$$

$$[A_{i,k}, A_{j,k} + B_{i,j}] = [A_{j,k}, B_{i,j} + A_{i,k}] = [B_{i,j}, A_{i,k} + A_{j,k}] = O$$
$$(1 \le i, j \le n, i \ne j; 1 \le k \le p),$$

$$[A_{i,k}, B_{j,l}] = O \quad (1 \le i, j, l \le n, i \ne j, l \ne i, j; 1 \le k \le p),$$

$$[B_{i,l}, B_{j,m}] = O \quad (1 \le i, j, l, m \le n, i \ne j, l \ne m, l \ne i, j, m \ne i, j),$$

$$[B_{i,l}, B_{j,l} + B_{i,j}] = [B_{j,l}, B_{i,j} + B_{i,l}] = [B_{i,j}, B_{i,l} + B_{j,l}] = O$$
$$(1 \le i, j, l \le n, i \ne j, l \ne i, j).$$

$$(11.21)$$

Chapter 12
Regular Singularity

12.1 Local Analysis at a Regular Singular Point

In this section, we study the behavior of solutions of a linear Pfaffian system in a neighborhood of a singular point of the coefficients, where we restrict singular points to ones corresponding to regular singular points in ordinary differential equations. Note that a singular point of analytic functions in several variables is not a point but a set of codimension one (a hypersurface).

Let $U \subset \mathbb{C}^n$ be a domain, and $\varphi(x) = \varphi(x_1, x_2, \ldots, x_n)$ a holomorphic function on U. We set

$$E = \{x \in U \; ; \; \varphi(x) = 0\}.$$

A point x^0 in E is called a *critical point* if

$$\operatorname{grad} \varphi(x^0) = (\varphi_{x_1}(x^0), \varphi_{x_2}(x^0), \ldots, \varphi_{x_n}(x^0)) = 0 \tag{12.1}$$

holds. A point in E that is not a critical point is called an *ordinary point*. We denote by E^0 the set of ordinary points in E. Namely we set

$$E^0 = \{x \in E \; ; \; \operatorname{grad} \varphi(x) \neq 0\}.$$

We are going to study the behavior in a neighborhood of E^0 of a linear Pfaffian system having a logarithmic singularity along E. Explicitly, we consider the linear Pfaffian system of the form

$$du = (A(x)d \log \varphi(x) + \Omega_1(x))u, \tag{12.2}$$

© The Editor(s) (if applicable) and The Author(s), under exclusive license to Springer Nature Switzerland AG 2020
Y. Haraoka, *Linear Differential Equations in the Complex Domain*, Lecture Notes in Mathematics 2271, https://doi.org/10.1007/978-3-030-54663-2_12

where $A(x)$ is an $N \times N$ matrix with entries holomorphic on U and $\Omega_1(x)$ a 1-form of $N \times N$ matrix with entries holomorphic on U.

Take any point $a = (a_1, a_2, \ldots, a_n) \in E^0$. Since we have $\operatorname{grad} \varphi(a) \neq 0$, by the inverse mapping theorem, there exists a biholomorphic map

$$
\begin{array}{rccc}
\Phi: & U_a & \rightarrow & V_0 \\
& x = (x_1, x_2, \ldots, x_n) & \mapsto & \xi = (\xi_1, \xi_2, \ldots, \xi_n)
\end{array}
$$

from a neighborhood U_a of a to a neighborhood V_0 of 0 which sends a to 0 and $U_a \cap E^0$ to $V_0 \cap \{\xi_1 = 0\}$. For example, we take i such that $\varphi_{x_i}(a) \neq 0$, and may assume that $i = 1$. Then we can obtain Φ by defining

$$
\xi_1 = \varphi(x), \ \xi_2 = x_2 - a_2, \ \ldots, \ \xi_n = x_n - a_n.
$$

We shall write the linear Pfaffian system (12.2) in the new variables. Set $\xi = (\xi_1, \xi')$, $\xi' = (\xi_2, \xi_3, \ldots, \xi_n)$, and

$$
A(\Phi^{-1}(\xi)) = B(\xi).
$$

Since $B(\xi)$ is holomorphic at $\xi = 0$, the Taylor expansion with respect to ξ_1 gives

$$
B(\xi) = B_0(\xi') + \xi_1 B_1(\xi),
$$

where $B_0(\xi')$ is holomorphic at $\xi' = 0$ and $B_1(\xi)$ is holomorphic at $\xi = 0$. Note that

$$
d \log \varphi(x) = \frac{d\xi_1}{\xi_1} + \omega,
$$

where ω is a 1-form holomorphic at $\xi = 0$. Then there exist $N \times N$ matrices $C_1(\xi), A_2(\xi), \ldots, A_n(\xi)$ holomorphic at $\xi = 0$ such that the Pfaffian system (12.2) can be written as

$$
du = \left(\left(\frac{B_0(\xi')}{\xi_1} + C_1(\xi) \right) d\xi_1 + A_2(\xi) d\xi_2 + \cdots + A_n(\xi) d\xi_n \right) u.
$$

Thus in the following we consider the linear Pfaffian system

$$
\frac{\partial u}{\partial x_k} = A_k(x) u \quad (1 \leq k \leq n), \tag{12.3}
$$

where $A_2(x), A_3(x), \ldots, A_n(x)$ are $N \times N$ matrices with entries holomorphic in a neighborhood U_0 of 0, and $A_1(x)$ is given by

$$
A_1(x) = \frac{B_1(x')}{x_1} + C_1(x) \tag{12.4}
$$

with $N \times N$ matrix $B_1(x')$ holomorphic at $x' = 0$ (we set $x = (x_1, x')$, $x' = (x_2, x_3, \ldots, x_n)$) and $N \times N$ matrix $C_1(x)$ holomorphic in U_0.

The following theorem is fundamental.

Theorem 12.1 *Assume that the linear Pfaffian system (12.3) is completely integrable. Then the Jordan canonical form of $B_1(x')$ is a constant matrix, i.e. independent of x'.*

Proof By the complete integrability condition (11.4), we have

$$\frac{\partial}{\partial x_k}\left(\frac{B_1}{x_1} + C_1\right) + \left(\frac{B_1}{x_1} + C_1\right) A_k = \frac{\partial A_k}{\partial x_1} + A_k\left(\frac{B_1}{x_1} + C_1\right)$$

for $k > 1$. We set $D_k(x') = A_k(0, x')$. Comparing the coefficients of $1/x_1$ in both sides, we get

$$\frac{\partial B_1}{\partial x_k} = [D_k, B_1] \quad (2 \le k \le n). \tag{12.5}$$

Also by the complete integrability condition (11.4) with $2 \le k, l \le n$, we get

$$\frac{\partial D_k}{\partial x_l} + D_k D_l = \frac{\partial D_l}{\partial x_k} + D_l D_k \quad (2 \le k, l \le n) \tag{12.6}$$

by putting $x_1 = 0$. These conditions are the complete integrability condition of the linear Pfaffian system

$$\frac{\partial p}{\partial x_k} = D_k(x')p \quad (2 \le k \le n)$$

in variable x', and hence there exists a fundamental matrix solution $P(x')$ for this Pfaffian system. For $2 \le k \le n$, we have

$$\frac{\partial}{\partial x_k}(P^{-1}B_1 P) = -P^{-1}\frac{\partial P}{\partial x_k}P^{-1}B_1 P + P^{-1}\frac{\partial B_1}{\partial x_k}P + P^{-1}B_1\frac{\partial P}{\partial x_k}$$

$$= -P^{-1}D_k B_1 P + P^{-1}[D_k, B_1]P + P^{-1}B_1 D_k P$$

$$= O,$$

which shows that $P^{-1}B_1 P$ does not depend on x'. Of course it does not depend on x_1, and hence is a constant matrix. The Jordan canonical form of B_1 is similar to $P^{-1}B_1 P$, and then is also a constant matrix.   □

Now we give the main theorem in this section.

Theorem 12.2 *Consider the linear Pfaffian system (12.3) with $A_1(x)$ given by (12.4) and $A_2(x), A_3(x), \ldots, A_n(x)$ holomorphic on U_0. If the Pfaffian system (12.3) is completely integrable, there exists a fundamental system of solutions consisting of solutions of the form*

$$u(x) = x_1^\rho \sum_{j=0}^q (\log x_1)^j \sum_{m=0}^\infty u_{jm}(x') x_1^m, \tag{12.7}$$

where ρ is an eigenvalue of $B_1(x')$ whose real part is minimum among the eigenvalues having integral difference with ρ, q is a non-negative integer determined for each ρ by the Jordan canonical form of $B_1(x')$, and $u_{jm}(x')$ are vectors holomorphic at $x' = 0$.

Proof By using the similar transformation by P in the proof of Theorem 12.1, we may assume that $B_1(x')$ is already the Jordan canonical form J. For the differential equation in x_1 direction

$$\frac{\partial u}{\partial x_1} = \left(\frac{J}{x_1} + C_1(x_1, x') \right) u, \tag{12.8}$$

we have already constructed a fundamental system of solutions of the form (12.7) in Theorem 4.5. In order to quote the result of Theorem 4.5, we define several notation. Let

$$\{ \rho + k_0, \rho + k_1, \ldots, \rho + k_d \}$$

be the set of the eigenvalues of J having integral difference with ρ, where $0 = k_0 < k_1 < \cdots < k_d$ is a sequence of integers. We denote by J_i the direct sum of the Jordan cells with the eigenvalue $\rho + k_i$. Then J can be written as the direct sum

$$J = \bigoplus_{i=0}^d J_i \oplus J', \tag{12.9}$$

where J' is the direct sum of the Jordan cells with the eigenvalues having no integral difference with ρ. We write any vector $v \in \mathbb{C}^N$ as

$$v = {}^t([v]_0, [v]_1, \ldots, [v]_d, [v]')$$

according to this direct sum decomposition. Note that the non-negative integer q is determined by sizes of the Jordan cells in $J_0, \ldots, J_d$ as explained in Theorem 4.5.

As is shown in Theorem 4.5, the solution of the form (12.7) is uniquely determined by $[u_{00}]_0, [u_{0k_1}]_1, \ldots, [u_{0k_d}]_d$. Then we shall determine these datum so that u satisfies the differential equations in $x_2, \ldots, x_n$ directions. Let u be a solution of the form (12.7) of the differential equation (12.8) in x_1 direction (12.7), and set

$$v_k(x) = \frac{\partial u}{\partial x_k} - A_k(x)u$$

for $k > 1$. By using the complete integrability condition, we obtain

$$\frac{\partial v_k}{\partial x_1} = \frac{\partial}{\partial x_k}\left(\frac{\partial u}{\partial x_1}\right) - \frac{\partial A_k}{\partial x_1}u - A_k\frac{\partial u}{\partial x_1}$$

$$= A_1\frac{\partial u}{\partial x_k} + \left(\frac{\partial A_1}{\partial x_k} - \frac{\partial A_k}{\partial x_1} - A_kA_1\right)u$$

$$= A_1 v_k.$$

Also we derive from (12.7)

$$v_k(x) = x_1{}^\rho \sum_{j=0}^q (\log x_1)^j \sum_{m=0}^\infty \left(\frac{\partial u_{jm}}{\partial x_k} - \sum_{s+t=m} A_{ks}u_{jt}\right)x_1{}^m$$

$$=: x_1{}^\rho \sum_{j=0}^q (\log x_1)^j \sum_{m=0}^\infty v_{jm}^k x_1{}^m. \tag{12.10}$$

Thus $v_k(x)$ is also a solution of (12.8) of the form (12.7). Here we denoted the Taylor expansion of $A_k(x)$ $(2 \le k \le n)$ in x_1 by

$$A_k(x_1, x') = \sum_{m=0}^\infty A_{km}(x')x_1{}^m.$$

In terms of the symbol used in the proof of Theorem 12.1, we have $A_{k0}(x') = D_k(x')$. We have (12.5) by the complete integrability condition, and assumed $B_1 = J$, which is a constant matrix. Then $[D_k, J] = O$ holds. In the direct sum decomposition (12.9) of J, the eigenvalues of each component are mutually distinct. Then owing to Lemma 4.7, D_k can be decomposed into a direct sum of the same form. Namely we have

$$D_k = \bigoplus_{i=0}^d D_k^i \oplus D_k',$$

where D_k^i and J_i have the same size, and so do D_k' and J'.

From (12.10) we obtain

$$[v_{00}^k]_0 = \left[\frac{\partial u_{00}}{\partial x_k} - D_k u_{00}\right]_0 = \frac{\partial[u_{00}]_0}{\partial x_k} - D_k^0[u_{00}]_0.$$

Then we define $[u_{00}]_0$ so that

$$\frac{\partial[u_{00}]_0}{\partial x_k} = D_k^0[u_{00}]_0 \tag{12.11}$$

holds. By the proof of Theorem 4.5, for the coefficients of the expansion (12.10) we see that

$$v_{jm}^k = \frac{\partial u_{jm}}{\partial x_k} - \sum_{s+t=m} A_{ks} u_{jt} = 0 \quad (m < k_1) \tag{12.12}$$

holds. Next we have

$$[v_{0k_1}^k]_1 = \left[\frac{\partial u_{0k_1}}{\partial x_k} - \sum_{s+t=k_1} A_{ks} u_{0t}\right]_1 = \frac{\partial[u_{0k_1}]_1}{\partial x_k} - D_k^1[u_{0k_1}]_1 - \left[\sum_{t=0}^{k_1-1} A_{k,k_1-t} u_{0t}\right]_1.$$

Then we take as $[u_{0k_1}]_1$ a solution of the inhomogeneous differential equation

$$\frac{\partial[u_{0k_1}]_1}{\partial x_k} = D_k^1[u_{0k_1}]_1 + \left[\sum_{t=0}^{k_1-1} A_{k,k_1-t} u_{0t}\right]_1, \tag{12.13}$$

where the inhomogeneous term in the right hand side is determined by $[u_{00}]_0$ taken so far. We continue similar procedures to determine $[u_{0k_i}]_i$ so that $[v_{0k_i}^k]_i = 0$ holds. Then all datum $[v_{00}^k]_0, [v_{0k_1}^k]_1, \ldots, [v_{0k_d}^k]_d$ which determine (12.10) become zero, which implies $v_k(x) = 0$ so that

$$\frac{\partial u}{\partial x_k} = A_k(x)u$$

holds.

In the above we fixed one $k > 1$, and determined $[u_{0k_i}]_i$ by solving differential equations in x_k, so that $u(x)$ satisfies the differential equation in x_k direction. Hence, in order that $u(x)$ satisfies all differential equations in x_k $(2 \le k \le n)$ directions, the differential equations used to determine $[u_{0k_i}]_i$ should be compatible for all k. We shall show this.

For the differential equation (12.11), we see the compatibility as a linear Pfaffian system in x' by the condition (12.6) and the direct sum decomposition of D_k. Then we can take full set of linearly independent solutions $[u_{00}]_0$. We consider the next inhomogeneous differential equation (12.13). The associated homogeneous linear Pfaffian system

$$\frac{\partial [u_{0k_1}]_1}{\partial x_k} = D_k^1 [u_{0k_1}]_1 \quad (2 \le k \le n)$$

is completely integrable by the same reason as above. Then we have only to check the condition (11.13) in Theorem 11.7 on the inhomogeneous terms. We shall compute the following for $k, l > 1$. By using the complete integrability condition (11.4) and the condition (12.12), we have

$$\frac{\partial}{\partial x_l} \left(\sum_{t=0}^{k_1-1} A_{k,k_1-t} u_{0t} \right) - \frac{\partial}{\partial x_k} \left(\sum_{t=0}^{k_1-1} A_{l,k_1-t} u_{0t} \right)$$

$$= \sum_{t=0}^{k_1-1} \left(\frac{\partial A_{k,k_1-t}}{\partial x_l} u_{0t} + A_{k,k_1-t} \frac{\partial u_{0t}}{\partial x_l} - \frac{\partial A_{l,k_1-t}}{\partial x_k} u_{0t} - A_{l,k_1-t} \frac{\partial u_{0t}}{\partial x_k}' \right)$$

$$= \sum_{t=0}^{k_1-1} \Big(\sum_{i+j=k_1-t} (A_{li} A_{kj} - A_{ki} A_{lj}) u_{0t}$$

$$+ A_{k,k_1-t} \sum_{i+j=t} A_{li} u_{0j} - A_{l,k_1-t} \sum_{i+j=t} A_{ki} u_{0j} \Big)$$

$$= \sum_{t=0}^{k_1-1} \Big(\sum_{i=0}^{k_1-t-1} (A_{li} A_{k,k_1-t-i} - A_{ki} A_{l,k_1-t-i}) u_{0t}$$

$$+ A_{k,k_1-t} \sum_{j=0}^{t-1} A_{l,t-j} u_{0j} - A_{l,k_1-t} \sum_{j=0}^{t-1} A_{k,t-j} u_{0j} \Big)$$

$$= \sum_{t=0}^{k_1-1} \sum_{i=0}^{k_1-t-1} (A_{li} A_{k,k_1-t-i} - A_{ki} A_{l,k_1-t-i}) u_{0t}$$

$$+ \sum_{j=0}^{k_1-2} \sum_{t=j+1}^{k_1-1} (A_{k,k_1-t} A_{l,l-j} - A_{l,k_1-t} A_{k,t-j}) u_{0j}$$

$$= \sum_{t=0}^{k_1-1} \sum_{i=0}^{k_1-t-1} (A_{li} A_{k,k_1-t-i} - A_{ki} A_{l,k_1-t-i}) u_{0t}$$

$$- \sum_{t=0}^{k_1-2} \sum_{i=1}^{k_1-t-1} (A_{li} A_{k,k_1-t-i} - A_{ki} A_{l,k_1-t-i}) u_{0t}$$

$$= \sum_{t=0}^{k_1-1} (A_{l0} A_{k,k_1-t} - A_{k0} A_{l,k_1-t}) u_{0t}.$$

Taking the $[\ ]_1$ components of the both sides, we have the identity

$$\frac{\partial}{\partial x_l}\left[\sum_{t=0}^{k_1-1} A_{k,k_1-t}u_{0t}\right]_1 - \frac{\partial}{\partial x_k}\left[\sum_{t=0}^{k_1-1} A_{l,k_1-t}u_{0t}\right]_1$$

$$=\left[A_{l0}\sum_{t=0}^{k_1-1} A_{k,k_1-t}u_{0t}\right]_1 - \left[A_{k0}\sum_{t=0}^{k_1-1} A_{l,k_1^t}u_{0t}\right]_1$$

$$=D_l^1\left[\sum_{t=0}^{k_1-1} A_{k,k_1-t}u_{0t}\right]_1 - D_k^1\left[\sum_{t=0}^{k_1-1} A_{l,k_1^t}u_{0t}\right]_1$$

for the inhomogeneous parts of the differential equation (12.13). Then, thanks to Theorem 11.7, there exists $[u_{0k_1}]_1$ satisfying (12.13) for every k ($2 \le k \le n$) for any initial data. The assertions for $[u_{0k_2}]_2$ and so on can be shown similarly. Thus we can show that, for each step, $[u_{0k_i}]_i$ exists for any initial data. Consequently, we have linearly independent solutions of the linear Pfaffian system (12.3) with exponent ρ of the same number as sum of the dimensions of the generalized eigenspaces of J for $\rho + k_i$ ($i = 0, 1, \ldots, d$).

This shows that we can construct a fundamental system of solutions of the form (12.7). $\square$

Corollary 12.3 *Consider the linear Pfaffian system (12.3) with $A_1(x)$ given by (12.4) and $A_2(x), A_3(x), \ldots, A_n(x)$ holomorphic on U_0. If the Pfaffian system (12.3) is completely integrable and $B_1(x')$ is non-resonant, there exists a fundamental matrix solution of the form*

$$\mathcal{U}(x) = F(x)x_1^{\ J}, \tag{12.14}$$

where J is the Jordan canonical form of $B_1(x')$ and $F(x)$ is an $N \times N$ matrix holomorphic and invertible at $x = 0$.

Proof Thanks to Corollary 4.6, there exists a fundamental matrix solution of the form (12.14) of the differential equation (12.8) in x_1 direction. Each column of the fundamental matrix solution is of the form (12.7), and hence, by Theorem 12.2, we can determine the x' dependence so that it becomes a solution of the linear Pfaffian system (12.3). $\square$

We explain the construction of the above fundamental matrix solution more explicitly. When $B_1(x')$ is non-resonant, d in the proof of Theorem 12.2 becomes 0. Then we can determine a basis of the solutions of exponent ρ only by determining $[u_{00}]_0$. We expand $F(x)$ in the fundamental matrix solution (12.14) with respect to

x_1 at $x_1 = 0$ as

$$F(x) = \sum_{m=0}^{\infty} F_m(x')x_1{}^m.$$

Then the initial term $F_0(x')$ is a block diagonal matrix each of whose diagonal blocks is obtained from the linearly independent columns $[u_{00}]_0$, and the other terms $F_m(x')$ $(m > 0)$ are uniquely determined by $F_0(x')$ as we have seen in the proof of Theorem 4.5 (i).

We write the conclusions of Theorem 12.2 and Corollary 12.3 in the coordinate before the coordinate change. For the completely integrable linear Pfaffian system (12.2), we have a fundamental system of solutions consisting of solutions of the form

$$u(x) = \varphi(x)^{\rho} \sum_{j=0}^{q} (\log \varphi(x))^j \sum_{m=0}^{\infty} u_{jm}(x)\varphi(x)^m, \qquad (12.15)$$

where ρ is a constant and $u_{jm}(x)$ are vector functions in the variables corresponding to x' in the new coordinate. Moreover, in the case where the matrix corresponding to $B_1(x')$ is non-resonant, J being its Jordan canonical form, we have the fundamental matrix solution

$$\mathcal{U}(x) = F(x)\varphi(x)^J,$$

where $F(x)$ is a matrix function holomorphic and invertible on a neighborhood of a point in E^0.

Example 12.1 We consider the case $n = 2$, and set $(x_1, x_2) = (x, y)$. We take

$$\varphi(x, y) = xy(x - 1)(y - 1)(x - y),$$

and consider a linear Pfaffian system

$$du = \left(A_1 \frac{dx}{x} + A_2 \frac{dy}{y - 1} + A_3 \frac{d(x - y)}{x - y} + A_4 \frac{dx}{x - 1} + A_5 \frac{dy}{y} \right) u \qquad (12.16)$$

with logarithmic singularity along $E = \{\varphi(x, y) = 0\}$, where $A_1, A_2, \ldots, A_5$ are $N \times N$ constant matrices. We have already studied this Pfaffian system in Example 11.1, where the complete integrability condition (11.15) is obtained. We assume it. It can be easily shown that this Pfaffian system has logarithmic singularities along E by writing the 1-form in the right hand side as

$$A_1 d \log x + A_2 d \log(y - 1) + A_3 \log(x - y) + A_4 \log(x - 1) + A_5 d \log y.$$

Since we have

$$\varphi_x(x, y) = y(y - 1)(3x^2 - 2(y + 1)x + y),$$

$$\varphi_y(x, y) = -x(x - 1)(3y^2 - 2(x + 1)y + x),$$

at the points $(x, y) = (0, 0), (0, 1), (1, 0), (1, 1)$ on E, we get $\operatorname{grad} \varphi(x, y) = 0$, so that Theorems 12.1 and 12.2 and Corollary 12.3 cannot be applied. For the other points we can apply these results, and obtain a local solution. For example, we take $p \neq 0, 1$. Then, if A_3 is non-resonant, we have a local fundamental matrix solution

$$u(x, y) = F(x, y)(x - y)^{A_3}$$

at the point (p, p), where $F(x, y)$ is an $N \times N$ matrix function holomorphic and invertible at (p, p).

In the above, we studied the Pfaffian system locally at the ordinary points E^0 in the singular locus $E = \{\varphi(x) = 0\}$. Afterwards in Chap. 13, we will proceed to the global analysis such as the monodromy and the rigidity, and in these global analysis we can obtain enough information from the local analysis at the ordinary points of E. On the other hand, the local analysis at critical points of E, namely at the points $x \in E$ satisfying $\operatorname{grad}\varphi(x) = 0$, is a natural and important problem by itself.

We consider the case where $\varphi(x)$ is a polynomial. Some polynomials $\varphi(x)$ are irreducible and have critical points. For example, the polynomial

$$\varphi(x_1, x_2) = x_1{}^2 - x_2{}^3$$

is irreducible with the critical point $(x_1, x_2) = (0, 0)$ in E. A standard way of the local analysis at such points is to use the blowing-up, which makes the critical point an ordinary point. For reducible polynomials $\varphi(x)$, another kind of critical points appear. Let $\varphi(x)$ be reducible and reduced, and take its irreducible decomposition

$$\varphi(x) = \prod_{i=1}^{m} \varphi_i(x).$$

Then any common zeros of at least two irreducible components are critical points of E. Let a be a common zero of $\varphi_1(x), \varphi_2(x), \ldots, \varphi_k(x)$. (We assume that a is not a zero of the other irreducible components.) If we can extend $(\varphi_1(x), \varphi_2(x), \ldots, \varphi_k(x))$ to a coordinate at a, we say that E is *normally crossing* at a. For a point $a \in \mathbb{C}^n$, E is normally crossing at a only if $k \leq n$ holds. If E is not normally crossing at a point, we usually use the blowing-up for the local analysis.

Now we consider the case that E is normally crossing at a. There exists a coordinate change

$$(x_1, x_2, \ldots, x_n) \to (\xi_1, \xi_2, \ldots, \xi_n)$$

in a neighborhood of a such that $\varphi_i(x) = 0$ is sent to $\xi_i = 0$ for each i ($1 \le i \le k$) and a is sent to 0. Therefore in this case, a completely integrable linear Pfaffian system with logarithmic singularity along E can be reduced to the linear Pfaffian system of the form

$$\begin{cases} \dfrac{\partial u}{\partial x_i} = \left(\dfrac{B_i(x)}{x_i} + C_i(x) \right) u & (1 \le i \le k), \\[2mm] \dfrac{\partial u}{\partial x_j} = A_j(x)u & (k+1 \le j \le n), \end{cases} \tag{12.17}$$

where $A_j(x), B_i(x), C_i(x)$ are matrix functions holomorphic at $x = 0$, and $B_i(x)$ are matrix functions in the variables $(x_1, \ldots, x_{i-1}, x_{i+1}, \ldots, x_n)$. For the linear Pfaffian systems of this form, Gérard [41] and Yoshida-Takano [192] constructed local solutions.

Theorem 12.4 *Assume that the linear Pfaffian system (12.17) is completely integrable. Then there exists a fundamental matrix solution*

$$U(x){x_1}^{L_1} \cdots {x_k}^{L_k}{x_1}^{G_1} \cdots {x_k}^{G_k}$$

of (12.17), where $U(x)$ is a matrix function holomorphic and invertible at $x = 0$, L_i ($1 \le i \le k$) diagonal matrices with diagonal entries in non-negative integers, and $G_1, \ldots, G_k$ mutually commutative constant matrices.

We remark that Theorem 12.4 contains Theorem 12.2 and Corollary 12.3 as particular cases.

12.2 Restriction to a Singular Locus

For completely integrable systems in two or more variables, the singular locus becomes a variety (hypersurface) having a positive dimension. Then we may define the operation that restricts the integrable system to the singular locus. This is a distinguished feature for the completely integrable systems in several variables, which cannot be defined for ordinary differential equations where the singular points are isolated points. Actually we have many remarkable phenomena by taking the restriction to a singular locus.

Here we understand that the restriction to a singular locus means the differential equation satisfied by the restriction to the singular locus of solutions of the original differential equation which are holomorphic at the singular locus. Of course there exists solutions having singularities at the singular locus, and then we cannot define the restriction of such solutions. While there may exist solutions holomorphic at a singular point. We take such solutions to make the restriction.

We consider a completely integrable linear Pfaffian system (12.2) with loga-rithmic singularity along E. By taking a coordinate change, we have a Pfaffian system (12.3) singular along $x_1 = 0$, where $A_1(x)$ is given by (12.4), and $A_k(x)$ $(2 \leq k \leq n)$ are holomorphic at $x = 0$. Moreover we assume that $B_1(x')$ is non-resonant.

By Theorem 12.2, the Pfaffian system (12.3) has a solution of the form (12.7). In general the solution has singularity on $x_1 = 0$, however, if $\rho \in \mathbb{Z}_{\geq 0}$ and there appear no logarithmic term $\log x_1$, the solution becomes holomorphic at $x_1 = 0$, and hence we can input $x_1 = 0$. Note that, if $\rho \in \mathbb{Z}_{>0}$, the result of the restriction of $u(x)$ becomes $u(0, x') = 0$ and we have no meaningful result. Then we assume $\rho = 0$. Therefore we consider the restriction when $B_1(x')$ has the eigenvalue 0.

By a gauge transformation by $P(x')$ which appeared in the proof of Theo-rem 12.1, we may assume that $B_1(x')$ is a constant matrix C_0. As C_0 we may take the Jordan canonical form of $B_1(x')$, however, we do not need to assume that C_0 is the Jordan canonical form. We assume that C_0 is non-resonant and has the eigenvalue 0. The linear Pfaffian system becomes

$$du = \left(\left(\frac{C_0}{x_1} + C_1(x_1, x') \right) dx_1 + \sum_{k=2}^{n} A_k(x)\, dx_k \right) u. \tag{12.18}$$

We take a solution $u(x)$ of the Pfaffian system of exponent 0 and holomorphic at $x_1 = 0$. Expanding $u(x)$ in the Taylor series with respect to x_1, we get

$$u(x) = \sum_{m=0}^{\infty} u_m(x') x_1^{m}.$$

Put this into the differential equation in x_1 direction of the Pfaffian system (12.18). Then we have

$$\sum_{m=0}^{\infty} m u_m x_1^{m-1} = \sum_{m=0}^{\infty} C_0 u_m x_1^{m-1} + C_1(x_1, x') u(x).$$

The second term of the right hand side is holomorphic at $x_1 = 0$. Comparing the coefficients of x_1^{-1} in both sides, we get

$$C_0 u_0 = 0.$$

Thus we should take $u_0(x')$ as an eigenvector of C_0 for the eigenvalue 0. On the other hand, putting $x_1 = 0$ into the differential equation in x_k direction

$$\frac{\partial u}{\partial x_k} = A_k(x) u$$

of (12.18), we have

$$\frac{\partial u_0}{\partial x_k} = D_k(x')u_0 \quad (2 \leq k \leq n), \tag{12.19}$$

where $A_k(0, x') = D_k(x')$ as in the proofs of Theorems 12.1 and 12.2. As we have seen in (12.6), the linear Pfaffian system (12.19) in variable x' is completely integrable, and hence we have a solution. We want to know whether there exists a solution $u_0(x')$ of (12.19) such that it remains to be eigenvectors of C_0 for the eigenvalue 0 for any x'. In the condition (12.5) which is obtained from the complete integrability condition, we take $B_1 = C_0$, a constant matrix, so that we have

$$[D_k, C_0] = O \quad (2 \leq k \leq n). \tag{12.20}$$

Let u_{00} be a constant eigenvector of C_0 for the eigenvalue 0, and $u_0(x')$ the solution of the completely integrable linear Pfaffian system (12.19) satisfying the initial condition

$$u_0(0) = u_{00}.$$

Owing to the relation (12.20), we have

$$\frac{\partial}{\partial x_k}(C_0 u_0) = C_0 \frac{\partial u_0}{\partial x_k} = C_0 D_k(x')u_0 = D_k(x')(C_0 u_0),$$

and hence $C_0 u_0(x')$ is also a solution of the Pfaffian system (12.19). This solution satisfies the initial condition

$$C_0 u_0(0) = C_0 u_{00} = 0,$$

so that we have

$$C_0 u_0(x') = 0$$

by the uniqueness of solutions shown in Theorem 11.1. Thus $u_0(x')$ is a 0-eigenvector of C_0 for any x'.

According to the definition of the restriction which we gave at the beginning of this section, the differential equation (12.19) satisfied by the function u_0 which remains to be a 0-eigenvector of C_0 is the restriction of (12.18). Thus the differential equation (12.19) itself is the restriction equation, however, we need the additional condition that u_0 is a 0-eigenvector of C_0. We want to get a differential equation without any additional condition.

Let W be the 0-eigenspace of C_0 in $\mathbb{C}^N$. The relation (12.20) implies that W is $(D_2(x'), D_3(x'), \ldots, D_n(x'))$-invariant. Then denoting the action of $D_k(x')$ $(2 \leq k \leq n)$ on W by $\bar{D}_k(x')$, we see that the linear Pfaffian system

$$\frac{\partial v}{\partial x_k} = \bar{D}_k(x')v \quad (2 \leq k \leq n) \tag{12.21}$$

becomes completely integrable. The rank of this Pfaffian system is $\dim W$. We define (12.21) as the *restriction* of the linear Pfaffian system (12.18) to the singular locus $x_1 = 0$. The restriction is an operation which reduces the number of the independent variables by one and the rank by at least one.

In the above argument, we put a special meaning to the eigenvalue 0 among the eigenvalues of C_0. However, by the following procedure, we may define the restriction for the other eigenvalues of C_0. Assume that C_0 is non-resonant, and let ρ be an eigenvalue of C_0. By operating the gauge transformation

$$u = x_1{}^{\rho} \tilde{u}$$

to the linear Pfaffian system (12.18), it is transformed to

$$d\tilde{u} = \left(\left(\frac{C_0 - \rho}{x_1} + C_1(x_1, x') \right) dx_1 + \sum_{k=2}^{n} A_k(x)\, dx_k \right) \tilde{u},$$

where the residue matrix $C_0 - \rho$ has the eigenvalue 0. Then we can take the restriction of this Pfaffian system to the singular locus $x_1 = 0$. We call this restriction the restriction of the linear Pfaffian system (12.18) to the singular locus $x_1 = 0$ with respect to the eigenvalue ρ. The rank of this restriction becomes the dimension of the ρ-eigenspace of C_0.

Example 12.2 Appell's hypergeometric series $F_1(\alpha, \beta, \beta', \gamma; x, y)$ in two variables satisfies a system of partial differential equations, which can be reduced to a linear Pfaffian system (12.16) of rank three with singular locus given by $\varphi(x, y) = xy(x - 1)(y - 1)(x - y) = 0$. Explicitly the Pfaffian system is given by

$$du = \left(A_1 \frac{dx}{x} + A_2 \frac{dy}{y-1} + A_3 \frac{d(x-y)}{x-y} + A_4 \frac{dx}{x-1} + A_5 \frac{dy}{y} \right) u \qquad (12.22)$$

with the residue matrices

$$A_1 = \begin{pmatrix} 0 & 1 & 0 \\ 0 & \beta' - \gamma + 1 & 0 \\ 0 & -\beta' & 0 \end{pmatrix}, \quad A_2 = \begin{pmatrix} 0 & 0 & 0 \\ 0 & 0 & 0 \\ -\alpha\beta' & -\beta' & \gamma - \alpha - \beta' - 1 \end{pmatrix},$$

$$A_3 = \begin{pmatrix} 0 & 0 & 0 \\ 0 & -\beta' & \beta \\ 0 & \beta' & -\beta \end{pmatrix}, \quad A_4 = \begin{pmatrix} 0 & 0 & 0 \\ -\alpha\beta & \gamma - \alpha - \beta - 1 & -\beta \\ 0 & 0 & 0 \end{pmatrix},$$

$$A_5 = \begin{pmatrix} 0 & 0 & 1 \\ 0 & 0 & -\beta \\ 0 & 0 & \beta - \gamma + 1 \end{pmatrix}.$$

We compute two restrictions of this Pfaffian system.

First we compute the restriction to $y = 0$. The Pfaffian system (12.22) is already of the form (12.18), and then we have only to study the 0-eigenspace of the residue matrix A_5. We see that the 0-eigenspace of A_5 is of dimension two, and that A_5 is already of block triangle form. Then the differential equation on the 0-eigenspace can be obtained by taking the principal 2×2 part. Namely, for the differential equation in x direction of (12.22)

$$\frac{\partial u}{\partial x} = \left(\frac{A_1}{x} + \frac{A_3}{x - y} + \frac{A_4}{x - 1} \right) u,$$

we put $y = 0$, and then take the principal 2×2 part. Thus we get the restriction

$$\frac{dv}{dx} = \left(\frac{B_0}{x} + \frac{B_1}{x - 1} \right) v, \tag{12.23}$$

where B_0 and B_1 are principal 2×2 parts of $A_1 + A_3$ and A_4, respectively, and given by

$$B_0 = \begin{pmatrix} 0 & 1 \\ 0 & 1 - \gamma \end{pmatrix}, \quad B_1 = \begin{pmatrix} 0 & 0 \\ -\alpha\beta & \gamma - \alpha - \beta - 1 \end{pmatrix}.$$

The restriction (12.23) to the singular locus $y = 0$ is equivalent to the Gauss hypergeometric differential equation.

Next we compute the restriction to $x = y$. We operate the change of variables $(\xi, \eta) = (x, x - y)$, and then compute the restriction to $\eta = 0$. By this change of variables, the Pfaffian system (12.22) is transformed to

$$du = \left(A_1 \frac{d\xi}{\xi} + A_2 \frac{d(\xi - \eta)}{\xi - \eta - 1} + A_3 \frac{d\eta}{\eta} + A_4 \frac{d\xi}{\xi - 1} + A_5 \frac{d(\xi - \eta)}{\xi - \eta} \right) u.$$

The residue matrix at $\eta = 0$ is A_3, which has two dimensional 0-eigenspace. We diagonalize the residue matrix A_3. Take

$$Q^{-1} A_3 Q = \begin{pmatrix} 0 & 0 & 0 \\ 0 & 0 & 0 \\ 0 & 0 & -\beta - \beta' \end{pmatrix},$$

and operate the gauge transformation $u = Q\tilde{u}$. Then we get a linear Pfaffian system where each residue matrix A_i is transformed to $A_i' = Q^{-1} A_i Q$ for each i. The differential equation in ξ direction is given by

$$\frac{\partial \tilde{u}}{\partial \xi} = \left(\frac{A_1'}{\xi} + \frac{A_2'}{\xi - \eta - 1} + \frac{A_4'}{\xi - 1} + \frac{A_5'}{\xi - \eta} \right) \tilde{u}.$$

We put $\eta = 0$ into this equation, and take the principal 2×2 part, which corresponds to the 0-eigenspace of A'_3. Thus we get the restriction to $x = y$

$$\frac{dv}{dx} = \left(\frac{C_0}{x} + \frac{C_1}{x-1} \right) v, \tag{12.24}$$

where C_0, C_1 are the principal 2×2 parts of $A'_1 + A'_5$, $A'_2 + A'_4$, respectively, and given explicitly by

$$C_0 = \begin{pmatrix} 1 - \gamma & 0 \\ \beta + \beta' & 0 \end{pmatrix}, \quad C_1 = \begin{pmatrix} \gamma - \alpha - \beta - \beta' - 1 & -\alpha \\ 0 & 0 \end{pmatrix}.$$

This restriction (12.24) to $x = y$ is also the Gauss hypergeometric differential equation.

Chapter 13
Monodromy Representations

We have shown in Theorem 11.2 that, for a completely integrable linear Pfaffian system, any solution can be analytically continued to the full domain X of definition of the coefficients. Then, taking any point $b \in X$ and a fundamental system of solutions $\mathcal{U}(x)$ in its simply connected neighborhood, we have the monodromy representation

$$\rho : \pi_1(X, b) \to \mathrm{GL}(N, \mathbb{C})$$

by the analytic continuations of $\mathcal{U}(x)$. This aspect seems to be very similar to the case of ordinary differential equations, however, in the case of more than one variables we have various fundamental groups according to the geometric nature of X, and hence we will find new and interesting aspects.

In the following of this chapter, we assume that X is a complement of a hypersurface in $\mathbb{C}^n$, where a hypersurface means the zero locus of a polynomial in n variables. Then a hypersurface S is given by a polynomial $\varphi(x) = \varphi(x_1, x_2, \ldots, x_n) \in \mathbb{C}[x_1, x_2, \ldots, x_n]$ as

$$S = \{x \in \mathbb{C}^n \,;\, \varphi(x) = 0\}.$$

We always assume that the polynomial $\varphi(x)$ is reduced. We call $\varphi(x)$ the *defining polynomial* for S. Let

$$\varphi(x) = \prod_i \varphi_i(x) \tag{13.1}$$

be the irreducible decomposition of $\varphi(x)$ in $\mathbb{C}[x]$, where each $\varphi_i(x) \in \mathbb{C}[x]$ is irreducible. Since we assumed that $\varphi(x)$ is reduced, there is no multiple factor in

© The Editor(s) (if applicable) and The Author(s), under exclusive license to Springer Nature Switzerland AG 2020
Y. Haraoka, *Linear Differential Equations in the Complex Domain*, Lecture Notes in Mathematics 2271, https://doi.org/10.1007/978-3-030-54663-2_13

the irreducible decomposition. We set

$$S_i = \{x \in \mathbb{C}^n \,;\, \varphi_i(x) = 0\} \tag{13.2}$$

for each i. Then we have

$$S = \bigcup_i S_i, \tag{13.3}$$

which we call the irreducible decomposition of S. Each S_i is called an *irreducible component* of S.

In studying a linear Pfaffian system defined on the complement $X = \mathbb{C}^n \setminus S$ of a hypersurface S in $\mathbb{C}^n$, the behavior at infinity plays an important role, which is similar as in the case of ordinary differential equations. Thus we understand that the singular locus of a linear Pfaffian system is a union of the hypersurface S and the infinity. However, different from the one dimensional case, there is no canonical compactification $\mathbb{C}^n$, and hence the infinity depends on the compactification. As a compactification of $\mathbb{C}^n$, we take a union of $\mathbb{C}^n$ and several hyperplanes H_∞^j at infinity, and denote it by $\overline{\mathbb{C}^n}$. Then the singular locus becomes

$$\bar{S} = \bigcup_i S_i \cup \bigcup_j H_\infty^j,$$

and we regard X as the complement $X = \overline{\mathbb{C}^n} \setminus \bar{S}$. We often take $\mathbb{P}^n$ and $(\mathbb{P}^1)^n$ as compactifications of $\mathbb{C}^n$; $\mathbb{P}^n$ is obtained by adding a hyperplane H_∞ at infinity, and $(\mathbb{P}^1)^n$ is obtained by adding n hyperplanes

$$\bigcup_{i=1}^n \{x_i = \infty\}.$$

The choice of a compactification does not make any change of the domain of definition of the Pfaffian system, however, it will make a change of the rigidity.

In the following, we quote several results on the topology of a complement space of a hypersurface. We recommend the book [175] as a fundamental reference.

13.1   Local Monodromy

In Part I, we described the local monodromy of an ordinary differential equation by using the expression of a local solution in Sect. 4.3, and also by using a topological property of the fundamental group in Sect. 5.2 (Theorem 5.3). For a completely integrable linear Pfaffian system, the local monodromy can be formulated similarly in two ways. We first explain the formulation by topology.

Definition 13.1 For a hypersurface S in $\mathbb{C}^n$ and for an element γ in the fundamental group $\pi_1(\mathbb{C}^n \setminus S)$, the *winding number* $n(\gamma, S)$ of γ with respect to S is defined by

$$n(\gamma, S) = \frac{1}{2\pi\sqrt{-1}} \int_{\varphi \circ \gamma} \frac{dx}{x},$$

where φ is a defining polynomial of S.[1]

Namely we define the winding number $n(\gamma, S)$ by the winding number of the loop $\varphi \circ \gamma$ in $\mathbb{C}$ with respect to 0. If we take another defining polynomial of S, it becomes a scalar multiple of φ, and hence the value $n(\gamma, S)$ does not change. It is also easy to see that the integral in the definition of $n(\gamma, S)$ does not depend on the choice of a representative of γ.

Decompose φ into irreducible factors as in (13.1), and take the corresponding irreducible decomposition (13.3) of S. Then we can define the winding number

$$n(\gamma, S_i) = \frac{1}{2\pi\sqrt{-1}} \int_{\varphi_i \circ \gamma} \frac{dx}{x}$$

for each irreducible component S_i.

Let a be an ordinary point in S on an irreducible component S_i. Take a complex line H passing through a transversally with S. Take a path L in $\mathbb{C}^n \setminus S$ from the base point $b \in \mathbb{C}^n \setminus S$ to a point c in $H \setminus (S \cap H)$, and a $(+1)$-loop K in $H \setminus (S \cap H)$ for a with the base point c. We call the loop LKL^{-1} and the element in the fundamental group represented by LKL^{-1} a $(+1)$-*loop* (or a *monodromy*) for S_i (Fig. 13.1). For a $(+1)$-loop γ for S_i, we have $n(\gamma, S_i) = 1$ and $n(\gamma, S_j) = 0$ $(j \neq i)$. The following theorem is fundamental.

Theorem 13.1 *Let S_i be an irreducible component of a hypersurface S. Any two $(+1)$-loops for S_i are conjugate in $\pi_1(\mathbb{C}^n \setminus S)$.*

This theorem can be shown by applying [175, Proposition 1.31], where we use that the intersection of S_i and the set of the ordinary points of S is path-connected.

Example 13.1 Let S be the irreducible curve $x^2 - y = 0$ in $\mathbb{C}^2$. As a line intersecting transversally with S, we take $H : y = 1$. Then the intersection $S \cap H$ becomes two points $p = (-1, 1)$ and $q = (1, 1)$. Take a base point $b = (-\sqrt{-1}, 1) \in H$, and define $(+1)$-loops γ_-, γ_+ in H for p, q, respectively, as in Fig. 13.2a, where we regard H as complex x-plane. By the definition, γ_-, γ_+ become $(+1)$-loops for S. We shall show that these are conjugate (and moreover identical) in $\pi_1(\mathbb{C}^2 \setminus S, b)$.

For $\theta \in [0, 2\pi]$, we define a line H_θ by

$$y - 1 = m(\theta)(x + \sqrt{-1}),$$

---

[1] I do not find any definition of the winding number for a hypersurface in references. This definition is due to Professor Takeshi Abe in Kumamoto University.

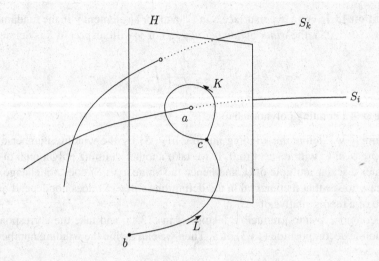

Fig. 13.1 (+1)-loop for $a \in S_i$

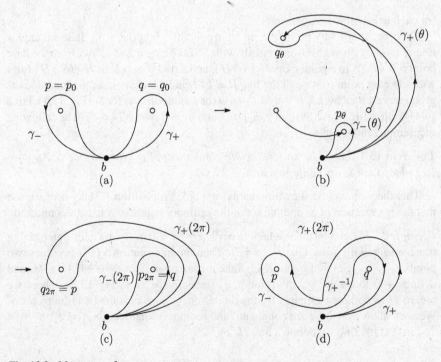

Fig. 13.2 Movement of γ_-, γ_+

where

$$m(\theta) = (\sqrt{2} - 1)\sqrt{-1}(1 - e^{\sqrt{-1}\theta}).$$

We see that H_θ passes through b for all θ, and coincides with H for $\theta = 0, 2\pi$. Since H_θ avoids the singular point $(0, 0)$ of S, the intersection $S \cap H_\theta$ remains to be two points p_θ and q_θ, which are the continuous image of p and q, respectively. Let $\gamma_-(\theta), \gamma_+(\theta)$ be loops in $H_\theta \setminus (S \cap H_\theta) = H_\theta \setminus \{p_\theta, q_\theta\}$ with base point b which are the continuous images of γ_-, γ_+, respectively. The movement of $p_\theta, q_\theta, \gamma_-(\theta)$ and $\gamma_+(\theta)$ are illustrated in Fig. 13.2b, where we also regard H_θ as complex x-plane. We have $p_{2\pi} = q$ and $q_{2\pi} = p$ (Fig. 13.2c). Note that $\gamma_-(2\pi), \gamma_+(2\pi)$ are continuous deformations of γ_-, γ_+, respectively, in the space $\mathbb{C}^2 \setminus S$, and hence identical to γ_-, γ_+, respectively, in $\pi_1(\mathbb{C}^2 \setminus S)$. On the other hand, we have $\gamma_-(2\pi) = \gamma_+$ and $\gamma_+(2\pi) = \gamma_+\gamma_-\gamma_+^{-1}$ in $\pi_1(H \setminus \{p, q\}, b)$ as illustrated in Fig. 13.2c and d. Therefore we get

$$\gamma_- = \gamma_-(2\pi) = \gamma_+$$

and

$$\gamma_+ = \gamma_+(2\pi) = \gamma_+\gamma_-\gamma_+^{-1},$$

from both of which we derive $\gamma_- = \gamma_+$. Thus we explicitly get the conclusion of Theorem 13.1 in this case.

Theorem 13.1 enables us to define the local monodromy. Let

$$\rho : \pi_1(X, b) \to \mathrm{GL}(N, \mathbb{C})$$

be an anti-representation of the fundamental group of the complement space $X = \mathbb{C}^n \setminus S$ of a hypersurface S. We do not necessarily assume that ρ is a monodromy representation of a linear Pfaffian system. Let (13.3) be the irreducible decomposition of S. The *local monodromy* at S_j is defined as the conjugacy class $[\rho(\gamma)]$ of the image of a $(+1)$-loop $\gamma \in \pi_1(X, b)$ for S_j. Thanks to Theorem 13.1, the local monodromy does not depend on the choice of a $(+1)$-loop γ, and is determined only by S_j.

Next we formulate the local monodromy by using a local solution of a linear Pfaffian system. Let S be a hypersurface. We consider a completely integrable linear Pfaffian system

$$du = \Omega u$$

on $X = \mathbb{C}^n \setminus S$ with logarithmic singularity along S. We have the irreducible decomposition (13.3) of $S = \{\varphi(x) = 0\}$. Take an irreducible component $S_i = \{\varphi_i(x) = 0\}$ and a point $a \in S_i$ which is an ordinary point of S. Then we have grad $\varphi(a) \neq 0$, and hence, as discussed in Chap. 12, we have a coordinate change

$$\Phi : x = (x_1, x_2, \ldots, x_n) \mapsto \xi = (\xi_1, \xi_2, \ldots, \xi_n)$$

such that a is sent to $\xi = 0$ and a neighborhood of a in S_i is sent to $\{\xi_1 = 0\}$. The linear Pfaffian system is transformed to

$$du = \left(\left(\frac{B_0(\xi')}{\xi_1} + C_1(\xi) \right) d\xi_1 + \sum_{k=2}^{n} A_k(\xi) \, d\xi_k \right) u,$$

where $C_1(\xi)$, $A_k(\xi)$ $(2 \leq k \leq n)$ are holomorphic at $\xi = 0$. The matrix function $B_0(\xi')$ is determined by

$$A(\Phi^{-1}(\xi)) = B_0(\xi') + \xi_1 B_1(\xi),$$

where we write

$$\Omega = (A(x) d \log \varphi_i(x) + \Omega_1).$$

As stated in Theorem 12.2, we have a fundamental matrix solution $\mathcal{U}(\xi)$ in a neighborhood of $\xi = 0$. In Sect. 4.3, we obtained a circuit matrix with respect to a similar fundamental matrix solution. Also in this case, for a $(+1)$-loop γ on the complex ξ_1 plane obtained by $\xi' = 0$, the analytic continuation of $\mathcal{U}(\xi)$ along γ is given by

$$\gamma_* \mathcal{U}(\xi) = \mathcal{U}(\xi) M$$

with some matrix $M \in \mathrm{GL}(N, \mathbb{C})$. When $B_0(\xi')$ is non-resonant, as stated in Corollary 12.3, we have a fundamental matrix solution

$$\mathcal{U}(\xi) = F(\xi) \xi_1{}^J$$

with a matrix function $F(\xi)$ holomorphic at $\xi = 0$ and the Jordan canonical form J of $B_0(\xi')$. In this case, we have

$$M = e^{2\pi \sqrt{-1} J}.$$

For a moment, we assume that $B_0(\xi')$ is non-resonant.

We shall show that the conjugacy class $[M]$ of M does not depend on the choice of the coordinate change Φ or the point $a \in S_i$. Let

$$\Psi : x = (x_1, x_2, \ldots, x_n) \mapsto \eta = (\eta_1, \eta_2, \ldots, \eta_n)$$

be another coordinate change which send a to $\eta = 0$ and a neighborhood of a in S_i to $\{\eta_1 = 0\}$. Then the composition

$$\Psi \circ \Phi^{-1} : \xi \mapsto \eta$$

becomes biholomorphic in a neighborhood of $\xi = 0$, and sends $\{\xi_1 = 0\}$ to $\{\eta_1 = 0\}$. Therefore the restriction of $\Psi \circ \Phi^{-1}$ to $\xi_1 = 0$ becomes a biholomorphic map $\xi' \mapsto \eta'|_{\xi_1=0}$. Now we set

$$A(\Phi^{-1}(\xi)) = B_0(\xi') + \xi_1 B_1(\xi),$$

$$A(\Psi^{-1}(\eta)) = D_0(\eta') + \eta_1 D_1(\eta).$$

Since $\xi_1 = 0$ corresponds to $\eta_1 = 0$, we get

$$D_0(\eta'(0, \xi')) = B_0(\xi').$$

By Theorem 12.1, we see that the Jordan canonical form of $D_0(\eta')$ does not depend on η', and hence coincides with the Jordan canonical form of $D_0(\eta'(0, \xi')) = B_0(\xi')$. Thus the Jordan canonical form of $D_0(\eta')$ is also J, so that M does not change. Next we perturb $a \in S_i$. If we perturb a in a domain where the coordinate change Φ is valid, the perturbation corresponds to the perturbation, in ξ space, of ξ' in a neighborhood of $\xi' = 0$ with $\xi_1 = 0$ being fixed. In this case, since the Jordan canonical form of $B_0(\xi')$ does not depend on ξ' and hence is J, the conjugacy class $[M]$ is invariant. The intersection of S_i and the regular points of S is path connected, and then $[M]$ is still invariant on the intersection.

When J is resonant, the circuit matrix may not be determined only by J, and then the above argument will not hold. In order to show that the conjugacy class $[M]$ does not depend on the choice of Φ or $a \in S_i$, we need to examine how the circuit matrix is determined in the way explained in Sect. 4.3. We do not go into detail in this book. Note that, even if J is resonant, the conjugacy class $[M]$ is determined only by S_i thanks to Theorem 13.1.

In conclusion, the conjugacy class $[M]$ of the circuit matrix M with respect to the fundamental matrix solution constructed in Theorem 12.2 is determined only by S_i. We call $[M]$ the *local monodromy* at S_i.

13.2 Analysis Using Monodromy Representation

In Sect. 5.4 in Chap. 5, we saw that the monodromy representation plays a fundamental role in the analysis of Fuchsian ordinary differential equations. For completely integrable Pfaffian systems in several variables with logarithmic singularities along hypersurfaces, we can develop a similar argument. For simplicity, we take $\mathbb{P}^n$ as a compactification of $\mathbb{C}^n$, and assume that the singular locus is the hypersurface $\bar{S} = S \cup H_\infty$ with the locus S of a reduced polynomial $\varphi(x_1, x_2, \ldots, x_n)$ and the hyperplane at infinity H_∞ in $\mathbb{P}^n$.

Set $X = \mathbb{P}^n \setminus \bar{S}$. We consider a completely integrable linear Pfaffian system

$$du = \Omega u \qquad (13.4)$$

on X of rank N with logarithmic singularity along $\bar{S}$. Let

$$\bar{S} = \bigcup_i S_i \qquad (13.5)$$

be the irreducible decomposition of $\bar{S}$. Note that H_∞ is among S_i's. If the local monodromy at an irreducible component S_i of $\bar{S}$ is trivial, we call S_i an *apparent singularity*. Each of the other irreducible components is called a *branching locus*.

The next theorem is fundamental.

Theorem 13.2 *Assume the linear Pfaffian system (13.4) on $X = \mathbb{P}^n \setminus \bar{S}$ is completely integrable and possesses a logarithmic singularity along the hypersurface $\bar{S}$. If a rational function in entries of solutions of (13.4) is holomorphic and single valued on X, it is a rational function in $x_1, x_2, \ldots, x_n$.*

Proof Let S_i be an irreducible component of S other than H_∞, and $\varphi_i(x)$ its defining polynomial. We set $\varphi(x) = \prod \varphi_i(x)$. Let $f(x)$ be a rational function in entries of solutions of (13.4) that is holomorphic and single valued on X. Take a coordinate change $(x_1, x_2, \ldots, x_n) \rightarrow (\xi_1, \xi_2, \ldots, \xi_n)$ on a neighborhood of a generic point of S_i such that S_i is sent to $\{\xi_1 = 0\}$. For each fixed $\xi' = (\xi_2, \ldots, \xi_n)$, f is single valued in a neighborhood of $\xi_1 = 0$ in ξ_1-plane except $\xi_1 = 0$, and at most regular singular at $\xi_1 = 0$. Then, owing to Theorem 4.1, $\xi_1 = 0$ is at most a pole of f. The order of the pole is globally bounded since S_i and X are path connected. This holds for every S_i, and hence there exists a non-negative integer m such that $\varphi(x)^m f(x)$ is holomorphic and single valued on $\mathbb{C}^n \setminus S^{\text{crit}}$, where S^{crit} denotes the set of critical points in S. The codimension of S^{crit} is at least two, and then, by the Hartogs principle, $\varphi(x)^m f(x)$ can be extended to a holomorphic and single valued function on $\mathbb{C}^n$. Let $(X_0 : X_1 : \cdots : X_n)$ be the homogeneous coordinate of $\mathbb{P}^n$ such that $x_i = X_i/X_0$ $(1 = 1, 2, \ldots, n)$ makes the coordinate of $\mathbb{C}^n$. We consider the Taylor expansion of $\varphi(x)^m f(x)$ at $x = 0$. Since $\varphi(x)^m f(x)$ is holomorphic in $\mathbb{C}^n$, the Taylor expansion gives the Laurent expansion at $X_0 = 0$. In a similar argument for S_i, we see that the principal part of the Laurent expansion consists of a finite number of terms. Thus $\varphi(x)^m f(x)$ is a polynomial in $x_1, x_2, \ldots, x_n$. □

Thanks to this theorem, we can derive similar assertions as in Chap. 5. First, as an analogous result to Theorem 5.5, we have some equivalence between completely integrable Pfaffian systems with logarithmic singularities and their monodromy representations.

Theorem 13.3 *Consider two completely integrable linear Pfaffian systems with logarithmic singularities along a common hypersurface in $\mathbb{P}^n$, and assume that the sets of their branching loci coincide. Then, the monodromy representations of these Pfaffian systems are equivalent if and only if these Pfaffian systems are transformed each other by gauge transformations with coefficients in rational functions.*

By Theorem 11.5, we see that the determinant of a fundamental matrix solution does not vanish on X. Then we can apply Theorem 13.2, and hence can prove Theorem 13.3 in a similar way as Theorem 5.5.

A zero of a polynomial in several variables is called an algebraic function. Namely, an algebraic function is a solution T of the equation

$$a_0(x)T^m + a_1(x)T^{m-1} + \cdots + a_m(x) = 0$$

with coefficients $a_0(x), a_1(x), \ldots, a_m(x) \in \mathbb{C}[x_1, x_2, \ldots, x_n]$.

Theorem 13.4 *For a completely integrable linear Pfaffian system with logarithmic singularity along a hypersurface in $\mathbb{P}^n$, any solution is an algebraic function if and only if the monodromy group is a finite group.*

This theorem can be shown in a similar way as Theorems 5.11 and 5.12. For several completely integrable systems containing the systems satisfied by Appell's F_1, F_2, F_3, F_4 and Lauricella's F_D, conditions for the finiteness of the monodromy groups are obtained [123, 150, 151]. On the other hand, the completely integrable system in [60] mentioned in Sect. 5.4.3 is constructed so that the monodromy group becomes finite.

In Sect. 5.4, we mentioned to the works of Schwarz and the differential Galois theory. Sasaki and Yoshida extended the Schwarz derivative to several variables case, which seems interesting [122, 152]. They also constructed several completely integrable systems [189]. The differential Galois theory can be extended to several variables case by using a partial differential field which has several differentials. However, in my personal opinion, the differential Galois theory seems to be a theory for ordinary differential equations, and then it may work better if we apply it to an ordinary differential equation which appears as a one dimensional section of a linear Pfaffian system than applying to the Pfaffian system directly.

So far we have explained, for linear Pfaffian systems, many analogous results to ones for Fuchsian ordinary differential equations hold. On the other hand, we can show a proper phenomenon for several variables case also by using the monodromy representation.

Theorem 13.5 *If the fundamental group $\pi_1(X) = \pi_1(\mathbb{P}^n \setminus \bar{S})$ is commutative, any solution of a completely integrable linear Pfaffian system with logarithmic singularity along $\bar{S}$ becomes elementary.*

Proof Let (13.5) be the irreducible decomposition of $\bar{S}$. First, thanks to Theorem 13.1, all $(+1)$-loops for an irreducible component S_i of $\bar{S}$ define a single element in $\pi_1(X)$ if it is commutative. Then, by Zariski-van Kampen theorem

(Theorem 13.9) which will be given later, $\pi_1(X)$ is generated by the $(+1)$-loops γ_i for S_i.

Take a fundamental system of solutions

$$\mathcal{U}(x) = (u^1(x), u^2(x), \ldots, u^N(x))$$

for the linear Pfaffian system, and let

$$\rho : \pi_1(X) \to \mathrm{GL}(N, \mathbb{C})$$

be the monodromy representation with respect to $\mathcal{U}(x)$. For each irreducible component S_i of S, we set

$$M_i = \rho(\gamma_i),$$

where γ_i is the $(+1)$-loop for S_i. Since $\pi_1(X)$ is commutative, M_i's are commutative each other. Let $\varphi_i(x)$ be the defining polynomial for S_i. Take a matrix A_i such that $e^{2\pi\sqrt{-1}A_i} = M_i$ and $[A_i, A_j] = O$ for any i, j. We show the existence of such matrices in Theorem 13.6 and Corollary 13.7 which will be given soon later. By using these, we define

$$\Phi(x) = \prod_i \varphi_i(x)^{A_i}.$$

Since A_i's are commutative, $\Phi(x)$ does not depend on the order of the product. Then we have

$$\gamma_{i*}\Phi(x) = \Phi(x)e^{2\pi\sqrt{-1}A_i} = \Phi(x)M_i.$$

Note that $\Phi(x)$ is holomorphic and invertible on X. Now we consider the matrix function

$$\mathcal{U}(x)\Phi(x)^{-1},$$

which is holomorphic and single valued on X. Then we apply Theorem 13.2 to show that each entry of the matrix function is a rational function in $x_1, x_2, \ldots, x_n$. Thus each entry of $\mathcal{U}(x)$ is elementary.   □

The above theorem suggests that, for completely integrable systems, solutions may become easier than solutions to ordinary differential equations. It is an important problem to construct a completely integrable system having non-elementary solutions.

Here we give results on the logarithm of matrices.

Theorem 13.6 *Let A be a non-singular matrix, and $A = S + N$ its Jordan decomposition. Here S is semi-simple, N is nilpotent and S and N are commutative. Take $m \in \mathbb{Z}_{\geq 0}$ such that $N^m = O$. Let*

$$S = \alpha_1 P_1 + \alpha_2 P_2 + \cdots + \alpha_r P_r$$

be the spectral decomposition of S; $\alpha_1, \alpha_2, \ldots, \alpha_r$ are the eigenvalues of A, and $P_1, P_2, \ldots, P_r$ are projections satisfying

$$P_1 + P_2 + \cdots + P_r = I, \quad P_i P_j = O \ (i \neq j).$$

Since A is non-singular, we have $\alpha_i \neq 0 \ (1 \leq i \leq r)$, and S is non-singular. Now we set

$$C = \sum_{i=1}^{r} (\log \alpha_i) P_i + \sum_{k=1}^{m-1} \frac{(-1)^{k-1}}{k} (S^{-1} N)^k.$$

Then we have

$$e^C = A.$$

A proof of this theorem will be found in textbooks on advanced linear algebra. I know a reference [163] which is written in Japanese. See also [40].

Corollary 13.7 *Let A, B be commutative non-singular matrices. Then there exist mutually commutative matrices C, D such that*

$$e^C = A, \ e^D = B.$$

Proof We see that the matrices P_i, S, N appeared in constructing C in Theorem 13.6 are polynomials in the matrix A. Then, if we define C and D for A and B by Theorem 13.6, C and D are polynomials in A and B, respectively, and hence become commutative owing to the commutativity of A and B. □

13.3 Presentation of Fundamental Groups

Monodromy representations are anti-representations of fundamental groups, and hence, in studying monodromy representations, it is substantial to know the fundamental groups. For the case of ordinary differential equations, i.e. the one dimensional case, we have a simple presentation (5.7) of the fundamental group (Sect. 5.3), and the group structure is completely determined only by the number of the subtracted points from $\mathbb{P}^1$. For two or more dimensional cases, there is no such uniform presentation. This diversity brings interesting natures for higher dimensional cases.

For the complement $X = \mathbb{C}^n \setminus S$ of a hypersurface S, we can in principle obtain a presentation of the fundamental group by combining two theorems—Zariski's hyperplane section theorem and Zariski-van Kampen theorem. We give the statements of these theorems here, however, the proofs are left to references.

Theorem 13.8 (Zariski's Hyperplane Section Theorem) *Set $n \geq 2$, and let S be a hypersurface in $\mathbb{P}^n$. If we take a plane $\mathbb{P}^2$ in $\mathbb{P}^n$ in general position with respect to S, we have an isomorphism*

$$\pi_1(\mathbb{P}^n \setminus S) \simeq \pi_1(\mathbb{P}^2 \setminus (\mathbb{P}^2 \cap S))$$

between the fundamental groups.

For the proof, please refer to the references [45, 160]. Thanks to this theorem, the problem of obtaining a presentation is reduced to the two dimensional case. For the two dimensional case, a presentation is given by the following Zariski-van Kampen theorem. Before stating the theorem, we prepare several symbols.

Let (x, y) be the coordinate of $\mathbb{C}^2$. We take $\mathbb{P}^2$ as a compactification of $\mathbb{C}^2$, and let H_∞ be the line at infinity. Suppose that the (hyper)surface S is given by the polynomial $\varphi(x, y)$ of degree d:

$$S = \{(x, y) \in \mathbb{C}^2 ; \varphi(x, y) = 0\}.$$

We may assume that $\varphi(x, y)$ is reduced. Take any point $b \in \mathbb{C}^2 \setminus S$. For a line F passing through b, we denote by $p(F)$ the intersection of F and H_∞ in $\mathbb{P}^2$. If we restrict $\varphi(x, y)$ to F, it becomes a polynomial in one variable of the coordinate of F of degree at most d, and hence we have

$$\#F \cap S \leq d.$$

Among the lines F, take a line F_0 such that this value is equal to d, and set $p(F_0) = p_0$. There are only finite many lines F passing through b such that

$$\#F \cap S < d.$$

We denote by $p_1, p_2, \ldots, p_e$ the points $p(F)$ for such lines F, where e denotes the number of such lines.

On the line F_0, which is two dimensional over $\mathbb{R}$, there is the point b, and also are d intersection points $a_1, a_2, \ldots, a_d$ of F_0 and S. The fundamental group $\pi_1(F_0 \setminus \{a_1, a_2, \ldots, a_d\}, b)$ is a free group generated by d elements $\gamma_1, \gamma_2, \ldots, \gamma_d$, where, for each i, γ_i is a $(+1)$-loop for a_i with the base point b. Note that, for any choice of the $(+1)$-loops γ_i, they define elements of $\pi_1(\mathbb{C}^2 \setminus S, b)$. Look at the line H_∞ at infinity, on which the intersection p_0 with F_0 and points $p_1, p_2, \ldots, p_e$ are. The fundamental group $\pi_1(H_\infty \setminus \{p_1, p_2, \ldots, p_e\}, p_0)$ is a free group generated by e

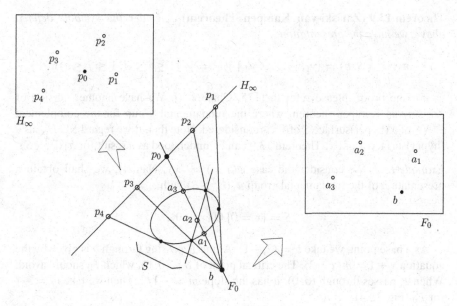

Fig. 13.3 F_0 and F's with $\#F \cap S < d$

elements $g_1, g_2, \ldots, g_e$ (Fig. 13.3). (We may take each g_j as a $(+1)$-loop for p_j with the base point p_0.)

The line F passing through b is uniquely determined by $p(F)$. Then, when $p \in H_\infty$ moves along g_j, the corresponding line F starts from F_0, moves continuously, and returns to F_0. According to this movement, the set of the points $F \cap S$ starts from $a_1, a_2, \ldots, a_d$, moves continuously, and returns to $a_1, a_2, \ldots, a_d$. This movement may bring a permutation of $a_1, a_2, \ldots, a_d$. Since g_j avoids the points $p_1, p_2, \ldots, p_e$, the number d of the set $F \cap S$ is kept invariant during this movement, and then no two points coalesce. Therefore the loops γ_i $(1 \le i \le d)$ on F_0 can be deformed continuously according to this movement so that they do not touch S. We denote by $\gamma_i{}^{g_j}$ the result of γ_i by this movement. Since $\gamma_i{}^{g_j}$ defines an element of $\pi_1(F_0 \setminus \{a_1, a_2, \ldots, a_d\}, b)$, it can be written by using the generators $\gamma_1, \gamma_2, \ldots, \gamma_d$. On the other hand, $\gamma_i{}^{g_j}$ is a result of a continuous deformation of γ_i in $\mathbb{C}^2 \setminus S$ with the base point b fixed, and hence we have $\gamma_i{}^{g_j} = \gamma_i$ in $\pi_1(\mathbb{C}^2 \setminus S, b)$. We replace the left hand side by the expression in terms of the generators $\gamma_1, \gamma_2, \ldots, \gamma_d$. Then this equality gives relation among $\gamma_1, \gamma_2, \ldots, \gamma_d$. Zariski-van Kampen theorem asserts that γ_i $(1 \le i \le d)$ are generators of the fundamental group $\pi_1(\mathbb{C}^2 \setminus S, b)$, and the relations $\gamma_i{}^{g_j} = \gamma_i$ generate the relations among the generators.

Theorem 13.9 (Zariski-van Kampen Theorem)   *Under the symbols defined above, we have the presentation*

$$\pi_1(\mathbb{C}^2 \setminus S, b) = \langle \gamma_1, \gamma_2, \ldots, \gamma_d \mid \gamma_i^{g_j} = \gamma_i \ (1 \le i \le d, 1 \le j \le e) \rangle.$$

For the proof, please refer to [175, Chapter 4]. We have another version of Zariski-van Kampen theorem where the fundamental group for the complement $\mathbb{P}^2 \setminus \bar{S}$ of a (hyper)surface $\bar{S}$ of $\mathbb{P}^2$ is considered. Note that, if we regard $S \cup H_\infty$ as a (hyper)surface $\bar{S}$ of $\mathbb{P}^2$, Theorem 13.9 can be understood as a result for $\pi_1(\mathbb{P}^2 \setminus \bar{S})$.

Example 13.2 We consider the case $\varphi(x, y) = xy$. Namely, we shall obtain a presentation of the fundamental group $\pi_1(\mathbb{C}^2 \setminus S)$ with

$$S = \{x = 0\} \cup \{y = 0\}.$$

As a base point, we take $b = (2, -1)$. A line F passing through b is given by the equation $y + 1 = m(x - 2)$. The critical point of S is $(0, 0)$, which F_0 should avoid. When F passes through $(0, 0)$, it has the slope $m = -1/2$. Then we take $m = -1$ for F_0 (Fig. 13.4):

$$y + 1 = -(x - 2).$$

In order to move the line F passing through b so that it does not pass through $(0, 0)$, it is enough to move the slope m in the complex m-plane so that it does not

Fig. 13.4 S and F_0 $y + 1 = -(x - 2).$

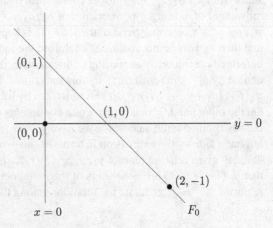

pass through $m = -1/2$. Then we set

$$m(\theta) = -\frac{2}{3} - \frac{1}{3}e^{\sqrt{-1}\theta} \quad (\theta \in [0, 2\pi])$$

and

$$F(\theta) : y + 1 = m(\theta)(x - 2).$$

The line F_0 coincides with $F(0)$. We use x as a coordinate on the complex line $F(\theta)$. Then the base point b is given by $x = 2$, and the intersection $F(\theta) \cap \{x = 0\}$ is always given by $x = 0$. Set

$$F(\theta) \cap \{y = 0\} = (y_0(\theta), 0).$$

Then the intersection $F(\theta) \cap \{y = 0\}$ is given by $x = y_0(\theta)$. Thus on the complex line F_0, we have the base point $x = 2$ and the intersection $x = 0$ and $x = 1(= y_0(0))$ with S. As a generator of $\pi_1(F_0 \setminus (F_0 \cap S), b)$, we take the loops γ_1, γ_2 illustrated in Fig. 13.5.

Since

$$y_0(\theta) = 2 + \frac{1}{m(\theta)},$$

the point $y_0(\theta)$ on the line $F(\theta)$ moves on a circle starting from 1 and encircles $x = 0$ once in the positive direction as θ moves from 0 to 2π. We show in Fig. 13.6 how γ_1, γ_2 are deformed according to the movement of $y_0(\theta)$.

We denote the deformed loops by $\gamma_1{}^g, \gamma_2{}^g$ as in Fig. 13.7. Figure 13.7 also shows how the deformed loops $\gamma_1{}^g, \gamma_2{}^g$ can be written in terms of γ_1 and γ_2. Consequently, we get the relations

$$\gamma_1{}^g = \gamma_1\gamma_2\gamma_1(\gamma_1\gamma_2)^{-1},$$

$$\gamma_2{}^g = \gamma_1\gamma_2\gamma_1{}^{-1}.$$

Fig. 13.5 Loops γ_1, γ_2

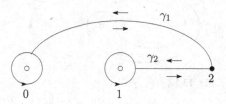

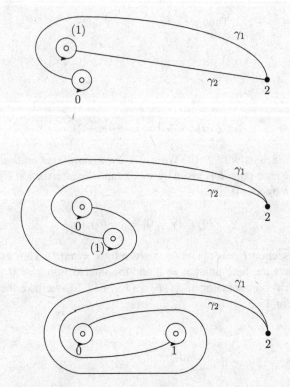

Fig. 13.6 Deformations of γ_1 and γ_2

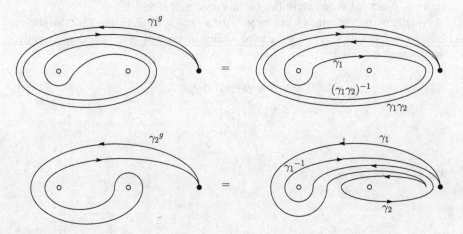

Fig. 13.7 $\gamma_1{}^g, \gamma_2{}^g$

By setting $\gamma_1{}^8 = \gamma_1, \gamma_2{}^8 = \gamma_2$ as in Theorem 13.9, both of these relations are reduced to

$$\gamma_1\gamma_2 = \gamma_2\gamma_1.$$

Then by Theorem 13.9, we get the presentation

$$\pi_1(\mathbb{C}^2 \setminus \{xy = 0\}, b) = \langle \gamma_1, \gamma_2 \mid \gamma_1\gamma_2 = \gamma_2\gamma_1 \rangle.$$

Thus the fundamental group is commutative and isomorphic to $\mathbb{Z}^2$.

In Example 13.2, S is a collection of two lines $x = 0, y = 0$ which intersect at a point $(0, 0)$. In this case, we can take (x, y) as a local coordinate at $(0, 0)$, and hence S is normally crossing at $(0, 0)$ (see Sect. 10.1). In general, if a hypersurface S is a collection of hyperplanes such that any intersection is normally crossing, the fundamental group $\pi_1(\mathbb{C}^n \setminus S)$ of the complement becomes abelian. This can be shown in a similar way as Example 13.2.

Example 13.3 As an example of non-normally crossing hyperplane arrangements, we consider S given by

$$\varphi(x, y) = xy(x - 1)(y - 1)(x - y)$$

in the case $n = 2$. Namely,

$$S = \{x = 0\} \cup \{y = 0\} \cup \{x = 1\} \cup \{y = 1\} \cup \{x = y\}$$

(Fig. 13.8).

Fig. 13.8 $S = \{xy(x - 1)(y - 1)(x - y) = 0\}$

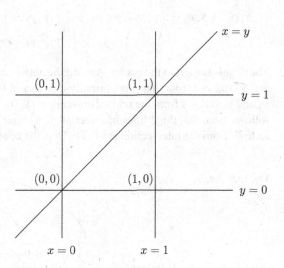

The five irreducible components intersect at four points. Among them, at two points $(1, 0)$, $(0, 1)$ S is normally crossing, while at $(0, 0)$, $(1, 1)$ three lines intersect and then S is not normally crossing. In order to describe the fundamental group of the complement of S, we take a base point $b = (2, -1/2)$ and a line F_0 given by

$$F_0 : x + y = \frac{3}{2}.$$

We use x as a coordinate of F_0. Then the five intersections with S are given by

$$F_0 \cap \{x = 0\} : x = 0,$$

$$F_0 \cap \{y = 1\} : x = \frac{1}{2},$$

$$F_0 \cap \{x = y\} : x = \frac{3}{4},$$

$$F_0 \cap \{x = 1\} : x = 1,$$

$$F_0 \cap \{y = 0\} : x = \frac{3}{2}.$$

We can take generators $\gamma_1, \gamma_2, \ldots, \gamma_5$ in F_0 as in Fig. 13.9.

We move F so that it encircles one of the four intersection points $(0, 0)$, $(0, 1)$, $(1, 0)$, $(1, 1)$, and then deform the generators $\gamma_1, \gamma_2, \ldots, \gamma_5$. Then we obtain $5 \times 4 = 20$ relations. From these relations, we obtain the following presentation:

$$\pi_1(\mathbb{C}^2 \setminus S, b) = \left\langle \gamma_1, \gamma_2, \gamma_3, \gamma_4, \gamma_5 \;\middle|\; \begin{matrix} \gamma_1\gamma_2 = \gamma_2\gamma_1, \ \gamma_4\gamma_5 = \gamma_5\gamma_4, \\ \gamma_1\gamma_3\gamma_5 = \gamma_3\gamma_5\gamma_1 = \gamma_5\gamma_1\gamma_3, \\ \gamma_2\gamma_3\gamma_4 = \gamma_3\gamma_4\gamma_2 = \gamma_4\gamma_2\gamma_3 \end{matrix} \right\rangle. \tag{13.6}$$

The details are left to the reader. Among the above relations, the commutativity of γ_1 and γ_2 follows from that S is normally crossing at $(0, 1)$, and the commutativity of γ_4 and γ_5 follows from the normal crossing at $(1, 0)$. The cyclic relation of $\gamma_1, \gamma_3, \gamma_5$ follows from that three lines intersect at $(0, 0)$, and the cyclic relation of $\gamma_2, \gamma_3, \gamma_4$ follows from the intersection at $(1, 1)$. Thus the relations come from the way of the

Fig. 13.9 $\gamma_1, \gamma_2, \ldots, \gamma_5$

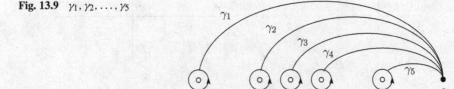

intersections of the lines. In general, when S is a collection of hyperplanes in $\mathbb{C}^n$, we call it a hyperplane arrangement, and the fundamental group of the complement of S can be described in a uniform way [135, §5.3]. For the fundamental groups of complements of algebraic curves in $\mathbb{C}^2$ or $\mathbb{P}^2$, there is a detailed description in the book [175, Chap. 5, 6].

Gérard-Levelt [42] is a pioneering work for completely integrable systems in several variables, which opened our modern understandings. There, it is described how the topological nature of the singular locus determines the analytic behavior of solutions. The following linear Pfaffian system is considered:

$$du = \left(\sum_i A_i d \log \varphi_i(x) \right) u, \tag{13.7}$$

where A_i are $N \times N$ constant matrices and $\varphi_i(x)$ is a linear polynomial in n variables. The singular locus is $S = \{x \in \mathbb{C}^n ; \prod_i \varphi_i(x) = 0\}$. Assume that the Pfaffian system is completely integrable. One of the main results of the paper is that, if S is normally crossing at any intersection point of irreducible components, then any solution is elementary. In proving the result, the authors obtain commutation relations among the matrices A_i from the complete integrability condition, and then show that (13.7) becomes reducible.

This result can be shown from our standpoint. If S is normally crossing at any intersection point, the fundamental group becomes abelian as stated after Example 13.2. Then we can apply Theorem 13.5 to show that any solution is elementary.

Therefore we do not get any non-trivial function in several variables as a solution in the normally crossing case, and then we will consider hyperplane arrangements of non-normally crossing type. Among such arrangements, S in Example 13.3 seems to be the simplest. The second main result of Gérard-Levelt is on the Pfaffian system (13.7) of rank 3 with S in Example 13.3. They showed that, if such Pfaffian system has a non-elementary function solution, it can be reduced to the Pfaffian system satisfied by Appell's hypergeometric series F_1. The condition that the Pfaffian system has a non-elementary solution seems as assuming that the monodromy representation is irreducible, and the conclusion says that the Pfaffian system is unique. In terms of the monodromy representation, this result can be regarded that the monodromy representation is unique if it is irreducible. This suggests that, also for completely integrable systems in several variables, the notion of *rigidity* may be defined, so that the second main result of Gérard-Levelt is reformulated in this viewpoint. In the next section, we realize this observation.

13.4   Rigidity

Definition 13.2 Let $\overline{\mathbb{C}^n}$ be a compactification of $\mathbb{C}^n$, and $\bar{S}$ a hypersurface in $\overline{\mathbb{C}^n}$. An anti-representation

$$\rho : \pi_1(\overline{\mathbb{C}^n} \setminus \bar{S}, b) \to \mathrm{GL}(N, \mathbb{C})$$

of the fundamental group is said to be *rigid* if ρ is uniquely determined by the local monodromies up to isomorphisms.

This definition is the same in appearance as in one dimensional case (Definition 7.2), where the index of rigidity is introduced after the definition. The index of rigidity gives a simple criterion for the rigidity (Theorem 7.8). In the proof of Theorem 7.8, the presentation (5.7) of the fundamental group is essentially used. Namely, the anti-representation ρ is given by a tuple $(M_0, M_1, \ldots, M_p)$ of $p + 1$ matrices satisfying $M_0 M_1 \cdots M_p = I$, which comes from the relation of the generators of the fundamental group. In higher dimensional case, there are various relations of generators of the fundamental groups depending on hypersurfaces, which induces various relations among the matrices representing ρ. Thus we can not expect a uniform definition of the index of rigidity.

If we fix a hypersurface, then a presentation of the fundamental group is fixed, and hence we may define the index of rigidity. However, by looking at various examples, it seems difficult to define the index even in such case.

Naively, there appear many relations among the generators of the fundamental group in several variables case as we have seen in Example 13.3. Then the number of relations among the matrices generating the monodromy representation becomes greater than the one variable case, and hence the monodromy representation tends to become rigid. On the other hand, if there are many relations among the tuple of the matrices, the tuple becomes hard to exist. In several cases, the tuple exists under some additional conditions not coming from the relations in a presentation of the fundamental group.

Moreover, further distinct feature in higher dimensional case is observed. In one dimensional case, if a monodromy representation is not rigid, there are infinitely many representations with the same local monodromies (such representations depends on several number of complex numbers). While in higher dimensional cases, for some collections of the local monodromies, there exist finitely many but not unique representations.

Thus the rigidity for higher dimensional cases seems quite different from the one dimensional case, and then seems to be quite interesting. The study may require various knowledges in topology, algebraic geometry, representation theory and so on.

We end this section by presenting an illustrative example.

Example 13.4 We consider the surface S studied in Example 13.3. Let H_∞ be the line at infinity in $\mathbb{P}^2$, and set $\bar{S} = S \cup H_\infty$. We shall consider how many irreducible anti-representations

$$\rho : \pi_1(\mathbb{P}^2 \setminus \bar{S}, b) \to \mathrm{GL}(3, \mathbb{C})$$

of dimension three exist. As we have seen in Sect. 7.4, the rigidity is determined by the spectral types of the local monodromies. Then we prescribe spectral types of the local monodromies, and study the existence and uniqueness of ρ.

Appell's hypergeometric series F_1 satisfies a linear Pfaffian system of rank three with the singular locus $\bar{S}$. For the Pfaffian system, the spectral type of the local monodromy at each irreducible component is (21). Then we assume that ρ has the same spectral type at each irreducible component.

Since $\mathbb{P}^2 \setminus \bar{S} = \mathbb{C}^2 \setminus S$, we can use the presentation (13.6) of the fundamental group given in Example 13.3. For each generator γ_j of the fundamental group, we set

$$\rho(\gamma_j) = M_j \quad (1 \le j \le 5).$$

Then the tuple $(M_1, M_2, \dots, M_5)$ of matrices determines ρ, and satisfies

$$M_1 M_2 = M_2 M_1, \quad M_4 M_5 = M_5 M_4,$$
$$M_5 M_3 M_1 = M_1 M_5 M_3 = M_3 M_1 M_5, \qquad (13.8)$$
$$M_4 M_3 M_2 = M_2 M_4 M_3 = M_3 M_2 M_4,$$

which come from the relations in (13.6). Now we assume

$$M_j^\natural = (21) \quad (1 \le j \le 5).$$

If we multiply each M_j by a scalar in $\mathbb{C}^\times$, the irreducibility, the spectral types and the rigidity do not change. Thus, without loss of generality, we may assume that the double eigenvalue of each M_j is 1. Therefore we assume that, for each j, there exists $e_j \in \mathbb{C} \setminus \{0, 1\}$ such that

$$M_j \sim \begin{pmatrix} 1 & & \\ & 1 & \\ & & e_j \end{pmatrix} \quad (1 \le j \le 5) \qquad (13.9)$$

holds.

The argument is similar to Examples 7.1 and 7.2. We consider the simultaneous conjugacy class $[(M_1, M_2, \dots, M_5)]$ of the tuple $(M_1, M_2, \dots, M_5)$ of the matrices. Then we may send one of $M_1, M_2, \dots, M_5$ to a diagonal matrix. In this case we have the relation $M_1 M_2 = M_2 M_1$, and hence we may send M_1 and M_2 to diagonal matrices simultaneously. There are two possibilities

$$M_1 = \begin{pmatrix} 1 & & \\ & 1 & \\ & & e_1 \end{pmatrix}, \quad M_2 = \begin{pmatrix} 1 & & \\ & 1 & \\ & & e_2 \end{pmatrix} \tag{13.10}$$

and

$$M_1 = \begin{pmatrix} 1 & & \\ & 1 & \\ & & e_1 \end{pmatrix}, \quad M_2 = \begin{pmatrix} 1 & & \\ & e_2 & \\ & & 1 \end{pmatrix}. \tag{13.11}$$

In the first case (13.10), M_1, M_2 do not change by a similar transformation by a matrix in $GL(2, \mathbb{C}) \times GL(1, \mathbb{C})$. Then we can normalize another matrix, say M_3. Thus we may assume that the principal 2×2 part of M_3 is of the Jordan canonical form:

$$M_3 = \begin{pmatrix} \alpha & * & \\ & \beta & * \\ * & * & * \end{pmatrix} \text{ or } \begin{pmatrix} \alpha & 1 & * \\ & \alpha & * \\ * & * & * \end{pmatrix}.$$

By using the condition $\operatorname{rank}(M_3 - 1) = 1$, we may set $\alpha = 1$, and can determine some other entries of M_3. Putting the normalized M_1, M_2, M_3 into the relations (13.8), we can determine M_4, M_5. It turns out that we obtain a reducible representation in this case. Therefore we should assume that M_1, M_2 have the second form (13.11).

By (13.9), for M_3, M_4, M_5 we can set

$$M_j = I_3 + \begin{pmatrix} x_j \\ y_j \\ z_j \end{pmatrix} (1 \ p_j \ q_j) \quad (j = 3, 4, 5),$$

where

$$x_j = e_j - 1 - y_j p_j - z_j q_j.$$

A similar transformation by a diagonal matrix does not change M_1, M_2, and then we may assume $p_3 = q_3 = 1$. Thus the unknowns are $y_3, z_3, y_4, z_4, p_4, q_4, y_5, z_5, p_5, q_5$, and M_j's are parametrized by these unknowns. Hence the relations (13.8) become algebraic equations in these unknowns.

We solve these algebraic equations under the condition that ρ is irreducible. Finally we get the following result. We first find that an irreducible representation exists only if the relation

$$e_1 e_2 = e_4 e_5$$

among the eigenvalues e_j of M_j holds. In this case, we have two sets of solutions to the relations (13.8). Let f_j be a square root of e_j ($j = 2, 3, 4$). Then, for each set of solutions, every entry of M_3, M_4, M_5 is a rational function in e_1, f_2, f_3, f_4. The ambiguity of the square roots induces the action by

$$\sigma_j : f_j \mapsto -f_j \quad (j = 2, 3, 4),$$

each of which replaces the two sets. The local monodromy at the line H_∞ at infinity is represented by $(M_5 M_4 M_3 M_2 M_1)^{-1}$, and then is the conjugacy class of the matrix

$$\begin{pmatrix} (e_1 e_2)^{-1} & & \\ & (e_1 e_2)^{-1} & \\ & & e_3{}^{-1} \end{pmatrix}.$$

Therefore we cannot distinguish the two sets of solutions by the local monodromy at H_∞.

Consequently, we find that the irreducible anti-representation ρ can be determined without using the local monodromy at the line at infinity. Then, if we change the compactification of $\mathbb{C}^2$, we get the same answer. Now we take $\mathbb{P}^1 \times \mathbb{P}^1$ as a compactification, and compute the local monodromies at the infinities $\{x = \infty\}, \{y = \infty\}$. The local monodromy at $\{x = \infty\}$ is represented by $(M_4 M_3 M_1)^{-1}$, and one at $\{y = \infty\}$ is represented by $(M_5 M_3 M_2)^{-1}$. For the first set of solutions, the local monodromies at $\{x = \infty\}$ and $\{y = \infty\}$ are given by

$$\begin{pmatrix} \frac{f_2}{f_3 f_4} & & \\ & \frac{f_2}{f_3 f_4} & \\ & & (e_1 e_2)^{-1} \end{pmatrix}, \begin{pmatrix} \frac{f_4}{f_2 f_3} & & \\ & \frac{f_4}{f_2 f_3} & \\ & & (e_1 e_2)^{-1} \end{pmatrix},$$

respectively, and for the second set of solutions, they are given by

$$\begin{pmatrix} -\frac{f_2}{f_3 f_4} & & \\ & -\frac{f_2}{f_3 f_4} & \\ & & (e_1 e_2)^{-1} \end{pmatrix}, \begin{pmatrix} -\frac{f_4}{f_2 f_3} & & \\ & -\frac{f_4}{f_2 f_3} & \\ & & (e_1 e_2)^{-1} \end{pmatrix}.$$

We notice that these local monodromies can be distinguished by the choices of the square roots f_j, and hence ρ is uniquely determined by the local monodromies. Namely, ρ is rigid if we take $\mathbb{P}^1 \times \mathbb{P}^1$ as a compactification of $\mathbb{C}^2$.

Thus you may find there are several distinct features from the one dimensional case. For the explicit computation to obtain (M_3, M_4, M_5), please refer to [61], where similar considerations for the other Appell's hypergeometric series are described.

This example can be regarded as another proof of the second main result in Gérard-Levelt from the standpoint of the monodromy.

Besides this example, we determined monodromy representations for several completely integrable systems by using rigidity [62, 64].

We may also consider the rigidity of completely integrable systems. However, it will be difficult to formulate the rigidity, because there is no canonical form of completely integrable systems. The work of Kato [93] is an illustrative and pioneering work in this direction.

Chapter 14
Middle Convolution

We have seen that the middle convolution plays a substantial role in the study of
Fuchsian ordinary differential equations. Then it is natural to extend the middle
convolution to completely integrable systems. In this chapter, we explain how
to extend the middle convolution to completely integrable systems, and how to
describe it. As an illustrative example, we give an explicit description of the middle
convolution for linear Pfaffian systems of KZ (Knizhnik-Zamolodchikov) type
appeared in Example 11.2 in Chap. 11. We also study the properties and applications
of our middle convolution.

14.1 Definition and Properties of Middle Convolution

In Sect. 7.5, we defined the middle convolution for Fuchsian systems of ordinary
differential equations of normal form as an algebraic operation for the residue
matrices. While for completely integrable systems, even if restricted to linear
Pfaffian systems, there is no such normal form. Then it seems difficult to define
the middle convolution by a formal extension of the algebraic operation. Thus we
look at the analytic realization of the middle convolution, and try to extend by using
it.

We consider a collection of hyperplanes in $\mathbb{C}^n$ (a hyperplane arrangement).
Namely, let $h_i(x) = h_i(x_1, x_2, \ldots, x_n)$ $(1 \leq i \leq g)$ be a polynomial of degree
one, and set

$$H_i = \{x \in \mathbb{C}^n \, ; \, h_i(x) = 0\} \quad (1 \leq i \leq g).$$

© The Editor(s) (if applicable) and The Author(s), under exclusive
license to Springer Nature Switzerland AG 2020
Y. Haraoka, *Linear Differential Equations in the Complex Domain*, Lecture Notes
in Mathematics 2271, https://doi.org/10.1007/978-3-030-54663-2_14

We set

$$S = \bigcup_{i=1}^{g} H_i,$$

and consider the linear Pfaffian system

$$du = \left(\sum_{i=1}^{g} A_i d \log h_i \right) u \qquad (14.1)$$

with logarithmic singularity along S, where A_i $(1 \le i \le g)$ is an $N \times N$ constant matrix. Assume that (14.1) is completely integrable. The complete integrability condition is described in terms of the matrices A_i. The middle convolution of the Pfaffian system (14.1) is defined and studied in [52], and in this section we explain the ideas and some results in this reference. The complete integrability condition is also found in this reference.

As we have seen in Sect. 7.5, the middle convolution is an operation to construct a differential equation satisfied by the Riemann-Liouville transform of solutions of the original differential equation. From this viewpoint, we formulate as follows. Choose one variable among $x_1, x_2, \ldots, x_n$. Without loss of generality, we may choose x_1. By exchanging the indices if necessary, we can assume

$$\frac{\partial h_i(x)}{\partial x_1} \ne 0 \quad (1 \le i \le g'),$$

$$\frac{\partial h_j(x)}{\partial x_1} = 0 \quad (g' < j \le g).$$

Then, for $1 \le i \le g'$, we can write as

$$h_i(x) = c_i(x_1 - a_i), \qquad (14.2)$$

where c_i is a non-zero constant and a_i is a polynomial in $x' = (x_2, x_3, \ldots, x_n)$ of degree at most one. Take $\lambda \in \mathbb{C}$. For a solution $u(x)$ of the linear Pfaffian system (14.1), we define the Riemann-Liouville integral

$$v_i(x) = \int_\Delta \frac{u(t, x')}{t - a_i} (t - x_1)^\lambda dt \quad (1 \le i \le g'),$$

and set

$$v(x) = {}^t(v_1(x), v_2(x), \ldots, v_{g'}(x)).$$

We shall obtain partial differential equations satisfied by $v(x)$.

We obtain from the Pfaffian system (14.1) and (14.2) the ordinary differential equation

$$\frac{\partial u}{\partial x_1} = \left(\sum_{i=1}^{g'} \frac{A_i}{x_1 - a_i} \right) u$$

in x_1 direction satisfied by u. Then $v(x)$ is just the Riemann-Liouville integral used in constructing the analytic realization of the middle convolution for $u(x)$ when we regard it as a function of one variable x_1. Therefore $v(x)$ becomes a solution of the Fuchsian system of ordinary differential equations of normal form with the residue matrices G_j given by (7.30), where $A_1, A_2, \ldots, A_{g'}$ are used in defining G_j. We define the subspaces $\mathcal{K}$ and $\mathcal{L}$ of $(\mathbb{C}^N)^{g'}$ by (7.31) with $(A_1, A_2, \ldots, A_{g'})$. Then we can define the matrix B_j as the action of G_j on the quotient space $(\mathbb{C}^N)^{g'}/(\mathcal{K}+\mathcal{L})$, and the Fuchsian system of normal form with the residue matrices $B_1, B_2, \ldots, B_{g'}$ becomes irreducible. This differential equation is the irreducible component of the partial differential equation in x_1 satisfied by $v(x)$. Thus, with respect to the chosen variable x_1, we get the same result as the middle convolution for ordinary differential equations.

We are going to obtain the partial differential equations in $x' = (x_2, x_3, \ldots, x_n)$ satisfied by $v(x)$. Note that, in the definition of $v(x)$, x' appears in $u(t, x')$ and in a_i. By the partial differentiations under the integral and by using the partial differential equations for u derived from the Pfaffian system (14.1), we get the partial differential equations in x' satisfied by $v(x)$. By using the complete integrability condition, we see that the partial differential equations contain a closed subsystem with values in $(\mathbb{C}^N)^{g'}/(\mathcal{K}+\mathcal{L})$. The system of partial differential equations thus obtained becomes a completely integrable Pfaffian system, which we define as the middle convolution of (14.1) in x_1 direction.

This is the outline of the definition of the middle convolution for the linear Pfaffian system (14.1). The details are written in [52]. In the following, we shall give an explicit description of the middle convolution for the Pfaffian systems of KZ type.

As we have introduced in Example 11.2, a linear Pfaffian system of KZ type is the system of the form

$$du = \left(\sum_{i=1}^{n} \sum_{k=1}^{p} A_{i,k} \frac{dx_i}{x_i - a_k} + \sum_{1 \le i < j \le n} B_{i,j} \frac{d(x_i - x_j)}{x_i - x_j} \right) u, \tag{14.3}$$

where $a_1, a_2, \ldots, a_p \in \mathbb{C}$ are distinct points and $A_{i,k}$ $(1 \le i \le n, 1 \le k \le p)$, $B_{i,j}$ $(1 \le i < j \le n)$ are $N \times N$ constant matrices. We define $B_{i,j} = B_{j,i}$ for $i > j$. We give the definition of the middle convolution of (14.3) in x_1 direction. The middle convolutions in other directions can be obtained in a similar way.

According to the above outline, first we consider the ordinary differential equation

$$\frac{\partial u}{\partial x_1} = \left(\sum_{k=1}^{p} \frac{A_{1,k}}{x_1 - a_k} + \sum_{l=2}^{n} \frac{B_{1,l}}{x_1 - x_l} \right) u \tag{14.4}$$

in x_1 direction of (14.3). Set $x' = (x_2, x_3, \ldots, x_n)$ and take $\lambda \in \mathbb{C}$. For $1 \leq k \leq p$, we set

$$v_k(x) = \int_\Delta \frac{u(t, x')}{t - a_k} (t - x_1)^\lambda \, dt, \tag{14.5}$$

and for $2 \leq j \leq n$, we set

$$w_j(x) = \int_\Delta \frac{u(t, x')}{t - x_j} (t - x_1)^\lambda \, dt. \tag{14.6}$$

We shall obtain the linear Pfaffian system satisfied by

$$V(x) = {}^t(v_1(x), v_2(x), \ldots, v_p(x), w_2(x), w_3(x), \ldots, w_n(x)). \tag{14.7}$$

As is explained in Sect. 7.5, the Fuchsian system of normal form with the residue matrices given by (7.30) using $(A_1, A_2, \ldots, A_p, B_{1,2}, B_{1,3}, \ldots, B_{1,n})$ becomes the ordinary differential equation in x_1 direction satisfied by V. The system is written for each entry as

$$\frac{\partial v_k}{\partial x_1} = \frac{1}{x_1 - a_k} \left(\sum_{m=1}^{p} (A_{1,m} + \delta_{km}\lambda)v_m + \sum_{l=2}^{n} B_{1,l}w_l \right) \quad (1 \leq k \leq p),$$

$$\frac{\partial w_j}{\partial x_1} = \frac{1}{x_1 - x_j} \left(\sum_{m=1}^{p} A_{1,m}v_m + \sum_{l=2}^{n} (B_{1,l} + \delta_{lj}\lambda)w_l \right) \quad (2 \leq j \leq n).$$

$$\tag{14.8}$$

Take $i > 1$. We shall obtain the differential equation in x_i direction satisfied by V. First we compute the partial derivative of v_k. We use the differential equation

$$\frac{\partial u}{\partial x_i} = \left(\sum_{m=1}^{p} \frac{A_{i,m}}{x_i - a_m} + \sum_{l \neq 1, i} \frac{B_{i,l}}{x_i - x_l} \right) u \tag{14.9}$$

in x_i direction for u to get

$$\frac{\partial v_k}{\partial x_i} = \int_\Delta \frac{1}{t - a_k} \frac{\partial u}{\partial x_i}(t, x') (t - x_1)^\lambda \, dt$$

$$= \int_\Delta \frac{1}{t - a_k} \left(\sum_{m=1}^p \frac{A_{i,m}}{x_i - a_m} + \frac{B_{i,1}}{x_i - t} + \sum_{l \neq 1,i} \frac{B_{i,l}}{x_i - x_l} \right) u(t, x') (t - x_1)^\lambda \, dt$$

$$= \left(\sum_{m=1}^p \frac{A_{i,m}}{x_i - a_m} + \sum_{l \neq 1,i} \frac{B_{i,l}}{x_i - x_l} \right) v_k$$

$$\quad + \int_\Delta \frac{B_{i,1}}{x_i - a_k} \left(\frac{1}{t - a_k} + \frac{1}{x_i - t} \right) u(t, x') (t - x_1)^\lambda \, dt$$

$$= \left(\sum_{m \neq k} \frac{A_{i,m}}{x_i - a_m} + \frac{A_{i,k} + B_{i,1}}{x_i - a_k} + \sum_{l \neq 1,i} \frac{B_{i,l}}{x_i - x_l} \right) v_k - \frac{B_{i,1}}{x_i - a_k} w_i.$$

$$(14.10)$$

Next we compute the partial derivatives of w_j for $2 \leq j \leq n$. For $j \neq i$, in a similar way as $\partial v_k / \partial x_i$ we get

$$\frac{\partial w_j}{\partial x_i} = \left(\sum_{k=1}^p \frac{A_{i,k}}{x_i - a_k} + \sum_{l \neq 1,i,j} \frac{B_{i,l}}{x_i - x_l} + \frac{B_{i,j} + B_{i,1}}{x_i - x_j} \right) w_j - \frac{B_{i,1}}{x_i - x_j} w_i.$$

$$(14.11)$$

For $j = i$, we need a slightly long consideration. Differentiating w_i with respect to x_i, we get

$$\frac{\partial w_i}{\partial x_i} = \int_\Delta \left(\frac{1}{t - x_i} \frac{\partial u}{\partial x_i}(t, x') + \frac{1}{(t - x_i)^2} u(t, x') \right) (t - x_1)^\lambda \, dt$$

$$= \int_\Delta \frac{1}{t - x_i} \left(\sum_{k=1}^p \frac{A_{i,k}}{x_i - a_k} + \sum_{l \neq 1,i} \frac{B_{i,l}}{x_i - x_l} + \frac{B_{i,1}}{x_i - t} \right) u(t, x') (t - x_1)^\lambda \, dt$$

$$\quad - \int_\Delta \frac{\partial}{\partial t} \left(\frac{1}{t - x_i} \right) u(t, x') (t - x_1)^\lambda \, dt$$

$$= \left(\sum_{k=1}^p \frac{A_{i,k}}{x_i - a_k} + \sum_{l \neq 1,i} \frac{B_{i,l}}{x_i - x_l} \right) w_i - B_{i,1} \int_\Delta \frac{u(t, x')}{(t - x_i)^2} (t - x_1)^\lambda \, dt$$

$$\quad + \int_\Delta \frac{1}{t - x_i} \left(\frac{\partial u}{\partial t}(t, x') + u(t, x') \frac{\lambda}{t - x_1} \right) (t - x_1)^\lambda \, dt,$$

where, in the last equality, we regarded Δ as a cycle to integrate by parts. We compute two terms in the last hand side:

$$\int_\Delta \frac{1}{t-x_i} \frac{\partial u}{\partial t}(t,x')(t-x_1)^\lambda dt$$

$$= \int_\Delta \frac{1}{t-x_i} \left(\sum_{k=1}^p \frac{A_{i,k}}{t-a_k} + \sum_{l=2}^n \frac{B_{1,l}}{t-x_l} \right) u(t,x')(t-x_1)^\lambda dt$$

$$= \int_\Delta \left(\sum_{k=1}^p \frac{A_{1,k}}{x_i-a_k} \left(\frac{1}{t-x_i} - \frac{1}{t-a_k} \right) + \sum_{l\neq 1,i} \frac{B_{1,l}}{x_i-x_l} \left(\frac{1}{t-x_i} - \frac{1}{t-x_l} \right) \right)$$

$$\times u(t,x')(t-x_1)^\lambda dt + B_{1,i} \int_\Delta \frac{u(t,x')}{(t-x_i)^2} (t-x_1)^\lambda dt$$

$$= \sum_{k=1}^p \frac{A_{1,k}}{x_i-a_k}(w_i-v_k) + \sum_{l\neq 1,i} \frac{B_{1,l}}{x_i-x_l}(w_i-w_l) + B_{1,i} \int_\Delta \frac{u(t,x')}{(t-x_i)^2}(t-x_1)^\lambda dt,$$

and

$$\int_\Delta \frac{u(t,x')}{t-x_i} \frac{\lambda}{t-x_1}(t-x_1)^\lambda dt = -\frac{\partial}{\partial x_1} \int_\Delta \frac{u(t,x')}{t-x_i}(t-x_1)^\lambda dt = -\frac{\partial w_i}{\partial x_1}.$$

Combining them and using (14.8), we obtain

$$\frac{\partial w_i}{\partial x_i} = \frac{1}{x_i-x_1} \left(\sum_{m=1}^p A_{1,m}v_m + \sum_{l=2}^n (B_{1,l}+\delta_{li}\lambda)w_l \right)$$

$$+ \left(\sum_{k=1}^p \frac{A_{i,k}}{x_i-a_k} + \sum_{l\neq 1,i} \frac{B_{i,l}}{x_i-x_l} \right) w_i \qquad (14.12)$$

$$+ \sum_{k=1}^p \frac{A_{1,k}}{x_i-a_k}(\dot{w}_i-v_k) + \sum_{l\neq 1,i} \frac{B_{1,l}}{x_i-x_l}(w_i-w_l).$$

Summing up the above, we get the following result.

Theorem 14.1 *Let $u(x)$ be a solution of the completely integrable linear Pfaffian system (14.3) of KZ type, and define $v_k(x)$ ($1 \leq k \leq p$) and $w_j(x)$ ($2 \leq j \leq n$) by (14.5), (14.6), respectively. We set $V(x)$ as (14.7). Then $V(x)$ satisfies the completely integrable linear Pfaffian system*

$$dV = \left(\sum_{i=1}^n \sum_{k=1}^p G_{i,k} \frac{dx_i}{x_i-a_k} + \sum_{1\leq i<j\leq n} H_{i,j} \frac{d(x_i-x_j)}{x_i-x_j} \right) V \qquad (14.13)$$

of KZ type, where $G_{i,k}$, $H_{i,j}$ are given as follows: For $1 \le k \le p$,

$$
G_{1,k} =
\begin{array}{c}
\\
1 \\
k \\
p \\
p+1 \\
\\
p+n-1
\end{array}
\begin{pmatrix}
\overset{1}{O} & \cdots & \overset{k}{\cdots} & \cdots & \overset{p}{O} & \overset{p+1}{O} & \cdots & \overset{p+n-1}{O} \\
& \cdots & \cdots & \cdots & & & \cdots & \\
A_{1,1} & \cdots & A_{1,k}+\lambda & \cdots & A_{1,p} & B_{1,2} & \cdots & B_{1,n} \\
& \cdots & \cdots & \cdots & & & \cdots & \\
O & \cdots & \cdots & \cdots & O & O & \cdots & O \\
O & \cdots & \cdots & \cdots & O & O & \cdots & O \\
& \cdots & \cdots & \cdots & & & \cdots & \\
O & \cdots & \cdots & \cdots & O & O & \cdots & O
\end{pmatrix},
$$

for $2 \le j \le n$,

$$
H_{1,j} =
\begin{array}{c}
\\
1 \\
p \\
p+1 \\
\\
p+j-1 \\
\\
p+n-1
\end{array}
\begin{pmatrix}
\overset{1}{O} & \cdots & \overset{p}{O} & \overset{p+1}{O} & \cdots & \overset{p+j-1}{\cdots} & \cdots & \overset{p+n-1}{O} \\
& \cdots & & & \cdots & \cdots & \cdots & \\
O & \cdots & O & O & \cdots & \cdots & \cdots & O \\
O & \cdots & O & O & \cdots & \cdots & \cdots & O \\
& & & & \cdots & \cdots & \cdots & \\
A_{1,1} & \cdots & A_{1,p} & B_{1,2} & \cdots & B_{1,j}+\lambda & \cdots & B_{1,n} \\
& & & & \cdots & \cdots & \cdots & \\
O & \cdots & O & O & \cdots & \cdots & \cdots & O
\end{pmatrix},
$$

for $2 \le i \le n$, $1 \le k \le p$,

$$
G_{i,k} =
\begin{array}{c}
1 \\
\\
k \\
\\
p \\
p+1 \\
\\
p+i-1 \\
\\
p+n-1
\end{array}
\begin{pmatrix}
A_{i,k} & & & & & & & \\
& \ddots & & & & & & \\
& & A_{i,k}+B_{i,1} & & & & -B_{i,1} & \\
& & & \ddots & & & & \\
& & & & A_{i,k} & & & \\
& & & & & A_{i,k} & & \\
& & & & & & \ddots & \\
& & -A_{1,k} & & & & A_{i,k}+A_{1,k} & \\
& & & & & & & \ddots \\
& & & & & & & A_{i,k}
\end{pmatrix},
$$

and for $2 \leq i < j \leq n$,

$$
H_{i,j} = \begin{pmatrix}
B_{i,j} & & & & & & \\
 & \ddots & & & & & \\
 & & B_{i,j} & & & & \\
 & & & B_{i,j} & & & \\
 & & & & \ddots & & \\
 & & & & & B_{i,j}+B_{j,1} & -B_{j,1} \\
 & & & & & & \ddots \\
 & & & & & -B_{1,i} & B_{i,j}+B_{1,i} \\
 & & & & & & & \ddots \\
 & & & & & & & & B_{i,j}
\end{pmatrix}.
$$

This assertion is an expression of the equalities (14.8), (14.10), (14.11) and (14.12) in terms of matrices. The linear Pfaffian system (14.13) is completely integrable, which can be examined directly, however, is evident because a solution $V(x)$ exists.

Remark 14.1 In the linear Pfaffian system (14.3) of KZ type, some of the residue matrices $A_{i,k}, B_{i,j}$ $(2 \leq i, j \leq n)$ can be zero matrices. In such case, the corresponding hyperplanes $x_i = a_k$ or $x_i = x_j$ are not included in the singular locus in (14.3). On the other hand, the residue matrices $G_{i,k}, H_{i,j}$ of the linear Pfaffian system (14.13) satisfied by the Riemann-Liouville transform are given by using $A_{1,m}$ and $B_{1,l}$, and hence are not zero. Therefore the singular locus increases in this case. This phenomenon does not occur in the case of one variable. Note that such phenomenon can happen not only in KZ type but in general (14.1).

Example 14.1 We have studied the linear Pfaffian system of KZ type for $n = 2$ in Examples 11.1 and 12.1. We shall write the result of Theorem 14.1 in this case. We use (x, y) as the coordinate of $\mathbb{C}^2$, and retain the symbols in Example 12.1. Then the Pfaffian system is given by

$$
du = \left(A_1 \frac{dx}{x} + A_2 \frac{dy}{y-1} + A_3 \frac{d(x-y)}{x-y} + A_4 \frac{dx}{x-1} + A_5 \frac{dy}{y} \right) u.
$$

We set $a_1 = 0, a_2 = 1$, and define $v_1(x, y), v_2(x, y)$ by (14.5) and $w_2(x, y)$ by (14.6). Set

$$
V(x, y) = {}^t(v_1(x, y), w_2(x, y), v_2(x, y)).
$$

Then the linear Pfaffian system (14.13) satisfied by $V(x, y)$ is given by

$$dV = \left(B_1 \frac{dx}{x} + B_2 \frac{dy}{y-1} + B_3 \frac{d(x-y)}{x-y} + B_4 \frac{dx}{x-1} + B_5 \frac{dy}{y} \right) V,$$

where the explicit forms of B_i $(1 \leq i \leq 5)$ are

$$B_1 = \begin{pmatrix} A_1 + \lambda \ A_3 \ A_4 \\ O \quad O \ O \\ O \quad O \ O \end{pmatrix}, \quad B_3 = \begin{pmatrix} O & O & O \\ A_1 \ A_3 + \lambda \ A_4 \\ O & O & O \end{pmatrix},$$

$$B_4 = \begin{pmatrix} O & O & O \\ O & O & O \\ A_1 \ A_3 \ A_4 + \lambda \end{pmatrix},$$

$$B_2 = \begin{pmatrix} A_2 & O & O \\ O & A_2 + A_4 & -A_4 \\ O & -A_3 & A_2 + A_3 \end{pmatrix}, \quad B_5 = \begin{pmatrix} A_5 + A_3 & -A_3 & O \\ -A_1 & A_5 + A_1 & O \\ O & O & A_5 \end{pmatrix}.$$

Now we define the invariant subspaces $\mathcal{K}$ and $\mathcal{L}$ in the same way as (7.31) in Sect. 7.5. Since the residue matrices in the ordinary differential equation (14.9) in x_1 direction are $(A_{1,1}, \ldots, A_{1,p}, B_{1,2}, \ldots, B_{1,n})$, the spaces $\mathcal{K}, \mathcal{L}$ are given by

$$\mathcal{K} = \{ {}^t(v_1, \ldots, v_p, w_2, \ldots, w_n) ; \ v_k \in \operatorname{Ker} A_{1,k}, \ w_j \in \operatorname{Ker} B_{1,j} \},$$

$$\mathcal{L} = \operatorname{Ker} \left(\sum_{k=1}^{p} G_{1,k} + \sum_{j=2}^{n} H_{1,j} \right). \tag{14.14}$$

As we have seen in Sect. 7.5, the spaces $\mathcal{K}, \mathcal{L}$ are $(G_{1,1}, \ldots, G_{1,p}, H_{1,2}, \ldots, H_{1,n})$-invariant. We shall see that $\mathcal{K}, \mathcal{L}$ are also invariant subspaces for the other $G_{i,k}, H_{i,j}$.

Take any $V = {}^t(v_1, \ldots, v_p, w_2, \ldots, w_n) \in \mathcal{K}$. For $2 \leq i \leq n, 1 \leq k \leq p$, we have

$$G_{i,k} V = \begin{array}{c} 1 \\ \\ k \\ \\ p \\ p+1 \\ \\ p+i-1 \\ \\ p+n-1 \end{array} \begin{pmatrix} A_{i,k} v_1 \\ \vdots \\ (A_{i,k} + B_{i,1}) v_k - B_{i,1} w_i \\ \vdots \\ A_{i,k} v_p \\ A_{i,k} w_2 \\ \vdots \\ -A_{1,k} v_k + (A_{i,k} + A_{1,k}) w_i \\ \vdots \\ A_{i,k} w_n \end{pmatrix} = \begin{pmatrix} A_{i,k} v_1 \\ \vdots \\ (A_{i,k} + B_{i,1}) v_k \\ \vdots \\ A_{i,k} v_p \\ A_{i,k} w_2 \\ \vdots \\ (A_{i,k} + A_{1,k}) w_i \\ \vdots \\ A_{i,k} w_n \end{pmatrix}.$$

By using the integrability condition (11.21) obtained in Example 11.2, we get

$$A_{1,m}(A_{i,k}v_m) = A_{i,k}A_{1,m}v_m = 0 \quad (m \neq k),$$

$$A_{1,k}((A_{i,k} + B_{i,1})v_k) = (A_{i,k} + B_{i,1})A_{1,k}v_k = 0,$$

$$B_{1,l}(A_{i,k}w_l) = A_{i,k}B_{1,l}w_l = 0 \quad (l \neq i),$$

$$B_{1,i}((A_{i,k} + A_{1,k})w_i) = (A_{i,k} + A_{1,k})B_{1,i}w_i = 0,$$

and hence we have $G_{i,k}V \in \mathcal{K}$. In a similar way, we can show $H_{i,j}V \in \mathcal{K}$. Therefore $\mathcal{K}$ is an invariant subspace for $G_{i,k}$ and $H_{i,j}$.

Next we examine the invariance of $\mathcal{L}$. By Lemma 7.16, we have

$$\mathcal{L} = \{{}^t(v, \ldots, v)\,;\ \Big(\sum_{k=1}^{p} A_{1,k} + \sum_{j=2}^{n} B_{1,j} + \lambda\Big)v = 0\}$$

for the case $\lambda \neq 0$, and

$$\mathcal{L} = \{{}^t(v_1, \ldots, v_p, w_2, \ldots, w_n)\,;\ \sum_{k=1}^{p} A_{1,k}v_k + \sum_{j=2}^{n} B_{1,j}w_j = 0\}$$

for the case $\lambda = 0$. Assume that $\lambda \neq 0$. Take any $V = {}^t(v, \ldots, v) \in \mathcal{L}$. For $2 \leq i \leq p, 1 \leq k \leq p$, it follows

$$G_{i,k}V = \begin{pmatrix} A_{i,k}v \\ \vdots \\ A_{i,k}v \end{pmatrix}$$

directly from the explicit forms of $G_{i,k}$ and V. Now we use the integrability condition (11.21) to get

$$\Big(\sum_{m=1}^{p} A_{1,m} + \sum_{j=2}^{n} B_{1,j} + \lambda\Big) A_{i,k}v$$

$$= \Big(\sum_{m \neq k} A_{1,m} + \sum_{j \neq 1,i} B_{1,j} + (A_{1,k} + B_{1,i}) + \lambda\Big) A_{i,k}v$$

$$= A_{i,k}\Big(\sum_{m \neq k} A_{1,m} + \sum_{j \neq 1,i} B_{1,j} + (A_{1,k} + B_{1,i}) + \lambda\Big) v$$

$$= 0.$$

Hence we have $G_{i,k}V \in \mathcal{L}$. In a similar way, we have $H_{i,j}V \in \mathcal{L}$, and therefore $\mathcal{L}$ is invariant under $G_{i,k}$ and $H_{i,j}$. Next assume that $\lambda = 0$. Take any $V = {}^t(v_1, \ldots, v_p, w_2, \ldots, w_n) \in \mathcal{L}$. For $2 \leq i \leq p$, $1 \leq k \leq p$, it follows

$$
G_{i,k}V = \begin{matrix} 1 \\ \\ k \\ \\ \\ p \\ p+1 \\ \\ \\ p+i-1 \\ \\ p+n-1 \end{matrix}
\begin{pmatrix}
A_{i,k}v_1 \\
\vdots \\
(A_{i,k} + B_{i,1})v_k - B_{i,1}w_i \\
\vdots \\
A_{i,k}v_p \\
A_{i,k}w_2 \\
\vdots \\
-A_{1,k}v_k + (A_{i,k} + A_{1,k})w_i \\
\vdots \\
A_{i,k}w_n
\end{pmatrix}.
$$

By using the integrability condition (11.21), we get

$$
\sum_{m \neq k} A_{1,m}(A_{i,k}v_m) + A_{1,k}((A_{i,k} + B_{i,1})v_k - B_{i,1}w_i)
$$

$$
+ \sum_{j \neq 1,i} B_{1,j}(A_{i,k}w_j) + B_{1,i}(-A_{1,k}v_k + (A_{i,k} + A_{1,k})w_i)
$$

$$
= A_{i,k} \sum_{m \neq k} A_{1,m}v_m + A_{i,k} \sum_{j \neq 1,i} B_{1,j}w_j
$$

$$
+ (A_{i,k} + B_{i,1})A_{1,k}v_k - A_{1,k}B_{i,1}w_i - B_{1,i}A_{1,k}v_k + (A_{i,k} + A_{1,k})B_{1,i}w_i
$$

$$
= A_{i,k} \left(\sum_{m=1}^{p} A_{1,m}v_m + \sum_{j=2}^{n} B_{1,j}w_j \right)
$$

$$
= 0,
$$

and hence we have $G_{i,k}V \in \mathcal{L}$. Similarly we have $H_{i,j}V \in \mathcal{L}$, and therefore $\mathcal{L}$ is invariant under $G_{i,k}$ and $H_{i,j}$ also in this case.

Thus we obtain the following theorem.

Theorem 14.2 *The linear Pfaffian system (14.13) is constructed from the completely integrable linear Pfaffian system (14.3) of KZ type in Theorem 14.1. Define linear spaces $\mathcal{K}$ and $\mathcal{L}$ by (14.14). Then $\mathcal{K}$ and $\mathcal{L}$ are invariant under the action of $G_{i,k}$ and $H_{i,j}$. Let $\bar{G}_{i,k}, \bar{H}_{i,j}$ be the actions of $G_{i,k}, H_{i,j}$, respectively, on the* *quotient space* $\mathcal{K} + \mathcal{L}$. *Then the linear Pfaffian system*

$$dv = \left(\sum_{i=1}^{n} \sum_{k=1}^{p} \bar{G}_{i,k} \frac{dx_i}{x_i - a_k} + \sum_{1 \le i < j \le n} \bar{H}_{i,j} \frac{d(x_i - x_j)}{x_i - x_j} \right) v \tag{14.15}$$

of KZ type becomes completely integrable. If the Eq. (14.4) in x_1 direction of (14.3) is irreducible, the Pfaffian system (14.15) is also irreducible.

The integrability of (14.15) follows directly from the integrability of (14.13). The irreducibility follows from Theorem 7.14, since the differential equation in x_1 direction of the Pfaffian system (14.15) is the middle convolution of (14.4) as and ordinary differential equation.

The completely integrable linear Pfaffian system (14.15) of KZ type is called the *middle convolution* of (14.3) with parameter λ in x_1 direction. The operation which construct (14.15) from (14.3) is also said to be the middle convolution.

In the case of ordinary differential equations, we have the additivity of the middle convolution: Symbolically we have

$$mc_0 = \mathrm{id.}, \quad mc_\lambda \circ mc_\mu = mc_{\lambda+\mu}$$

(Theorems 7.12 and 7.13). The middle convolution for linear Pfaffian systems possesses the corresponding property. The proof is given in [52, Theorem 3.1].

Example 14.2 Take $\alpha_1, \alpha_2, \ldots, \alpha_5 \in \mathbb{C}$. We consider the linear Pfaffian system

$$du = \left(\alpha_1 \frac{dx}{x} + \alpha_2 \frac{dy}{y-1} + \alpha_3 \frac{d(x-y)}{x-y} + \alpha_4 \frac{dx}{x-1} + \alpha_5 \frac{dy}{y} \right) u$$

of rank 1. By the middle convolution with parameter λ in x direction, we get the linear Pfaffian system

$$dv = \left(A_1 \frac{dx}{x} + A_2 \frac{dy}{y-1} + A_3 \frac{d(x-y)}{x-y} + A_4 \frac{dx}{x-1} + A_5 \frac{dy}{y} \right) v \tag{14.16}$$

of rank 3, where $A_1, A_2, \ldots, A_5$ are given by

$$A_1 = \begin{pmatrix} \alpha_1 + \lambda \, \alpha_3 \, \alpha_4 \\ 0 \quad 0 \; 0 \\ 0 \quad 0 \; 0 \end{pmatrix}, \quad A_2 = \begin{pmatrix} \alpha_2 & 0 & 0 \\ 0 & \alpha_2 + \alpha_4 & -\alpha_4 \\ 0 & -\alpha_3 & \alpha_2 + \alpha_3 \end{pmatrix},$$

$$A_3 = \begin{pmatrix} 0 & 0 & 0 \\ \alpha_1 \, \alpha_3 + \lambda \, \alpha_4 \\ 0 & 0 & 0 \end{pmatrix}, \quad A_4 = \begin{pmatrix} 0 & 0 & 0 \\ 0 & 0 & 0 \\ \alpha_1 \, \alpha_3 \, \alpha_4 + \lambda \end{pmatrix},$$

$$A_5 = \begin{pmatrix} \alpha_5 + \alpha_3 & -\alpha_3 & 0 \\ -\alpha_1 & \alpha_5 + \alpha_1 & 0 \\ 0 & 0 & \alpha_5 \end{pmatrix},$$

which we derive from Example 14.1. The Pfaffian system (14.16) is equivalent to a linear Pfaffian system satisfied by Appell's hypergeometric function F_1.

We continue to operate the addition

$$A_1' = A_1 - (\alpha_1 + \lambda), \ A_i' = A_i \ (2 \le i \le 5),$$

and then operate the middle convolution with parameter μ in x direction to the Pfaffian system obtained by replacing each A_i by A_i'. Since we have

$$\dim \operatorname{Ker} A_1' = 1, \ \dim \operatorname{Ker} A_3' = \dim \operatorname{Ker} A_4' = 2, \ \mathcal{L} = \{0\},$$

the rank of the resulting Pfaffian system becomes $3 \times 3 - (1 + 2 + 2) = 4$. The resulting Pfaffian system can be given by

$$dw = \left(B_1 \frac{dx}{x} + B_2 \frac{dy}{y-1} + B_3 \frac{d(x-y)}{x-y} + B_4 \frac{dx}{x-1} + B_5 \frac{dy}{y} \right) w, \qquad (14.17)$$

where

$$B_1 = \begin{pmatrix} \mu - \alpha_1 - \lambda & 0 & \alpha_1 & 0 \\ 0 & \mu - \alpha_1 - \lambda & 0 & \alpha_1 \\ 0 & 0 & 0 & 0 \\ 0 & 0 & 0 & 0 \end{pmatrix},$$

$$B_2 = \begin{pmatrix} \alpha_2 + \alpha_4 & -\alpha_4 & 0 & 0 \\ -\alpha_3 & \alpha_2 + \alpha_3 & 0 & 0 \\ 0 & 0 & \alpha_2 + \alpha_4 & -\alpha_4 \\ 0 & 0 & -\alpha_3 & \alpha_2 + \alpha_3 \end{pmatrix},$$

$$B_3 = \begin{pmatrix} 0 & 0 & 0 & 0 \\ 0 & 0 & 0 & 0 \\ -\frac{\lambda(\alpha_1 + \alpha_3 + \lambda)}{\alpha_1} & -\frac{\alpha_4 \lambda}{\alpha_1} & \alpha_3 + \lambda + \mu & \alpha_4 \\ 0 & 0 & 0 & 0 \end{pmatrix},$$

$$B_4 = \begin{pmatrix} 0 & 0 & 0 & 0 \\ 0 & 0 & 0 & 0 \\ 0 & 0 & 0 & 0 \\ -\frac{\alpha_3 \lambda}{\alpha_1} & -\frac{\lambda(\alpha_1 + \alpha_4 + \lambda)}{\alpha_1} & \alpha_3 & \alpha_4 + \lambda + \mu \end{pmatrix},$$

$$B_5 = \begin{pmatrix} \alpha_1 + \alpha_3 + \alpha_5 + \lambda & \alpha_4 & -\alpha_1 & 0 \\ 0 & \alpha_5 & 0 & 0 \\ \frac{\lambda(\alpha_1 + \alpha_3 + \lambda)}{\alpha_1} & \frac{\alpha_4 \lambda}{\alpha_1} & \alpha_5 - \lambda & 0 \\ 0 & 0 & 0 & \alpha_5 \end{pmatrix}.$$

This Pfaffian system (14.17) is equivalent to Pfaffian systems satisfied by Appell's hypergeometric series F_2, F_3. The spectral types of these residue matrices are given by

$$(B_1^\natural, B_2^\natural, B_3^\natural, B_4^\natural, B_5^\natural) = ((22), (22), (31), (31), (31)).$$

If we compactify $\mathbb{C}^2$ by $\mathbb{P}^1 \times \mathbb{P}^1$, the residue matrices at $x = \infty$ and $y = \infty$ are $-B_1 - B_3 - B_4$ and $-B_2 - B_3 - B_5$, respectively, and their spectral types are given by

$$((-B_1 - B_3 - B_4)^\natural, (-B_2 - B_3 - B_5)^\natural) = (211), (211)).$$

If we compactify by $\mathbb{P}^2$, the residue matrix at the line at infinity H_∞ is $-B_1 - B_2 - B_3 - B_4 - B_5$, whose spectral type is given by

$$(-B_1 - B_2 - B_3 - B_4 - B_5)^\natural = (31).$$

Example 14.3 We operate another addition and middle convolution to the linear Pfaffian system (14.16) in Example 14.2. First we operate the addition

$$A_1' = A_1 - (\alpha_1 + \lambda), \ A_4' = A_4 - (\alpha_4 + \lambda), \ A_i' = A_i \ (i = 2, 3, 5).$$

Then we operate the middle convolution with parameter $\mu = \lambda - \alpha_3$ in x direction to the Pfaffian system obtained by replacing each A_i by A_i'. Since

$$\dim \operatorname{Ker} A_1 = \dim \operatorname{Ker} A_4 = 1, \ \dim \operatorname{Ker} A_3 = 2, \ \dim \mathcal{L} = 1,$$

the rank of the resulting Pfaffian system becomes $3 \times 3 - (1 + 1 + 2 + 1) = 4$. We write the resulting Pfaffian system as

$$dz = \left(C_1 \frac{dx}{x} + C_2 \frac{dy}{y-1} + C_3 \frac{d(x-y)}{x-y} + C_4 \frac{dx}{x-1} + C_5 \frac{dy}{y} \right) z. \tag{14.18}$$

Here we do not give explicit forms of the residue matrices C_i. The spectral types are given by

$$(C_1^\natural, C_2^\natural, C_3^\natural, C_4^\natural, C_5^\natural) = ((22), (22), (31), (22), (22)).$$

The spectral types of the residue matrices at $x = \infty$ and $y = \infty$ are given by

$$((-C_1 - C_3 - C_4)^\natural, (-C_2 - C_3 - C_5)^\natural) = ((22), (22)),$$

and one at H_∞ is given by

$$(-C_1 - C_2 - C_3 - C_4 - C_5)^\natural = (211).$$

This Pfaffian system (14.18) is equivalent to the pull-back of the Pfaffian system satisfied by Appell's F_4 [94].

14.2 Application

In Sect. 8.6 in Chap. 8, we considered deformations of rigid ordinary differential equations. The deformations can be obtained by applying the middle convolution for Pfaffian systems of KZ type, since the middle convolution provides new completely integrable systems.

We consider a rigid Fuchsian differential equation. Since we are going to deform it, we assume that the number of the singular points is at least four. We know that a rigid Fuchsian ordinary differential equation can be given by a Fuchsian system of normal form, and then we assume that the equation is given by

$$\frac{du}{dx} = \left(\sum_{i=1}^{p} \frac{A_i}{x - t_i} \right) u \tag{14.19}$$

with $p \geq 3$. We understand that, apart from ∞, the singular points t_1, t_2 have been normalized by a fractional linear transformation. Thus we may assume, for example, $t_1 = 0, t_2 = 1$, and the other singular points $t_3, \ldots, t_p$ are the parameters of the deformation. The purpose of this section is to derive partial differential equations in the variables $t_3, \ldots, t_p$ for u compatible with (14.19).

Thanks to Theorem 7.22, the Fuchsian system (14.19) is obtained from a rank one equation by a finite iteration of additions and middle convolutions. Let

$$\frac{dy}{dx} = \left(\sum_{i=1}^{p} \frac{\alpha_i}{x - t_i} \right) y \tag{14.20}$$

be the differential equation of rank one, where $\alpha_1, \alpha_2, \ldots, \alpha_p \in \mathbb{C} \setminus \{0\}$. We may regard this ordinary differential equation as the equation in x direction of a linear Pfaffian system

$$dy = \left(\sum_{i=1}^{p} \alpha_i \frac{d(x - t_i)}{x - t_i} + \sum_{1 \leq i < j \leq p} \beta_{i,j} \frac{d(t_i - t_j)}{t_i - t_j} \right) y \tag{14.21}$$

of KZ type with variables $(x, t_3, t_4, \ldots, t_p)$. Here we understand $dt_1 = dt_2 = 0$, because t_1, t_2 are normalized (fixed). Since the Pfaffian system (14.21) is of rank one, it is completely integrable for any values of $\beta_{i,j}$. In particular, we may choose $\beta_{i,j} = 0$. We replace each middle convolution in constructing (14.19) from (14.20) by the middle convolution of linear Pfaffian systems of KZ type in x direction. Then we get a completely integrable linear Pfaffian system

$$du = \left(\sum_{i=1}^{p} A_i \frac{d(x - t_i)}{x - t_i} + \sum_{1 \le i < j \le p} B_{i,j} \frac{d(t_i - t_j)}{t_i - t_j} \right) u$$

whose equation in x direction coincides with (14.19). Thus we can construct the deformation of the rigid ordinary differential equation (14.19).

Example 14.4 The deformation we showed in Example 8.1 is obtained by the above procedure. For a differential equation

$$\frac{dy}{dx} = \left(\frac{\alpha_1}{x} + \frac{\alpha_2}{x - t} + \frac{\alpha_3}{x - 1} \right) y$$

of rank one, we regard it as the equation in x direction of the linear Pfaffian system

$$dy = \left(\alpha_1 \frac{dx}{x} + \alpha_2 \frac{d(x - t)}{x - t} + \alpha_3 \frac{dx}{x - 1} + 0 \frac{dt}{t} + 0 \frac{dt}{t - 1} \right) y,$$

and operate the middle convolution to this Pfaffian system with parameter λ in x direction. Then we obtain the completely integrable Pfaffian system

$$dY = \left(A_1 \frac{dx}{x} + A_3 \frac{dx}{x - 1} + A_2 \frac{d(x - t)}{x - t} + B_1 \frac{dt}{t} + B_3 \frac{dt}{t - 1} \right) Y$$

with residue matrices A_1, A_2, A_3, B_1, B_3 given in Example 8.1.

Thus the middle convolution for completely integrable systems will be useful for constructing new completely integrable systems, and we may expect many applications. On the other hand, the definition of the middle convolution depends on the coordinates. Then it will be a good problem to study the combination of the middle convolution and coordinate changes. Another problem is to study the relations of many known hypergeometric functions in several variables—Appell-Lauricella, GKZ, Heckman-Opdam, etc.—by using the middle convolution and other transformations.

Moreover, restriction to a singular locus is also an operation which changes the rank of the system and the number of the variables. It is a similar but different kind operation from the middle convolution, and then the combination of these two kinds of operations will be interesting.

Notes and References to This Book

Hypergeometric Differential Equations/Functions

As explained in Introduction, hypergeometric differential equations are illustrative
examples for Fuchsian differential equations. Then it will be helpful to the reader to
study hypergeometric differential equations. Hypergeometric differential equations
and hypergeometric functions are studied from various viewpoints. We recommend
recent literatures [5, 51, 106, 191] and classic one [182].

Connection Problem

The connection problem is a substantial and interesting problem also for completely
integrable systems with regular singularity (regular holonomic systems). In some
cases, such connection problems are solved by the help of the restriction to singular
locus [94, 141]. We can formulate the connection problem for regular holonomic
systems from our viewpoints in Chap. 6 and in Part II. The formulation and some
examples are given in [56].

Irregular Singularity

In Chap. 10, we explained the classical results of irregular singular points—the local
and global analysis for differential equations having irregular singular points. The
mainstream of the recent research is the Borel summability, which is a method of
summation of formal solutions according to their orders of divergence (the Gevrey

© The Editor(s) (if applicable) and The Author(s), under exclusive
license to Springer Nature Switzerland AG 2020
Y. Haraoka, *Linear Differential Equations in the Complex Domain*, Lecture Notes
in Mathematics 2271, https://doi.org/10.1007/978-3-030-54663-2

orders). Please refer to [8, 178]. Also the exact WKB analysis gives remarkable results, in particular for second order differential equations [101, 102, 112, 172].

Computation of Stokes multipliers is one of the central problems. As commented in Sect. 7.3.2, the Laplace transform may send the connection coefficients of Fuchsian differential equations of Okubo normal form to Stokes multipliers. Balser [9, 10] investigated the mechanism of determining the Stokes multipliers for confluent hypergeometric systems of Okubo normal form. Another way to observe the Stokes phenomenon is the exponential asymptotics. The idea is to truncate the formal series solution at an optimal place, and to study the behavior of the remainder. See [118].

Local analysis at irregular singularity of holonomic systems is a fundamental problem. This problem was opened by Gérard-Sibuya [43], and Majima [120] studied the problem by introducing the notion of strong asymptotics. From another viewpoint, Shimomura [161, 162] calculated the asymptotic behaviors and Stokes phenomena for some particular holonomic systems with irregular singularity.

Riemann-Hilbert Problem and Deformation Theory

An extension of Riemann-Hilbert problem for irregular singular case is considered by Birkhoff [20], and is called Riemann-Hilbert-Birkhoff problem. Sibuya made a substantial contribution to this problem [165].

Recently the deformation theory for resonant case is intensively studied [28, 44, 81, 88, 89]. This direction seems very important and fruitful.

Concerning the deformation theory in irregular singular case, moduli spaces of irregular connections are studied in [22, 71, 183–185].

Riemann-Hilbert problem for regular holonomic systems in several variables is solved by Kashiwara [91]. Majima [120] extended Sibuya's result to solve Riemann-Hilbert-Birkhoff problem for some irregular holonomic systems. Recently, D'Agnolo and Kashiwara [31] established the Riemann-Hilbert correspondence for general holonomic systems. The local analysis at irregular singular locus in several variables case, however, has not yet been completed. The work of Majima seems basic. See also the work of Mochizuki [127].

Katz Theory in One Variable

It seems that Katz constructed the theory of rigid local systems in order to solve Grothendieck's p-curvature conjecture. An affirmative answer to the conjecture for rigid Fuchsian ordinary differential equations is given in the book [100]. For the conjecture, please also refer to [18, 47, 99].

The extension of the Katz theory to the irregular singular case is a natural and important problem. The notions of rigidity and middle convolution are extended to the irregular singular case, and the analogous result to Theorem 7.22 is obtained [6, 68, 69, 103, 173]. As in regular singular case, Kac-Moody root systems are useful to study the structures of differential equations with irregular singular points. See [70] for this topic.

As expository literatures for the Katz theory, we give the references [54, 168] and [55].

The Katz theory is applied to various areas of mathematics and physics. As an application to the conformal field theory, we give a reference [15].

Katz Theory in Several Variables

Extending the Katz theory in several variables case is one of the main purposes of this book. We give two recent references in this direction. Oshima [140] determined the eigenvalues of residue matrices of the results of middle convolution for KZ equations. The middle convolution for the monodromy of holonomic systems is also fundamental. To describe the monodromy, we need the knowledge of the fundamental group. For the KZ type equation, the fundamental group becomes the pure braid group, and then we can describe the middle convolution for the monodromy [57].

On the other hand, constructing completely integrable systems is a fundamental problem, although there is no standard way. We find in [189, Chapter 12] a pioneering result. Recently, Kato and Sekiguchi [95] obtained a series of completely integrable systems by the help of finite complex reflection groups along the idea of Saito [146]. Saito's idea was developed by Sabbah [144], and Kato-Mano-Sekiguchi [96–98] applied it to get some explicit results using algebraic solutions of Painlevé equations.

Let a rational (algebraic) completely integrable system be given. If a one-dimensional section of the system becomes a rigid ordinary differential equation, we may obtain the system from the rigid ODE by operating middle convolutions and additions in several variables This is a way of the construction, which we explained in Chap. 14. If any one-dimensional section is not rigid, the system comes from an algebraic solution of some deformation equation. Then such system can be obtained from an algebraic solution of some deformation equation. This is the second way. Thus we are interested in algebraic solutions of deformation equations. For Painlevé VI equation, Lisovyy-Tykhyy [117] completed the list of all algebraic solutions. For the Garnier system, Diarra [36] obtained some class of algebraic solutions. These lists will become a good source of completely integrable systems.

References

1. D.V. Anosov, A.A. Bolibruch, The Riemann-Hilbert problem. Aspects Math. **E22** (1994). Friedr. Vieweg & Sohn, Braunschweig
2. K. Aomoto, Les équations aus différences linéaires et les intégrales des fonctions multiformes. U. Fac. Sci. Univ. Tokyo Sec. IA **22**, 271–297 (1975)
3. K. Aomoto, On vanishing of cohomology attached to certain many valued meromorphic functions. J. Math. Soc. Japan **27**, 248–255 (1975)
4. K. Aomoto, On the structure of integrals of power products of linear functions. Sci. Papers Coll. Gen. Education Univ. Tokyo **27**, 49–61 (1977)
5. K. Aomoto, M. Kita, *Theory of Hypergeometric Functions* (Springer, Tokyo, 2011)
6. D. Arinkin, Rigid irregular connections on $\mathbb{P}^1$. Compositio Math. **146**, 1323–1338 (2010)
7. E. Artin, Theory of braids. Ann. Math. **48**, 101–126 (1947)
8. W. Balser, *Formal Power Series and Linear Systems of Meromorphic Ordinary Differential Equations*. Universitext (Springer, New York, 2000)
9. W. Balser, Nonlinear difference equations and Stokes matrices. Adv. Dyn. Syst. Appl. **7**, 145–162 (2012)
10. W. Balser, Computation of the Stokes multipliers of Okubo's confluent hypergeometric system. Adv. Dyn. Syst. Appl. **9**, 53–74 (2014)
11. W. Balser, W.B. Jurkat, D.A. Lutz, Birkhoff invariants and Stokes' multipliers for meromorphic linear differential equations. J. Math. Anal. Appl. **71**, 48–94 (1979)
12. W. Balser, W.B. Jurkat, D.A. Lutz, A general theory of invariants for meromorphic differential equations; Part I, Formal invariants. Funkcial. Ekvac. **22**, 197–221 (1979)
13. W. Balser, W.B. Jurkat, D.A. Lutz, A general theory of invariants for meromorphic differential equations; Part II, Proper invariants. Funkcial. Ekvac. **22**, 257–283 (1979)
14. W. Balser, W.B. Jurkat, D.A. Lutz, On the reduction of connection problems for differential equations with an irregular singular point to ones with only regular singularities, I. SIAM J. Math. Anal. **12**, 691–721 (1981)
15. V. Belavin, Y. Haraoka, R. Santachiara, Rigid Fuchsian systems in 2-dimensional conformal field theory. Commun. Math. Phys. **365**, 17–60 (2019)
16. P. Belkale, Local systems on $\mathbb{P}^1 - S$ for S a finite set. Compositio Math. **129**, 67–86 (2001)
17. M.A. Bershtein, A.I. Shchechkin, Bilinear equations on Painlevé τ functions from CFT. Commun. Math. Phys. **339**, 1021–1061 (2015)
18. F. Beukers, G. Heckman, Monodromy for the hypergeometric function $_nF_{n-1}$. Invent. Math. **95**, 325–354 (1989)

© The Editor(s) (if applicable) and The Author(s), under exclusive
license to Springer Nature Switzerland AG 2020
Y. Haraoka, *Linear Differential Equations in the Complex Domain*, Lecture Notes
in Mathematics 2271, https://doi.org/10.1007/978-3-030-54663-2

19. G.D. Birkhoff, Singular points of ordinary linear differential equations. Trans. Amer. Math. Soc. **10**, 436–470 (1909)
20. G.D. Birkhoff, The generalized Riemann problem for linear differential equations and the allied problems for linear difference and q-difference equations. Proc. Amer. Acad. Arts and Sci. **49**, 521–568 (1913)
21. J.S. Birman, *Braids, Links, and Mapping Class Groups*. Annals of Mathematics Studies, No. 82 (Princeton University Press, Princeton, 1974)
22. P.P. Boalch, Geometry and braiding of Stokes data; fission and wild character varieties. Ann. Math. **179**, 301–365 (2014)
23. A.A. Bolibruch, Hilbert's twenty-first problem for Fuchsian linear systems, in *Developments in Mathematics: The Moscow School* (Chapman & Hall, London, 1993), pp. 54–99
24. A.A. Bolibruch, On isomonodromic deformations of Fuchsian systems. J. Dyn. Control Syst. **3**, 589–604 (1997)
25. A.A. Bolibruch, On isomonodromic confluence of Fuchsian singularities. Proc. Stek. Inst. Math. **221**, 117–132 (1998)
26. K. Cho, K. Matsumoto, Intersection theory for twisted cohomologies and twisted Riemann's period relations I. Nagoya Math. J. **139**, 67–86 (1995)
27. E.A. Coddington, N. Levinson, *Theory of Ordinary Differential Equations* (McGraw-Hill, New York, 1955)
28. G. Cotti, B. Dubrovin, D. Guzzetti, Isomonodromy deformations at an irregular singularity with coalescing eigenvalues. Duke Math. J. **168**, 967–1108 (2019)
29. W. Crawley-Boevey, Geometry of the moment map for representations of quivers. Compositio Math. **126**, 257–293 (2001)
30. W. Crawley-Boevey, On matrices in prescribed conjugacy classes with no common invariant subspace and sum zero. Duke Math. J. **118**, 339–352 (2003)
31. A. D'Agnolo, M. Kashiwara, Riemann-Hilbert correspondence for holonomic D-modules. Publ. Math. Inst. Hautes Études Sci. **123**, 69–197 (2016)
32. P. Deligne, *Équations différentielles à points singuliers réguliers*. Lecture Notes in Mathematics, vol. 163 (Springer-Verlag, 1970)
33. P. Deligne, G. D. Mostow, Monodromy of hypergeometric functions and non-lattice integral monodromy, Publ. Math. IHES **63**, 5–89 (1986)
34. M. Dettweiler, S. Reiter, An algorithm of Katz and its application to the inverse Galois problem. J. Symbolic Comput. **30**, 761–798 (2000)
35. M. Dettweiler, S. Reiter, Middle convolution of Fuchsian systems and the construction of rigid differential systems. J. Algebra **318**, 1–24 (2007)
36. K. Diarra, Construction et classification de certaines solutions algébriques des systèmes de Garnier, Bull. Braz. Math. Soc. **44**, 129–154 (2013)
37. H. Esnault, V. Schechtman, E. Viehweg, Cohomology of local systems on the complement of hyperplanes. Invent. Math. **109**, 557–561 (1992)
38. R. Fuchs, Über die analytische Natur der Lösungen von Differentialgleichungen zweiter Ordnung mit festen kritischen Punkten. Math. Ann. **75**, 469–496 (1914)
39. K. Fuji, T. Suzuki, Drinfeld-Sokolov hierarchies of type A and fourth order Painlevé systems. Funkcial. Ekvac. **53**, 143–167 (2010)
40. F.R. Gantmacher, *The Theory of Matrices, I* (Chelsea, New York, 1959)
41. R. Gérard, Théorie de Fuchs sur une variété analytique complexe. J. Math. Pures Appl. **47**, 321–404 (1968)
42. R. Gérard, A.H.M. Levelt, Étude d'une classe particulière de systèmes de Pfaff du type de Fuchs sur l'espace projectif complexe. J. Math. Pures Appl. **51**, 189–217 (1972)
43. R. Gérard, Y. Sibuya, *Etude de certains systèmes de Pfaff avec singularites*. Lecture Notes in Mathematics, vol. 712 (Springer, Berlin, 1970), pp. 131–288
44. D. Guzzetti, Notes on non-generic isomonodromy deformations. SIGMA **14**, 087, 34 pp. (2018)
45. H.A. Hamm, D.T. Lê, Un théorème de Zariski du type de Lefschetz. Ann. Sci. École Norm. Sup. **6**, 317–366 (1973)

46. V.L. Hansen, *Braids and Coverings*. London Mathematical Society Student Texts, vol. 18 (Cambridge University Press, Cambridge, 1989)

47. Y. Haraoka, Finite monodromy of Pochhammer equation. Ann. Inst. Fourier **44**, 767–810 (1994)

48. Y. Haraoka, Monodromy representations of systems of differential equations free from accessory parameters. SIAM J. Math. Anal. **25**, 1595–1621 (1994)

49. Y. Haraoka, Irreducibility of accessory parameter free systems. Kumamoto J. Math. **8**, 153–170 (1995)

50. Y. Haraoka, Integral representations of solutions of differential equations free from accessory parameters. Adv. Math. **169**, 187–240 (2002)

51. Y. Haraoka, *Hypergeometric Functions* (Asakura Publishing, Tokyo, 2002) (in Japanese)

52. Y. Haraoka, Middle convolution for completely integrable systems with logarithmic singularities along hyperplane arrangements. Adv. Stud. Pure Math. **62**, 109–136 (2012)

53. Y. Haraoka, Regular coordinates and reduction of deformation equations for Fuchsian systems. Banach Center Publ. **97**, 39–58 (2012)

54. Y. Haraoka, Globally analyzable Fuchsian differential equations. Sugaku Expositions **28**, 49–72 (2015)

55. Y. Haraoka, Holonomic systems, in *Analytic, Algebraic and Geometric Aspects of Differential Equations*. Trends in Mathematics (Birkhäuser/Springer, Cham, 2017), pp. 59–87

56. Y. Haraoka, Connection problem for regular holonomic systems in several variables, in *Analytic, Algebraic and Geometric Aspects of Differential Equations*. Trends in Mathematics (Birkhäuser/Springer, Cham, 2017), pp. 337–350

57. Y. Haraoka, Multiplicative middle convolution for KZ equations. Math. Z. **294**, 1787–1839 (2020)

58. Y. Haraoka, G. Filipuk, Middle convolution and deformation for Fuchsian systems. J. London Math. Soc. **76**, 438–450 (2007)

59. Y. Haraoka, S. Hamaguchi, Topological theory for Selberg type integral associated with rigid Fuchsian systems. Math. Ann. **353**, 1239–1271 (2012)

60. Y. Haraoka, M. Kato, Generating systems for finite irreducible complex reflection groups. Funkcial. Ekvac. **53**, 435–488 (2012)

61. Y. Haraoka, T. Kikukawa, Rigidity of monodromies for Appell's hypergeometric functions. Opuscula Math. **35**, 567–594 (2015)

62. Y. Haraoka, T. Matsumura, Three-dimensional representations of braid groups associated with some finite complex reflection groups. Int. J. Math. **28**(14), 1750109, 44 p. (2017)

63. Y.Haraoka, K. Mimachi, A connection problem for Simpson's even family of rank four. Funkcial. Ekvac. **54**, 495–515 (2011)

64. Y. Haraoka, Y. Ueno, Rigidity for Appell's hypergeometric series F_4. Funkcial. Ekvac. **51**, 149–164 (2008)

65. Y. Haraoka, T. Yokoyama, Construction of rigid local systems and integral representations of their sections. Math. Nachr. **279**, 255–271 (2006)

66. J. Harnad, Dual isomonodromic deformations and moment maps to loop algebras. Commun. Math. Phys. **166**, 337–365 (1994)

67. A. Hattori, *Topology I-III*. Iwanami Lectures on Fundamental Mathematics, vol. 19 (Iwanami Shoten, Tokyo, 1982) (in Japanese)

68. K. Hiroe, Linear differential equations on $\mathbb{P}^1$ and root systems. J. Algebra **382**, 1–38 (2013)

69. K. Hiroe, Linear differential equations on the Riemann sphere and representations of quivers. Duke Math. J. **166**, 855–935 (2017)

70. K. Hiroe, T. Oshima, A classification of roots of symmetric Kac-Moody root systems and its application, in *Symmetries, Integrable Systems and Representations* (Springer Science & Business Media, New York, 2012), pp. 195–241

71. K. Hiroe, D. Yamakawa, Moduli spaces of meromorphic connections and quiver varieties. Adv. Math. **266**(2–14), 120–151 (2014)

72. N. Hitchin, Geometrical aspects of Schlesinger's equation. J. Geom. Phys. **23**, 287–300 (1997)

73. R. Hotta, K. Takeuchi, T. Tanisaki, *D-modules, Perverse Sheaves, and Representation Theory*. Progress in Mathematics, vol. 236 (Birkhäuser, Basel, 2008)
74. M. Hukuhara, Sur les points singuliers des équations différentielles linéaires, Domaine réel. J. Fac. Sci. Hokkaido Imp. Univ. **2**, 13–88 (1934)
75. M. Hukuhara, Sur les points singuliers des équations différentielles linéaires, II. J. Fac. Sci. Hokkaido Imp. Univ. **5**, 123–166 (1937)
76. M. Hukuhara, Sur les points singuliers des équations différentielles linéaires, III. Mem. Fac. Sci. Kyūsyū Imp. Univ. A. **2**, 125–137 (1941)
77. M. Hukuhara, *Ordinary Differential Equations*, 2nd edn. (Iwanami Shoten, Tokyo, 1980) (in Japanese)
78. M. Inaba, K. Iwasaki, M.-H. Saito, Moduli of stable parabolic connections, Riemann-Hilbert correspondence and geometry of Painlevé equations of type VI, Part I. Publ. Res. Inst. Math. Sci. **42**, 987–1089 (2006); Part II, Adv. Stud. Pure Math. **45**, 387–432 (2006)
79. N. Iorgov, O. Lisovyy, Yu. Tykhyy, Painlevé VI connection problem and monodromy of $c = 1$ conformal blocks. J. High Energy Phys. **2013**, 29 (2013). https://doi.org/10.1007/JHEP12(2013)029
80. N. Iorgov, O. Lisovyy, J. Teschner, Isomonodromic tau-functions from Liouville conformal blocks. Commun. Math. Phys. **336**, 671–694 (2015)
81. K. Iwasaki, Finite branch solutions to Painlevé VI around a fixed singular point, Adv. Math. **217**, 1889–1934 (2008)
82. K. Iwasaki, H. Kimura, S. Shimomura, M. Yoshida, *From Gauss to Painlevé*. Aspects of Mathematics, vol. 16 (Friedr. Vieweg & Sohn, Braunschweig, 1991)
83. M. Jimbo, T. Miwa, Monodromy preserving deformations of linear ordinary differential equations with rational coefficients II. Physica D **2**, 407–448 (1981)
84. M. Jimbo, T. Miwa, Monodromy preserving deformations of linear ordinary differential equations with rational coefficients III. Physica D **4**, 26–46 (1981)
85. M. Jimbo, T. Miwa, Y. Mori, M. Sato, Density matrix of an impenetrable Bose gas and the fifth Painlevé transcendent. Physica D **1**, 80–158 (1980)
86. M. Jimbo, T. Miwa, K. Ueno, Monodromy preserving deformations of linear ordinary differential equations with rational coefficients I. Physica D **2**, 306–352 (1981)
87. W.B. Jurkat, D.A. Lutz, On the order of solutions of analytic linear differential equations. Proc. London Math. Soc. **22**, 465–482 (1971)
88. K. Kaneko, Painlevé VI transcendents with are meromorphic at a fixed singularity. Proc. Japan Acad. Ser. A Math. Sci. **82**, 71–76 (2006)
89. K. Kaneko, Y. Ohyama, Meromorphic Painlevé transcendents at a fixed singularity. Math. Nachr. **286**, 861–875 (2013)
90. I. Kaplansky, *An Introduction to Differential Algebra*, 2nd edn. (Hermann, Paris, 1976)
91. M. Kashiwara, The Riemann-Hilbert problem for holonomic systems. Publ. RIMS Kyoto Univ. **20**, 319–365 (1984)
92. M. Kashiwara, *D-modules and Microlocal Calculus*. Translations of Mathematical Monographs, vol. 217 (American Mathematical Society, Providence, 2003)
93. M. Kato, The Riemann problem for Appell's F_4. Mem. Fac. Sci. Kyushu Univ. Ser. A Math. **47**, 227–243 (1993)
94. M. Kato, Connection formulas for Appell's system F_4 and some applications. Funkcial. Ekvac. **38**, 243–266 (1995)
95. M. Kato, J. Sekiguchi, Uniformization systems of equations with singularities along the discriminant sets of complex reflection groups of rank three. Kyushu J. Math. **68**, 181–221 (2014)
96. M. Kato, T. Mano, J. Sekiguchi, Flat structures without potentials. Rev. Roumaine Math. Pures Appl. **60**, 481–505 (2015)
97. M. Kato, T. Mano, J. Sekiguchi, Flat structures and algebraic solutions to Painlevé VI equation, in *Analytic, Algebraic and Geometric Aspects of Differential Equations*. Trends in Mathematics (Birkhäuser/Springer, Cham, 2017), pp. 383–398

98. M. Kato, T. Mano, J. Sekiguchi, Flat structures and potential vector fields related with algebraic solutions to Painlevé VI equation. Opuscula Math. **38**, 201–252 (2018)
99. N.M. Katz, Algebraic solutions of differential equations (p-curvature and the Hodge filtration). Invent. Math. **18**, 1–118 (1972)
100. N.M. Katz, *Rigid Local Systems* (Princeton University Press, Princeton, 1996)
101. T. Kawai, Y. Takei, *Algebraic Analysis of Singular Perturbation Theory*. Translations of Mathematical Monographs, vol. 227 (AMS, Providence, 2005)
102. T. Kawai, T. Koike, Y. Takei, On the exxact WKB analysis of higher order simple-pole type operators. Adv. Math. **228**, 63–96 (2011)
103. H. Kawakami, Generalized Okubo systems and the middle convolution. Int. Math. Res. Not. IMRN **2010**(17), 3394–3421 (2010)
104. T. Kimura, On Riemann's equations which are solvable by quadratures. Funkcial. Ekvac. **12**, 269–281 (1969)
105. T. Kimura, *Analytic Theory of Linear Ordinary Differential Equations on a Riemann Surface – Riemann's problem*. Lecture Notes (University of Minnesota, Minneapolis, 1973)
106. H. Kimura, Introduction to hypergeometric functions. SGC Library 55, Saiensu-Sha Co. Ltd. (2007) (in Japanese)
107. H. Kimura, K. Okamoto, On the polynomial Hamiltonian structure of the Garnier systems. J. Math. Pures Appl. **63**, 129–146 (1984)
108. M. Kita, M. Yoshida, Intersection theory for twisted cycles. Math. Nachr. **166**, 287–304 (1994)
109. F. Klein, *Lectures on the Icosahedron and the Solution of Equations of the Fifth Degree* (Dover Publications, New York, 1956)
110. T. Kohno, Homology of a local system on the complement of hyperplanes. Proc. Japan Acad. Ser. A **62**, 144–147 (1986)
111. M. Kohno, *Global Analysis in Linear Differential Equations* (Kluwer Academic Publishers, Boston, 1999)
112. T. Koike, Y. Takei, On the Voros coefficient for the Whittaker equation with a large parameter – some progress around Sato's conjecture in exact WKB analysis. Publ. Res. Inst. Math. Sci. **47**, 375–395 (2011)
113. E.R. Kolchin, Algebraic matric groups and the Picard-Vessiot theory of homogeneous linear differential equations. Ann. Math. **49**, 1–42 (1948)
114. E.R. Kolchin, *Differential Algebra and Algebraic Groups* (Academic Press, New York, 1973)
115. V.P. Kostov, The Deligne-Simpson problem for zero index of rigidity, in *Perspective in Complex Analysis, Differential Geometry and Mathematical Physics* (World Scientific, Singapore, 2001), pp. 1–35
116. V.P. Kostov, On the Deligne-Simpson problem. Proc. Steklov Inst. Math. **238**, 148–185 (2002)
117. O. Lisovyy, Y. Tykhyy, Algebraic solutions of the sixth Painlevé equation. J. Geom. Phys. **85**, 124–163 (2914)
118. D. Lutz, R. Schäfke, On the remainder of the asymptotic expansion of solutions of differential equations near irregular singular points. Complex Var. Theory Appl. **26**, 202–213 (1994)
119. D. Lutz, R. Schäfke, Calculating connection coefficients for meromorphic differential equations. Complex Var. Theory Appl. **34**, 423–436 (1997)
120. H. Majima, *Asymptotic Analysis for Integrable Connections with Irregular Singular Points*. Lecture Notes in Mathematics, vol. 1075 (Springer, Berlin, 1984)
121. J. Martinet, J.P. Ramis, Théorie de Galois différentielle et resommation, in *Computer Algebra and Differential Equations* (Academic Press, New York, 1989), pp. 115–214
122. K. Matsumoto, T. Sasaki, M. Yoshida, Recent progress of Gauss-Schwarz theory and related geometric structures. Mem. Fac. Sci. Kyushu Univ. Ser. A **47**, 283–381 (1993)
123. K. Matsumoto, T. Sasaki, N. Takayama, M. Yoshida, Monodromy of the hypergeometric differential equation of type (3, 6) II, The unitary reflection group of order $2^9 \cdot 3^7 \cdot 5 \cdot 7$. Ann. Scuola Norm. Sup. Pisa Cl. Sci. Ser. IV **20**, 63–90 (1993)

124. K. Mimachi, Intersection numbers for twisted cycles and the connection problem associated with the generalized hypergeometric function $_{n+1}F_n$. Int. Math. Res. Not. **2011**, 1757–1781 (2011)

125. K. Mimachi, M. Yoshida, Intersection numbers of twisted cycles and the correlation functions of the conformal field theory. Commun. Math. Phys. **234**, 339–358 (2003)

126. K. Mimachi, M. Yoshida, Intersection numbers of twisted cycles associated with the Selberg integral and an application to the conformal field theory. Commun. Math. Phys. **250**, 23–45 (2004)

127. T. Mochizuki, Wild harmonic bundles and wild pure twister D-modules. Astérisque **340** (2011)

128. K. Nishioka, General solutions of algebraic differential equations, in *Introduction to Differential Algebra*. Seminar on Mathematical Sciences, No. 11 (Department of Mathematics, Keio University, Tokyo, 1987) (in Japanese)

129. M. Noumi, *Painlevé Equations Through Symmetry*. Translation of Mathematical Monographs, vol. 223 (American Mathematical Society, Providence, 2004)

130. K. Okamoto, On the τ-function of the Painlevé equation. Physica D **2**, 525–535 (1981)

131. K. Okamoto, *Introduction to Painlevé Equations*. Sophia Kokyuroku in Mathematics, vol. 19 (Sophia University, Tokyo, 1985) (in Japanese)

132. K. Okamoto, *Painlevé Equations* (Iwanami Shoten, Tokyo, 2009) (in Japanese)

133. K. Okubo, *Connection Problems for Systems of Linear Differential Equations*. Lecture Notes in Mathematics, vol. 243 (Springer, Berlin, 1971), pp. 238–248

134. K. Okubo, On the group of Fuchsian equations. Seminar Reports, Tokyo Metropolitan University (1987)

135. P. Orlik, H. Terao, *Arrangements of Hyperplanes* (Springer, Berlin, 1992)

136. P. Orlik, H. Terao, *Arrangements and Hypergeometric Integrals*. MSJ Memoirs, vol. 9 (The Mathematical Society of Japan, Tokyo, 2001)

137. T. Oshima, *Fractional Calculus of Weyl Algebra and Fuchsian Differential Equations*. MSJ Meomoirs, vol. 28 (Mathematical Society of Japan, Tokyo, 2012)

138. T. Oshima, Classification of Fuchsian systems and their connection problem. RIMS Kōkyūroku Bessatsu **B37**, 163–192 (2013)

139. T. Oshima, Katz's middle convolution and Yokoyama's extending operation. Opuscula Math. **35**, 665–688 (2015)

140. T. Oshima, Transformations of KZ type equations. RIMS Kōkyūroku Bessatsu **B61**, 141–161 (2017)

141. T. Oshima, N. Shimeno, Heckman-Opdam hypergeometric functions and their specializations. RIMS Kōkyūroku Bessatsu **B20**, 129–162 (2010)

142. É. Picard, Sur une extension aux fonctions de deux variables du problème de Riemann relatif aux fonctions hypergéométriques. Ann. Sci. École Norm. Sup. **10**, 305–322 (1881)

143. C. Procesi, The invariant theory of $n \times n$ matrices. Adv. Math. **19**, 306–381 (1976)

144. C. Sabbah, Frobenius manifolds: isomonodromic deformations and infinitesimal period mappings. Exposition Math. **16**, 1–57 (1998)

145. T. Saito, On a problem of Riemann. Sûgaku **12**, 145–159 (1960–1961) (in Japanese)

146. K. Saito, On the uniformization of complements of discriminant loci. RIMS Kôkyûroku **287**, 117–137 (1977)

147. T. Saito, Linear differential equations and Fuchs functions I/I/III: reading the works of Poincaré. Kawai Institute for Culture and Education, 1991/1994/1998 (in Japanese)

148. H. Sakai, Rational surfaces associated with affine root systems and geometry of the Painlevé equations. Commun. Math. Phys. **220**, 165–229 (2001)

149. H. Sakai, Isomonodromic deformation and 4-dimensional Painlevé type equations. MSJ Mem. **37**, 1–23 (2018)

150. T. Sasaki, On the finiteness of the monodromy group of the system of hypergeometric differential equations (F_D). J. Fac. Sci. Univ. Tokyo Sect. IA Math. **24**, 565–573 (1977)

151. T. Sasaki, Picard-Vessiot group of Appell's system of hypergeometric differential equations and infiniteness of monodromy group. Kumamoto J. Sci. (Math.) **14**, 85–100 (1980/1981)

152. T. Sasaki, M. Yoshida, Linear differential equations in two variables of rank four, I/II. Math. Ann. **282**, 69–93/95–111 (1988)
153. Y. Sasano, Coupled Painlevé VI systems in dimension four with affine Weyl group symmetry of type $D_6^{(1)}$. RIMS Kôkyûroku Bessatsu **B5**, 137–152 (2008)
154. R. Schäfke, Über das globale analytische Verhalten der Lösungen der über die Laplacetransformation zusammenhängenden Differentialgleichungen $tx' = (A + tB)x$ und $(s - B)v' = (\rho - A)v$. Dissertation, University of Essen, 1979
155. R. Schäfke, D. Schmidt, The connection problem for general linear ordinary differential equations at two regular singular points with applications in the theory of special functions. SIAM J. Math. Anal. **11**, 848–862 (1980)
156. R. Schäfke, D. Schmidt, The connection problem for two neighboring regular singular points of general complex ordinary differential equations. SIAM J. Math. Anal. **11**, 863–875 (1980)
157. L. Schlesinger, Über eine Klasse von Differntialsystemen beliebiger Ordnung mit festen kritischen Punkten. J. Reine Angew. Math. **141**, 96–145 (1912)
158. L.L. Scott, Matrices and cohomology. Ann. Math. **105**, 473–492 (1977)
159. G.C. Shephard, J.A. Todd, Finite unitary reflection groups. Canadian J. Math. **6**, 274–304 (1954)
160. I. Shimada, Fundamental groups of complements to hypersurfaces. RIMS Kôkyûroku **1033**, 27–33 (1998)
161. S. Shimomura, Asymptotic expansions and Stokes multipliers of the confluent hypergeometric function Φ_2. Proc. R. Soc. Edinbrgh **123**, 1165–1177 (1993)
162. S. Shimomura, On a generalized Bessel function of two variables, I. J. Math. Anal. Appl. **187**, 468–484 (1994)
163. J. Shin, T. Naito, *Introduction to Linear Differential Equations, I*. Makino Shoten (2007) (in Japanese)
164. Y. Sibuya, *Global Theory of a Second Order Linear Ordinary Differential Equation with a Polynomial Coefficient*. Mathematics Studies, vol. 18 (North-Holland, Amsterdam, 1975)
165. Y. Sibuya, *Linear Differential Equations in the Complex Domain: Problems of Analytic Continuation*. Translations of Mathematical Monographs (American Mathematical Society, Providence, 1990)
166. C.L. Siegel, *Topics in Complex Function Theory I*. Elliptic Functions and Uniformization Theory (Wiley, New York, 1969)
167. C. Simpson, *Products of Matrices*. Canadian Mathematical Society Conference Proceedings, vol. 12 (American Mathematical Society, Providence, 1992), pp. 157–185
168. C. Simpson, Katz's middle convolution algorithm. Pure Appl. Math. Q. **5**, 781–852 (2009)
169. N.E. Steenrod, *The Topology of Fibre Bundles* (Princeton University Press, Princeton, 1951)
170. T. Suzuki, Six-dimesional Painlevé systems and their particular solutions in terms of rigid systems. J. Math. Phys. **55**, 102902 (2014)
171. K. Takano, E. Bannai, Global study of Jordan-Pochhammer differential equations. Funkcial. Ekvac. **19**, 85–99 (1976)
172. Y. Takei, Exact WKB analysis, and exact steepest descent method – a sequel to: "Algebraic analysis of singular perturbations". Sugaku Expositions **20**, 169–189 (2007)
173. K. Takemura, Introduction to middle convolution for differential equations with irregular singularities, in *New Trends in Quantum Integrable Systems* (World Scientific Publishing, Hackensack 2011), pp. 393–420
174. T. Terada, Problème de Riemann et fonctions automorphes provenant des fonctions hypergéometriques de plusieurs variables. J. Math. Kyoto Univ. **13**, 557–578 (1973)
175. H. Tokunaga, I. Shimada, *Fundamental Groups and Singular Points* (Kyoritsu Shuppan, Tokyo, 2001) (in Japanese)
176. H.L. Turrittin, Convergent solutions of ordinary linear homogeneous differential equations in the neighborhood of an irregular singular point. Acta Math. **93**, 27–66 (1955)
177. H. Umemura, Irreducibility of the Painlevé transcendental functions. Sugaku Expositions **2**, 231–252 (1989)

178. M. van der Put, M.F. Singer, *Galois Theory of Linear Differential Equations* (Springer, Berlin, 2003)

179. H. Völklein, The braid group and linear rigidity. Geom. Dedicata **84**, 135–150 (2001)

180. W. Wasow, *Asymptotic Expansions for Ordinary Differential Equations* (Dover Publications, New York, 1987). Reprint of the 1976 edition

181. G.N. Watson, *A Treatise on the Theory of Bessel Functions* (Cambridge University Press, Cambridge, 1944)

182. E.T. Whittaker, G.N. Watson, *A Course of Modern Analysis* (Cambridge University Press, Cambridge, 1927)

183. D. Yamakawa, Geometry of multiplicative projective algebra. Int. Math. Res. Pap. IMRP 2008, Art. ID rpn008 (2008)

184. D. Yamakawa, Middle convolution and Harnad duality. Math. Ann. **349**, 215–262 (2011)

185. D. Yamakawa, Fourier-Laplace transform and isomonodromic deformations. Funkcial. Ekvac. **59**, 315–349 (2016)

186. T. Yokoyama, On an irreducibility condition for hypergeometric systems. Funkcial. Ekvac. **38**, 11–19 (1995)

187. T. Yokoyama, Construction of systems of differential equations of Okubo normal form with rigid monodromy. Math. Nachr. **279**, 327–348 (2006)

188. T. Yokoyama, Recursive calculation of connection formulas for systems of differential equations of Okubo normal form. J. Dyn. Control Syst. **20**, 241–292 (2014)

189. M. Yoshida, *Fuchsian Differential Equations*. Aspects of Mathematics, vol. E11 (Friedr. Vieweg & Sohn, Wiesbaden, 1987)

190. M. Yoshida, The Schwarz program. Sugaku Expositions **5**, 35–49 (1992)

191. M. Yoshida, *Hypergeometric Functions, My Love – Modular Interpretations of Configuration Spaces*. Aspects of Mathematics, vol. 32 (Vieweg+Teubner Verlag, Berlin, 1997)

192. M. Yoshida, K. Takano, On a linear system of Pfaffian equations with regular singular points. Funkcial. Ekvac. **19**, 175–189 (1976)

Index

(+1)-loop, 67, 343

accessory parameter, 159
addition, 159
algebraic function, 349
apparent, 70
apparent singularity, 348
Appell's hypergeometric series, 338
asymptotically developable, 289
asymptotic behavior, 296
asymptotic expansion, 290

basic, 208
Birkhoff canonical form, 142
braid group, 257
branching locus, 348

characteristic exponent, 37, 50, 284
circuit matrix, 64
complete integrability condition, 312
connection coefficient, 104, 107, 111
connection matrix, 109
connection problem, 110
critical point, 325
cyclic vector, 18

defining polynomial, 341
deformation equation, 222
differential, 93
differential field, 92

differential field extension, 94
differential Galois group, 95
differential isomorphism, 94
dominant term, 284

Fuchsian, 70
Fuchsian system of normal form,
     133
Fuchs relation, 123
Fuchs' Theorem, 31
fundamental matrix solution, 26
fundamental system of solutions, 26, 27

Garnier system, 227
Gauss hypergeometric differential equation,
     86, 253
Gauss-Kummer formula, 113
Gauss-Kummer identity, 254
generalized Liouvillian extension, 98

Hamiltonian system, 236
Hitchin system, 223
hypersurface, 341

index of rigidity, 147, 158
irreducible, 72
irreducible component, 342
irregular singular point, 29
isomonodromic, 213
isomonodromic deformation, 215

© The Editor(s) (if applicable) and The Author(s), under exclusive
license to Springer Nature Switzerland AG 2020
Y. Haraoka, *Linear Differential Equations in the Complex Domain*, Lecture Notes
in Mathematics 2271, https://doi.org/10.1007/978-3-030-54663-2

LECTURE NOTES IN MATHEMATICS

 Springer

Editors in Chief: J.-M. Morel, B. Teissier;

Editorial Policy

1. Lecture Notes aim to report new developments in all areas of mathematics and their applications – quickly, informally and at a high level. Mathematical texts analysing new developments in modelling and numerical simulation are welcome.

   Manuscripts should be reasonably self-contained and rounded off. Thus they may, and often will, present not only results of the author but also related work by other people. They may be based on specialised lecture courses. Furthermore, the manuscripts should provide sufficient motivation, examples and applications. This clearly distinguishes Lecture Notes from journal articles or technical reports which normally are very concise. Articles intended for a journal but too long to be accepted by most journals, usually do not have this "lecture notes" character. For similar reasons it is unusual for doctoral theses to be accepted for the Lecture Notes series, though habilitation theses may be appropriate.

2. Besides monographs, multi-author manuscripts resulting from SUMMER SCHOOLS or similar INTENSIVE COURSES are welcome, provided their objective was held to present an active mathematical topic to an audience at the beginning or intermediate graduate level (a list of participants should be provided).

   The resulting manuscript should not be just a collection of course notes, but should require advance planning and coordination among the main lecturers. The subject matter should dictate the structure of the book. This structure should be motivated and explained in a scientific introduction, and the notation, references, index and formulation of results should be, if possible, unified by the editors. Each contribution should have an abstract and an introduction referring to the other contributions. In other words, more preparatory work must go into a multi-authored volume than simply assembling a disparate collection of papers, communicated at the event.

3. Manuscripts should be submitted either online at www.editorialmanager.com/lnm to Springer's mathematics editorial in Heidelberg, or electronically to one of the series editors. Authors should be aware that incomplete or insufficiently close-to-final manuscripts almost always result in longer refereeing times and nevertheless unclear referees' recommendations, making further refereeing of a final draft necessary. The strict minimum amount of material that will be considered should include a detailed outline describing the planned contents of each chapter, a bibliography and several sample chapters. Parallel submission of a manuscript to another publisher while under consideration for LNM is not acceptable and can lead to rejection.

4. In general, **monographs** will be sent out to at least 2 external referees for evaluation.

   A final decision to publish can be made only on the basis of the complete manuscript, however a refereeing process leading to a preliminary decision can be based on a pre-final or incomplete manuscript.

   Volume Editors of **multi-author works** are expected to arrange for the refereeing, to the usual scientific standards, of the individual contributions. If the resulting reports can be forwarded to the LNM Editorial Board, this is very helpful. If no reports are forwarded or if other questions remain unclear in respect of homogeneity etc, the series editors may wish to consult external referees for an overall evaluation of the volume.

5. Manuscripts should in general be submitted in English. Final manuscripts should contain at least 100 pages of mathematical text and should always include

   – a table of contents;
   – an informative introduction, with adequate motivation and perhaps some historical remarks: it should be accessible to a reader not intimately familiar with the topic treated;
   – a subject index: as a rule this is genuinely helpful for the reader.
   – For evaluation purposes, manuscripts should be submitted as pdf files.

6. Careful preparation of the manuscripts will help keep production time short besides ensuring satisfactory appearance of the finished book in print and online. After acceptance of the manuscript authors will be asked to prepare the final LaTeX source files (see LaTeX templates online: https://www.springer.com/gb/authors-editors/book-authors-editors/manuscriptpreparation/5636) plus the corresponding pdf- or zipped ps-file. The LaTeX source files are essential for producing the full-text online version of the book, see http://link.springer.com/bookseries/304 for the existing online volumes of LNM). The technical production of a Lecture Notes volume takes approximately 12 weeks. Additional instructions, if necessary, are available on request from lnm@springer.com.

7. Authors receive a total of 30 free copies of their volume and free access to their book on SpringerLink, but no royalties. They are entitled to a discount of 33.3 % on the price of Springer books purchased for their personal use, if ordering directly from Springer.

8. Commitment to publish is made by a *Publishing Agreement*; contributing authors of multiauthor books are requested to sign a *Consent to Publish form*. Springer-Verlag registers the copyright for each volume. Authors are free to reuse material contained in their LNM volumes in later publications: a brief written (or e-mail) request for formal permission is sufficient.

Addresses:
Professor Jean-Michel Morel, CMLA, École Normale Supérieure de Cachan, France
E-mail: moreljeanmichel@gmail.com

Professor Bernard Teissier, Equipe Géométrie et Dynamique,
Institut de Mathématiques de Jussieu – Paris Rive Gauche, Paris, France
E-mail: bernard.teissier@imj-prg.fr

Springer: Ute McCrory, Mathematics, Heidelberg, Germany,
E-mail: lnm@springer.com

Printed in the United States
By Bookmasters